Shah Alam Khan
Md. Jawaid Akhtar
Editors

Drug Repurposing

Novel Therapeutic Avenues and Innovations in Healthcare

nova
Medicine & Health
New York

DOI: https://doi.org/10.52305/ZKEA0671

Library of Congress Cataloging-in-Publication Data

ISBN: 979-8-89530-422-8 (Hardcover)
ISBN: 979-8-89530-478-5 (eBook)

Published by Nova Science Publishers, Inc. † New York

Contents

Preface

Drug repurposing involves identifying new therapeutic uses for existing medications. The guiding principle behind this book is to present drug repurposing as a promising avenue in modern healthcare. It not only minimizes the time and cost of the development of the drugs, but it also assures safety and minimizes the risk of pharmaceutical resistance. We have organized the books around new therapeutic strategies for the treatment of various diseases including fungal infections, diabetes, neurological disorders, breast cancer, non-small cell lung cancer (NSCLC), and kidney diseases. For centuries, natural compounds have been used to treat different diseases with minimal to no side-effects. Therefore, many clinically approved natural products are now being explored for their potential in treating new diseases.

The book presents a variety of drug repurposing approaches for various therapeutic applications. The first chapter focuses on the advancement in identifying the antifungal drugs against the resistant strains. Given the limited availability of the antifungal drugs and growing resistant fungal strains to the conventional therapies necessitates the search for new therapeutics options. The chapter highlights the treatment options using repurposed drugs against the strains of *Candida* spp., *Cryptococcus* spp., *Aspergillus* spp., and *Scedosporium/Lomentospora* spp.

Chapter 2 explores the potential of repurposed drugs in the treatment of neurodegenerative diseases such as Alzheimer's disease (AD) and Parkinson's disease (PD). These drugs target the mechanisms to remove the aggregate deposition in the cells and slow down the progression of the diseases. It discusses the use of effective strategy such as AI, bioinformatics, computer drug repurposing etc., to address the treatment gaps for these untreatable conditions. Similarly, in Chapter 3, antidiabetic drugs having the potential to be repurposed to cure various illnesses such as cardiovascular, cancer and neurological disorders have been discussed. The authors have supported the write up with numerous drugs that showed effectiveness in preclinical, clinical and observational studies. The diabetes disease is associated with additional comorbidities thus the drugs effective against diabetes also exhibit promising therapeutic effects against its associated health issues.

Further book chapters discuss drug repurposing techniques and mode of actions of the repurposed candidates to treat NSCLC and breast cancer. The quick development of the therapeutics resistance in cancer emphasizes the need to identify the novel medicines to offer additional therapy choices for treating the cancer. The natural products are distinctive with significant structural diversity and possesses diverse pharmacological properties. They are highly favourable in the drug discovery and carry several advantages over synthetic medicines in terms of efficacy and toxicity. The discovery of the natural drugs now includes more advanced and newer technologies including data mining, computer and *in silico* screening, proteomics and metabolomics.

The final chapter focuses on repurposing FDA approved drugs or investigational compounds against kidney diseases. The chapter provides an overview of the kidney diseases, drug repurposing strategies including target based, computer methodologies and personalized medicine development approaches. It also highlights the potential of both naturally occurring compounds and FDA-approved drugs in treating kidney conditions.

We would like to thank all the contributors who shared their valuable insights on drug repurposing against various diseases. Recent successes in repurposing drugs for COVID-19 illustrates the vast potential of drug repurposing in discovering the therapeutics for many other chronic diseases. The traditional methods of the drug discovery and approval takes long time than drug repurposing strategy. Consequently, drug repurposing strategies applied on the existing approved drugs with the focus on processes and the pathways involved in the diseases needs to be considered for finding the new indication from the existing drugs as evidenced by the recent literature presented in this book. The repurposing approach offers new opportunities to deliver the new medicines and are anticipated to be described in the near future.

Acknowledgments

We are deeply grateful to all the authors who made the completion of this book, *Drug Repurposing: Novel Therapeutic Avenues and Innovations in Healthcare*, possible. Our sincere thanks go to the distinguished authors whose expertise and research have enriched each of the eight chapters, focusing on drug repurposing strategies for challenging diseases such as neurodegenerative disorders, non-small cell lung cancer (NSCLC), breast cancer, fungal infections, diabetes, and other chronic diseases. Their insights have provided valuable perspectives on innovative therapeutic approaches.

We extend our heartfelt appreciation to Nova Science Publishers, Inc. for their trust in this project, providing us with the platform to disseminate cutting-edge research and ideas to the scientific community and beyond. They were very helpful in the various stages of developing and publishing this book.

Special thanks are due to our families for their unwavering encouragement and understanding during the demanding phases of book editing. Their patience and support have been invaluable.

We are deeply indebted to Professor Khalid Al Balushi, Dean of the College of Pharmacy, National University of Science and Technology, for his continuous support and encouragement.

Lastly, we acknowledge the readers whose interest and engagement drive our pursuit of knowledge and innovation in healthcare. Your enthusiasm for advancing therapeutic avenues through drug repurposing inspires our continued dedication to this field.

Thank you all for being part of this transformative journey.

Professor Shah Alam Khan
Dr. Md Jawaid Akhtar

About the Editors

Dr. Shah Alam Khan

Professor of Pharmaceutical Medicinal Chemistry
College of Pharmacy, National University of Science and Technology,
Bowsher, Muscat, Sultanate of Oman
Email: shahalam@nu.edu.om

Dr. Shah Alam Khan is currently working as a Professor of Pharmaceutical Medicinal Chemistry at the College of Pharmacy, National University of Science and Technology, Muscat, Oman (In academic partnership with West Virginia University, USA). He earned his Ph.D. from Jamia Hamdard University, New Delhi, India in 2003 for studying the chemistry and role of *Silybum marianum* (Milk thistle), silymarin and its synthetic analogues in the hepatoprotection. Dr. Khan is a recipient of the Junior Research Fellowship (JRF-UGC, GATE qualified) and Senior Research fellowship (SRF, CSIR) from the Government of India for pursuing master's and doctorate degrees. He has published 190+ full length scientific articles in the Scopus/WoS/ISI indexed national and international journals. In addition to this, he has contributed over 16 book chapters in publications by renowned publishers such as Elsevier, CRC Press, Springer Nature, Taylor and Francis, Nova, and Wiley. He is serving as an editorial board member of several international journals and has reviewed scientific papers for over 70 journals in the field of biomedical and pharmaceutical sciences. He has successfully completed eight funded research projects and a few more are ongoing. He has been invited as a guest speaker at numerous national and international conferences. His achievements have been recognized with the 'NU Best Researcher award' in 2020-21, 'NU Best academic excellence award' in 2021-22 and the 'Best Professor in Pharmacy' in July 2023. He is a life member of professional bodies including the IPA, ATINER, IPGA and APTI.

Dr. Md. Jawaid Akhtar
Assistant Professor of Pharmaceutical Medicinal Chemistry
College of Pharmacy, National University of Science and Technology,
Bowsher, Muscat, Sultanate of Oman
Email: mjawaid@nu.edu.om

Dr. Md Jawaid Akhtar is currently working as an Assistant Professor at the National University of Science and Technology, Muscat Oman. His areas of research interest are synthesis of heterocyclic compounds of medicinal importance and computer aided drug design. He is a recipient of a prestigious DST INSPIRE Fellowship from Govt of India to pursue a Ph.D. He qualified national level GATE 2009 and GPAT in 2010 and 2011. He has a total of 7 years of teaching and research experience with 66 publications and 3 books chapters in Scopus and WoS indexed journals to his credit. Recently, he secured research funding from the Ministry of Higher Education, Oman to design and develop synthetic anticancer agents. He serves as a reviewer for numerous prestigious international journals.

Chapter 1

Fungal Infection Treatment: Spotlight on Drug Repurposing as a New Therapeutic Strategy

Pedro Fernandes Barbosa[1,2]
Lucas Giovanini[1,2]
Ana Lúcia Casemiro[1,2]
Carolline Margot Albanez Lorentino[1,2]
Raquel Vicente de Lima[1,2]
Antonio Lima Braga[1,2]
Nahyara Louize Miranda Siqueira[1,2]
Márcia de Matos Fragata[1,3]
Julia Scaffo[1,4]
Larissa dos Santos Corrêa[1]
Lorena Rezende Franchini Affonso[1]
Gabriel Campos da Silva[1,5]
Camila Ferreira Miranda[1]
Matheus Mendes da Silva[1]
Lucieri Olegario Souza[1]
Lívia de Souza Ramos[1]
Marta Helena Branquinha[1,2,6,*]
André Luis Souza dos Santos[1,2,3,6,†]

[*] Corresponding Author's Email: mbranquinha@micro.ufrj.br
[†] Corresponding Author's Email: andre@micro.ufrj.br

In: Drug Repurposing
Editors: Shah Alam Khan and Md. Jawaid Akhtar
ISBN: 979-8-89530-422-8

and Thaís Pereira de Mello[1,2]

[1]Laboratório de Estudos Avançados de Microrganismos Emergentes e Resistentes (LEAMER), Departamento de Microbiologia Geral, Instituto de Microbiologia Paulo de Góes (IMPG), Universidade Federal do Rio de Janeiro (UFRJ), Rio de Janeiro, Brasil
[2]Programa de Pós-Graduação em Ciências (Microbiologia), Instituto de Microbiologia Paulo de Góes (IMPG), Universidade Federal do Rio de Janeiro (UFRJ), Rio de Janeiro, Brasil
[3]Programa de Pós-Graduação em Bioquímica, Instituto de Química, Universidade Federal do Rio de Janeiro (UFRJ), Rio de Janeiro, Brasil
[4]Programa de Pós-Graduação em Ciências Aplicadas a Produtos para a Saúde, Faculdade de Farmácia, Universidade Federal Fluminense (UFF), Rio de Janeiro, Brasil
[5]Graduação em Ciências Biológicas, Universidade Estadual do Rio de Janeiro (UERJ), Rio de Janeiro, Brasil
[6]Rede Micologia RJ – Fundação de Amparo à Pesquisa do Estado do Rio de Janeiro (FAPERJ), Rio de Janeiro, Brasil

Abstract

Fungal infections pose a significant and escalating threat to global health, particularly among individuals with compromised immune systems. Compounding this issue is the alarming rise in antifungal resistance observed over recent decades. This trend not only diminishes the effectiveness of current treatments but also increases the likelihood of fungal pathogens emerging as serious opportunistic threats to human health. The arsenal of antifungal medications available to treat severe infections primarily comprises three major classes: polyenes, azoles, and echinocandins. These drugs target critical fungal cell components, such as ergosterol or the enzymes involved in its synthesis. However, their clinical application is frequently limited by severe side effects, which restrict their use and emphasize the urgent need for innovative therapeutic strategies to combat fungal infections and protect human health. Traditional drug discovery methods are often constrained by lengthy timelines and substantial costs. In light of these challenges, drug repurposing has emerged as a compelling alternative. By leveraging drugs with established safety profiles, this approach bypasses many obstacles associated with conventional drug development, expediting the discovery process while significantly reducing financial and time-related investments. This chapter explores the current research efforts focused on drug repurposing as a strategy to counter fungal pathogens designated as priorities by the World Health Organization (WHO, 2022). Particular attention is given to yeast species such as *Candida albicans*, non-*albicans Candida* species (*Candida auris*, *Candida haemulonii*, *Candida glabrata*, *Candida parapsilosis*, *Candida krusei*, and *Candida tropicalis*), and *Cryptococcus* species (*C. neoformans* and *C. gattii*). Additionally, we examine filamentous fungi from the *Aspergillus* and *Scedosporium* genera.

Keywords: alternative therapy, drug repositioning, fungal pathogens, new strategies, virulence

1. Introduction

Over the past decade, research groups worldwide have reported a concerning increase in the incidence of fungal infections, accompanied by a gradual rise in associated mortality rates (Vanzolini and Magnani, 2024). Recent estimates suggest that fungal diseases affect over 6.55 million individuals annually, posing a substantial threat to global health and contributing to approximately 3.75 million deaths. Of these fatalities, it is estimated that 2.55 million are directly attributable to fungal infections (Denning, 2024). In addition to the significant human toll, the economic burden of fungal infections is considerable, driven by prolonged hospital stays and high treatment costs. For instance, in the United States of America (USA), annual healthcare expenses related to fungal infections range from approximately $22,000 per case to over $2 billion in total (Hamdy, Zaoutis, and Seo, 2016; Strollo et al., 2016).

Advancements in healthcare practices over the past century have significantly improved patient survival across various diseases (Arendrup and Patterson, 2017). However, this progress has also contributed to a rise in the number of immunosuppressed individuals, particularly those undergoing immunosuppressive therapies, making them more susceptible to infectious diseases (Kollef et al., 2021). Several factors have contributed to the growing population of immunosuppressed patients, including the prolonged use of antibiotics, coinfections with other microbial pathogens, extremes of age, the presence of immunosuppressive conditions such as neutropenia and diabetes, cancer chemotherapy, and invasive medical procedures (Chen et al., 2018; von Lilienfeld-Toal et al., 2019; Granato et al., 2021). As a result, the incidence and severity of fungal infections are strongly influenced by the host's immune status, the virulence arsenal of the pathogen, and environmental factors (Vanzolini and Magnani, 2024).

Current estimates indicate the existence of between 1.5 and 5 million fungal species, with at least 300 known to cause infections in humans (Bazzi, 2016). In 2022, the World Health Organization (WHO) published a landmark report categorizing 19 fungal pathogens into three priority levels based on their threat to global public health (WHO FPPL, 2022) (Figure 1). This classification was introduced in response to the increasing frequency of fungal outbreaks and infections reported each year. While this initiative was both pioneering and necessary, many clinically significant fungal species were excluded from the list, underscoring the need for an urgent revision. Such revisions should particularly address endemic fungal infections that disproportionately affect developing countries. Research groups worldwide have highlighted the dynamic and evolving nature of the mycological landscape, shaped by multiple factors. Among these, the rising resistance of fungal pathogens to commonly used antifungal treatments is particularly concerning. This resistance not only compromises the efficacy of existing therapies but also exacerbates the already significant challenges associated with fungal infections, including difficulties in achieving accurate diagnoses and selecting appropriate antifungal regimens (Wiederhold, 2021).

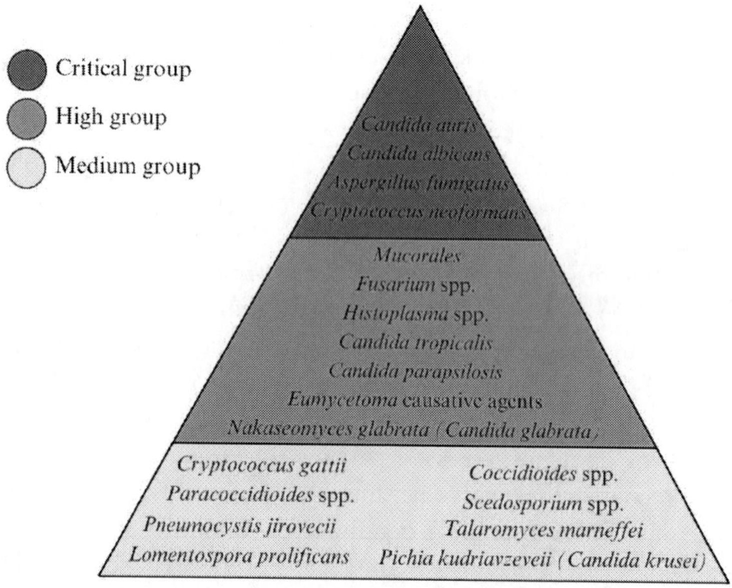

Figure 1. Fungal priority pathogens list to guide research, development and public health action, according to the World Health Organization (WHO, 2022).

Antifungals are a class of medications specifically developed to treat fungal infections by targeting key cellular structures in both yeasts and filamentous fungi (Nett and Andes, 2016). The primary categories of antifungal agents used in clinical practice include polyenes, azoles, echinocandins, and flucytosine (Figure 2) (Perfect, 2017). However, pathogens such as *Candida* spp., *Cryptococcus* spp., and *Aspergillus* spp. have evolved significant resistance mechanisms to these antifungal classes, posing a growing threat to global public health (Perlroth, Choi, and Spellberg, 2007; Geddes-McAlister and Shapiro, 2019).

The development of resistance mechanisms in fungal pathogens is largely attributed to the inappropriate use of antifungals in clinical settings. Additionally, environmental exposure to azole compounds, extensively employed as agricultural fungicides, exacerbates this issue. Due to the shared mechanism of action of azoles used in both healthcare and agriculture, saprophytic fungi such as *Aspergillus* spp. have developed widespread resistance. This occurs because the ergosterol biosynthesis pathway targeted by azoles is identical in both plant and animal fungal pathogens (Bowyer and Denning, 2014). Moreover, accelerated climate change introduces thermal stress to environmental microorganisms, driving significant alterations in their transcriptomes and proteomes. These adaptations enable certain heat-tolerant fungal species to survive, invade, and colonize human hosts more effectively (Garcia-Solache and Casadevall, 2010; Chow et al., 2017). Together, these factors promote the proliferation and selection of fungal strains with resistance mechanisms against currently available antifungal treatments (Cowen et al., 2014).

Azole resistance mechanisms are multifaceted and involve several key factors. One major mechanism is the overexpression of the *ERG11* gene in yeasts (or *CYP51* in filamentous fungi), which leads to elevated levels of lanosterol 14α-demethylase, which is the target of azoles. This overproduction requires higher antifungal doses to inhibit all target enzymes effectively. Additionally, the increased expression of membrane-associated efflux transporters accelerates the removal of azoles from fungal cells. Mutations or depletion of the *ERG3* gene, essential for producing toxic sterols in response to lanosterol 14α-demethylase inhibition, further reduces azole efficacy (Scorzoni et al., 2017).

Resistance to polyenes, while less common due to their fungicidal properties, is associated with specific adaptations. Fungal strains with reduced susceptibility to amphotericin B often exhibit overproduction of antioxidant enzymes like catalase, reduced respiration rates, and alterations in ergosterol concentrations within the plasma membrane (Lee, Robbins, and Cowen, 2023).

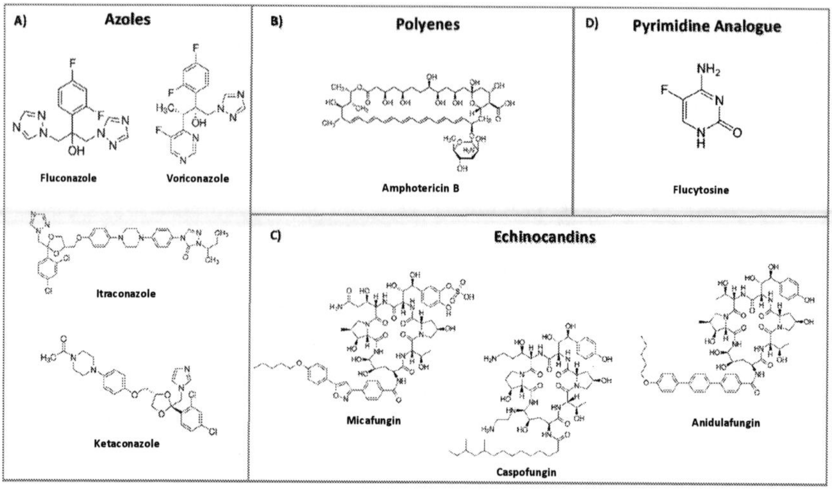

Figure 2. Chemical structures of commonly used antifungal drugs in medical practice. A) Azole antifungals inhibit the enzyme lanosterol 14α-demethylase, which is crucial for the conversion of lanosterol to ergosterol, an essential component of the fungal cell membrane; B) Polyene antifungals exert their effects by binding to ergosterol, creating pores in the fungal cell membrane, and inducing the production of reactive oxygen species, resulting in oxidative stress and futher damage to cellular components; C) Echinocandins inhibit the enzyme β-1,3-D-glucan synthase, which is essential for the synthesis of β-1,3-D-glucan, a critical component of the fungal cell wal; D) Pyrimidine analogues, such as flucytosine, interfere with fungal RNA and DNA synthesis, disrupting their functions and blocking protein synthesis.

Echinocandin resistance primarily stems from mutations in specific "hot spot" regions of the *FSK1* gene, which encodes the enzyme β-1,3-glucan synthase, the target of this antifungal class. These mutations reduce the enzyme's affinity for echinocandins, thereby diminishing the antifungal's effectiveness (Scorzoni et al., 2017; Kanafani and Perfect, 2008).

The formation of biofilms by clinically significant fungi, such as *Candida* spp. and *Aspergillus* spp., poses a substantial challenge in the context of antifungal resistance. Biofilms are highly organized microbial communities embedded in a self-produced extracellular polymeric matrix, capable of adhering to both biotic and abiotic surfaces (Scorzoni et al., 2017). This complex structure significantly enhances biofilm resilience to antifungal agents through multifaceted mechanisms. Key resistance factors include the high cell density within biofilms, increased efflux pump activity, and downregulation of genes involved in ergosterol biosynthesis (Cowen et al., 2014; Scorzoni et al., 2017). Additionally, the extracellular matrix acts as a physical barrier, hindering the effective penetration of antifungal drugs. For example, fluconazole and amphotericin B are known to bind to β-1,3-glucan within the extracellular matrix of *Candida albicans* biofilms, reducing their efficacy (Nett et al., 2007). These combined features contribute to the pronounced resistance of fungal biofilms, complicating antifungal therapy and highlighting the need for innovative treatment approaches.

Given the complexities in treating fungal infections, there has been a significant shift towards exploring innovative therapeutic approaches to enable faster and more effective treatments. Among these strategies, drug repositioning has emerged as a particularly promising avenue, offering the potential to accelerate the development of medications with potent antifungal properties against emerging fungal pathogens (Peyclit et al., 2021). Drug repositioning involves repurposing an existing pharmaceutical ingredient – originally developed and marketed for a specific disease – to target a new indication. This new application may require modifications in formulation, dosage, or administration routes, demanding thorough and innovative clinical investigations to elucidate its pharmacological mechanisms (Jourdan et al., 2020). One of the primary advantages of drug repositioning is its ability to streamline regulatory processes by leveraging existing safety and toxicity data, significantly reducing development time and cost. For instance, the development of a novel chemical entity typically takes about a decade and costs approximately USD 1 billion (Wouters, McKee, and Luyten, 2020; Donlin and Meyers, 2022). In contrast, repositioning existing drugs accelerates the timeline, lowers expenses, and improves the likelihood of market approval (Zhang et al., 2021; Barratt and Frail, 2015; Nosengo, 2016; Donlin and Meyers, 2022; Sangenito et al., 2022). This approach represents a pragmatic and efficient solution, bridging the gap between urgent clinical needs and the high cost and time demands of traditional drug development.

The repositioning approach has demonstrated significant potential, with numerous drugs successfully repurposed over the years, yielding promising outcomes. For example, polyene compounds, a well-established class of antifungal agents particularly amphotericin B, were discovered in 1953 and initially employed for the treatment of leishmaniasis. Similarly, azidothymidine (AZT), originally developed as a chemotherapeutic agent, was later repurposed as an antiretroviral drug in the 1980s. These examples highlight the effectiveness and versatility of this strategy, underscoring its value in addressing diverse therapeutic challenges (Nosengo, 2016; Ashburn and Thor, 2004; Bustamante et al., 2019; Ramos et al., 2022).

Given that drug repositioning can substantially reduce costs and development timelines, research in this area offers a highly attractive avenue for the discovery of new antimicrobial agents. In the following sections, we will explore approved drugs currently used for various medical conditions that have shown antifungal activity against clinically significant pathogens, including *Candida* spp., *Cryptococcus* spp., *Aspergillus* spp., and *Scedosporium/Lomentospora* spp. (Figure 3).

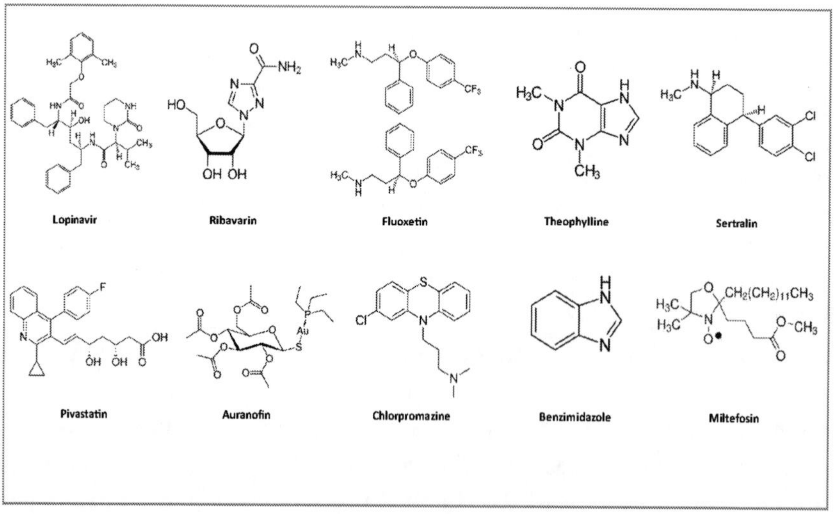

Figure 3. Chemical structures of approved drugs demonstrating antifungal effects against clinically resistant fungal pathogens.

2. *Candida albicans*

Candida species are globally recognized as the leading cause of invasive fungal infections and are a major concern in healthcare settings. In hospital environments, they are the primary etiological agents of nosocomial fungal infections (Du et al., 2019). These opportunistic pathogens can cause a wide range of diseases, from superficial infections to severe invasive candidiasis. The latter poses a significant challenge, particularly in immunosuppressed individuals with compromised immune defenses (Dadar et al., 2018). Among *Candida* species, *C. albicans* remains the most prevalent causative agent of invasive candidiasis in healthcare settings in the USA, with a crude mortality rate of approximately 40%, even with antifungal treatment (Lee et al., 2021).

This high mortality rate is attributed to the remarkable adaptability of *C. albicans*, which enables it to infect nearly every anatomical site in the human body, including the gastrointestinal, respiratory, and genitourinary tracts (Calgin and Cetinkol, 2018).

Drug resistance mechanisms in *C. albicans* are constantly evolving, posing a growing challenge in clinical settings (Padder et al., 2022). Among these, azole resistance is particularly concerning, as it significantly complicates the management of *C. albicans* infections. Reports of azole-resistant clinical isolates are increasingly common among hospitalized patients worldwide, further highlighting the need for alternative therapeutic strategies (Shi et al., 2021). In this context, drug repurposing has emerged as a promising approach for the discovery of novel antifungal agents. By identifying antifungal potential in existing drugs, this strategy offers a cost-effective and time-efficient pathway to address the rising threat of resistance. Numerous molecules are currently being investigated for their efficacy against *C. albicans* and other fungal pathogens (Miró-Canturri et al., 2019).

The search for antifungal molecules has seen a notable rise in studies focusing on protease inhibitors (PIs) due to the significant role of proteases in various pathologies (Ćwilichowska et al., 2022). PIs are compounds that inhibit proteolytic activity by binding either reversibly or irreversibly to the active site of target enzymes (Ćwilichowska et al., 2022). For instance, the use of HIV PIs in antiretroviral therapies has been shown to reduce the incidence of candidiasis in HIV-positive individuals (Patton et al., 2000). This has spurred interest in exploring the interactions between PIs and *Candida* species, leading to highly promising findings (Chakraborty et al., 2023). One example is lopinavir, a widely used HIV PI in clinical medicine, which has demonstrated the ability to disrupt several key biological processes in *C. albicans*, including nutrition, proliferation, and morphogenesis. At a concentration of 50 μM, lopinavir significantly reduced the viability of *C. albicans* across yeast densities ranging from 10^2 to 10^4 cells (Santos et al., 2021). Further investigations have revealed that lopinavir interferes with multiple processes essential for fungal virulence and infection. It inhibits the hydrolytic activity of secreted aspartyl proteases (Saps), the primary virulence factors of *C. albicans*, and disrupts the synthesis of lipases and mannoproteins. Lopinavir also prevents adhesion, biofilm formation, and interaction with epithelial cells. These inhibitory effects have been corroborated in vivo, where lopinavir effectively controlled disseminated fungal infections in animal models, including both immunocompetent and immunocompromised mice (Santos et al., 2021). These findings underscore the potential of PIs like lopinavir as innovative antifungal agents, warranting further exploration in the development of therapies against fungal infections.

Theophylline, a naturally occurring 1,3-dimethylxanthine, is found in tea leaves and shares pharmacological similarities with caffeine (Franzke et al., 1968). For over 70 years, it has been a cornerstone in the treatment of asthma and chronic obstructive pulmonary disease (COPD), historically ranking among the most prescribed drugs for respiratory conditions (Barnes, 2013). In addition to its bronchodilator effects, methylxanthines, including theophylline, are known to exhibit anti-chitinase activity, suggesting potential as antifungal agents (Tsirilakis et al., 2012). Recent research has focused on theophylline's antifungal properties, particularly against human pathogenic fungi, including *C. albicans*. Broth microdilution assays have shown theophylline's inhibitory effect against clinical isolates of *C. albicans* and non-albicans *Candida* species, with minimum inhibitory concentrations (MICs) ranging from 1.4 mg/mL to 1.8 mg/mL (Singh et al., 2019). Mechanistically, theophylline has been found to induce membrane damage, disrupt ionic balance, and deplete ergosterol levels, leading to increased membrane fluidity through elevated flippase activity. These effects culminate in significant membrane damage. Additionally, theophylline interferes with the metabolic flexibility of *C. albicans* by inhibiting key enzymes in the glyoxylate cycle, such as malate synthase and isocitrate lyase. In vivo studies using a *Caenorhabditis elegans* model of *C. albicans* infection have demonstrated that theophylline improves survival rates (Singh et al., 2019). Furthermore, theophylline's non-toxicity has been affirmed, as its hemolytic activity in erythrocytes has been classified as negligible.

Aripiprazole, an atypical antipsychotic, is primarily prescribed for psychiatric conditions such as bipolar disorder, schizophrenia, and treatment-resistant depression. Recent studies have explored its repurposing as an antifungal agent, particularly against species within the *Candida* genus, yielding promising results. Aripiprazole's antifungal activity appears to center on inhibiting the formation of biofilms and hyphae in *C. albicans*. Notably, at a concentration of 10 μg/mL, aripiprazole reduced biofilm formation by 60%. Higher concentrations showed more potent effects, with reductions of 80% and 90% at 25 μg/mL and 50 μg/mL, respectively. These findings highlight its effectiveness, as aripiprazole outperformed ketoconazole, a commonly used antifungal. Ketoconazole showed minimal impact on biofilm formation at similar concentrations, with only a slight reduction observed at 50 μg/mL. Additionally, aripiprazole was found to significantly attenuate the virulence of *C. albicans*. At concentrations ranging from 5 to 10 μg/mL, it effectively inhibited hyphal formation in *C. albicans* strains, whereas ketoconazole required a minimum concentration of 50 μg/mL to achieve a comparable effect. At higher concentrations (250 μg/mL), aripiprazole disrupted ergosterol synthesis, leading to approximately 70% damage to the plasma membrane of *C. albicans* cells. Computational studies further suggest that, at high concentrations, aripiprazole may exert antifungal effects similar to those of azole compounds. It acts as an inhibitor of the CYP51 enzyme, disrupting its catalytic cycle and interfering with the synthesis of ergosterol. This mechanism mirrors that of traditional azole antifungals, which inhibit enzymes critical to ergosterol biosynthesis (Rajasekharan et al., 2019).

In the search for novel antifungal therapies, researchers often investigate combinations of existing antimicrobials with repurposed drugs to identify synergistic effects that could enhance efficacy while potentially reducing the risk of resistance (Rossato et al., 2021). Artemisinin, a sesquiterpene lactone derived from the traditional Chinese herb *Artemisia annua* L., is the first-line treatment for malaria due to its high efficacy and low toxicity (Zhu et al., 2021). While artemisinin itself does not exhibit direct antifungal activity, recent studies have explored its potential to augment the efficacy of the polyene antifungal amphotericin B through a synergistic mechanism. This synergy occurs as artemisinin upregulates genes involved in the ergosterol biosynthesis pathway, thereby increasing the ergosterol content in fungal membranes. This enhancement improves the binding of amphotericin B to fungal cells, boosting its antifungal activity (Zhu et al., 2021). The combination of artemisinin and amphotericin B has demonstrated dose-dependent synergistic effects, with a decrease in the MICs and minimum fungicidal concentrations (MFCs) of amphotericin B as the concentration of artemisinin increased (from 25 to 100 mg/L). The fractional inhibitory concentration index (FICI) of this combination against clinical strains was below 0.5, indicating a potent antifungal effect against various *C. albicans* strains. Moreover, in vivo assays in murine models of oropharyngeal candidiasis caused by *C. albicans* confirmed the enhanced antifungal activity of the combination. While low doses of amphotericin B (0.5 mg/L) or artemisinin (50 mg/L) alone were ineffective at controlling the infection, their combined administration significantly reduced both the infection and fungal load (Zhu et al., 2021).

In a study by Chen et al. (2024), a screening of 624 compounds from the MCE Clinical Compound Library (MedChemExpress) was conducted against *C. albicans* strains, including the standard SC5314 strain and a voriconazole-resistant clinical isolate, designated as C5. Among the compounds tested, firsocostat emerged as the most potent antifungal, with a MIC of 4 μg/mL against the SC5314 strain and 8 μg/mL against the C5 strain. Notably, when voriconazole was added at a concentration of ¼×MIC, the MIC of firsocostat decreased by 8-fold against SC5314 and 32-fold against C5, indicating a strong synergistic effect between firsocostat and voriconazole. Firsocostat is an inhibitor of human acetyl-CoA carboxylase (ACC), specifically targeting the ACC1 and ACC2 isoenzymes. It acts by binding to the phosphopeptide-acceptor and dimerization sites of these enzymes, preventing their dimerization. The compound has been shown to synergize with a variety of clinical antifungals, including azoles such as fluconazole, itraconazole, and voriconazole, as well as with the polyene amphotericin B. *In vivo* studies using mouse models of systemic and local *C. albicans* infections revealed that incorporating firsocostat into treatment regimens significantly enhanced the activity of both voriconazole and amphotericin B. A current hypothesis linking firsocostat's antifungal activity is based on the structural similarities between human and *C. albicans* ACC enzymes. ACC plays a critical role in fatty acid biosynthesis, catalyzing the carboxylation of acetyl-CoA to malonyl-CoA, which is essential for the synthesis of fatty acids and certain secondary metabolites. Given that the ACC enzymes in humans and *C. albicans* share 46% amino acid identity, firsocostat may target fatty acid biosynthesis in *C. albicans*. This suggests the potential for developing small-molecule compounds that specifically target *C. albicans* ACC, exploiting differences in binding regions between fungal and human ACC enzymes.

Losartan, a widely prescribed medication for hypertension, functions as an angiotensin II receptor blocker (ARB). It promotes vasodilation, reduces blood pressure, and enhances blood and oxygen supply to the heart (Al-Majed et al., 2015). Beyond its cardiovascular benefits, inhibitors of the renin-angiotensin system, such as losartan, have been shown to influence fungal virulence factors, including the suppression of secreted aspartic proteases in C. albicans. Building on this knowledge, Lara et al. (2023) investigated the antimicrobial effects of losartan on *C. albicans* biofilms. Both losartan and aliskiren were found to reduce fungal viability at all tested concentrations. Specifically, losartan inhibited *C. albicans* biofilm growth by 47% to 88.5%, while aliskiren showed a broader range of inhibition, from 16% to 97.6%, at concentrations ranging from 1 to 10 mg/mL. Importantly, these drugs were also found to preserve the viability of human cells at certain concentrations. These results suggest that losartan and aliskiren exhibit both fungistatic and fungicidal activities against *C. albicans* biofilms, while remaining compatible with human cell viability.

3. *Candida auris*

Candida auris is a globally emerging, multidrug-resistant fungal pathogen, first identified in 2009 from the external ear canal of a Japanese patient (Satoh et al., 2009). Since its discovery, *C. auris* has gained significant attention due to its role in widespread outbreaks in healthcare settings across the world (Ahmad and Alfouzan, 2021). Genetic analyses have classified *C. auris* into four major clades, each corresponding to a distinct geographic region: South-Asian (Clade I), East Asian (Clade II), African (Clade III), and South American (Clade IV) (Ahmad and Alfouzan, 2021). More recently, the identification of a fifth clade, Clade V, was reported in Iranian isolates (Spruijtenburg et al., 2022).

The multidrug-resistant profile of *C. auris* presents significant challenges in treating infections, contributing to increased morbidity and mortality, with rates that can reach as high as 72% (Ahmad and Alfouzan, 2021). In addition to its resistance to multiple antifungal classes, *C. auris* exhibits remarkable environmental persistence, enabling it to survive for months on hospital surfaces despite standard disinfection efforts. This durability facilitates nosocomial transmission, making *C. auris* a major concern in healthcare settings. The pathogen frequently colonizes various body sites in hospitalized patients, including the axilla, groin, nares, respiratory, and urinary tracts. Notably, *C. auris* has been found to persist on reusable skin-surface axillary temperature probes, which correlates with a higher frequency of isolation from the axilla compared to other body sites (Cristina et al., 2023).

Miltefosine, an alkyl-phospholipid analogue, is clinically approved for treating leishmaniasis in many countries but was originally developed as a cancer treatment, making it a promising candidate for drug repurposing (Barreto et al., 2020; Wu et al., 2020). In recent years, miltefosine has shown broad-spectrum antifungal activity, including against *C. auris*. *In vitro* studies have demonstrated that miltefosine effectively inhibits the planktonic growth of *C. auris* isolates, with MIC values ranging from 1 to 4 μg/mL. It also exhibited fungicidal activity at concentrations of 2 and 4 μg/mL (Barreto et al., 2020). Furthermore, miltefosine was found to disrupt *C. auris* biofilm formation at concentrations as low as 0.25 μg/mL and up to 4 μg/mL, while higher concentrations (16 and 32 μg/mL) were effective at disrupting preformed biofilms. Notably, miltefosine also inhibited the dispersal of biofilm cells, which could prevent fungal dissemination and reinfection (Barreto et al., 2020). In addition to its antifungal effects, miltefosine, both in its free-form and encapsulated in alginate nanoparticles, has shown efficacy in controlling *C. auris* infections in the *Galleria mellonella* larvae model (Barreto et al., 2020). A synergistic effect between miltefosine and amphotericin B was observed in 25% (3/12) of the *C. auris* isolates tested, with a FICI of 0.5. However, the combination of miltefosine with fluconazole resulted in indifferent interactions across all isolates tested (Wu et al., 2020). These findings underscore miltefosine's potential as an effective antifungal agent, either alone or in combination with other treatments.

The *in vitro* activity of sertraline, a commonly used antidepressant, against *C. auris* has been documented in recent studies. Sertraline demonstrated inhibitory effects on the growth of planktonic *C. auris* cells, with MIC values ranging from 4-8 μg/mL (Alanís-Ríos et al., 2022) and 20-40 μg/mL (Gowri et al., 2020). Additionally, sertraline was found to inhibit the conversion of yeast cells into hyphae and disrupt preformed biofilms of *C. auris* (Gowri et al., 2020). The proposed mechanism of action for sertraline's antifungal effect is linked to damage to the fungal cell membrane. Molecular docking studies indicated that sertraline binds to sterol 14-α-demethylase, an enzyme involved in ergosterol biosynthesis (Gowri et al., 2020). Consistent with this, sertraline-treated *C. auris* cells showed a 5.5-fold reduction in ergosterol production (Gowri et al., 2020). Furthermore, the combination of sertraline with voriconazole exhibited a synergistic effect against 25% (3/12) of the *C. auris* isolates tested, with FICI values ranging from 0.37 to 0.49. A partial synergistic effect was observed in 8.3% (1/12) of isolates, with a FICI of 0.54. In contrast, the combination of sertraline with micafungin resulted in partial synergy for 8.3% (1/12) of isolates (FICI = 0.53) but showed indifferent effects for the remaining 91.7% (11/12) (Alanís-Ríos et al., 2022). These findings highlight sertraline's potential as an adjunctive treatment in *C. auris* infections, especially when combined with other antifungals.

The antimalarial drug artemether, derived from artemisinin, has shown broad-spectrum antifungal activity against various yeast species, including *C. auris* (Zhou et al., 2021). Artemether effectively inhibited the growth of *C. auris* at a concentration of 100 μM, and when combined with fluconazole (1 μg/mL), its antifungal effect was significantly enhanced. This synergistic interaction was attributed to artemether's ability to disrupt the multidrug efflux pump Pdr5, leading to an increased intracellular accumulation of fluconazole, thereby enhancing its antifungal activity (Zhou et al., 2021). In addition, myriocin, a serine palmitoyl transferase inhibitor used in treating inflammatory demyelinating diseases like multiple sclerosis, has demonstrated potent inhibitory effects against *C. auris*, with MIC_{80} values ranging from 0.25 to 2 μg/mL (Cheng et al., 2021). When combined with flucytosine, myriocin exhibited a synergistic effect in 35.7% (5/14) of *C. auris* isolates, with FICI values ranging from 0.49 to 0.50. In the remaining 64.3% (9/14) of isolates, the combination displayed an additive effect, with FICI values ranging from 0.51 to 0.75 (Cheng et al., 2021). These findings highlight the potential of artemether and myriocin, both individually and in combination with other antifungals, as effective therapeutic options against *C. auris* infections.

The antiemetic agent aprepitant has demonstrated chemosensitizing activity against *C. auris* isolates in a study by Eldesouky and colleagues (2020a). Aprepitant enhanced the antifungal activity of fluconazole against 70% (7/10) of *C. auris* isolates, showing a synergistic effect against four isolates (FICI ranging from 0.09 to 0.50) and an additive effect against three isolates (FICI ranging from 0.51 to 0.63). When combined with voriconazole, aprepitant exhibited a synergistic effect against two isolates (FICI = 0.19) and an additive effect against six isolates (FICI ranging from 0.53 to 0.56). However, the most pronounced effects were observed with the combination of aprepitant and itraconazole, which exhibited synergistic effects against eight isolates (FICI ranging from 0.14 to 0.31) and additive effects against the remaining two isolates (FICI = 0.52) (Eldesouky et al., 2020a). This combination significantly impaired *C. auris* biofilm formation by 95%, disrupted preformed biofilms by 52%, and notably increased the survival of *Caenorhabditis elegans* infected with *C. auris* by 90% after five days of infection, reducing the fungal burden by approximately 92%. Mechanistic studies suggested that the aprepitant/itraconazole combination impacts metal ion homeostasis in *C. auris*, compromising reactive oxygen species (ROS) detoxification mechanisms, thus enhancing the antifungal effect (Eldesouky et al., 2020a). In a related study, Eldesouky and colleagues (2020b) also demonstrated that pitavastatin, a drug used to treat hypercholesterolemia and dyslipidemia, exhibited potent azole chemosensitizing activity against *Candida* species, including *C. auris*. While pitavastatin alone showed high MICs against *C. auris* (256 μg/mL), its combination with fluconazole produced a synergistic effect against the five *C. auris* isolates tested (FICI ranging from 0.25 to 0.50). This combination reduced *C. auris* biofilm formation by approximately 41%, although it did not disrupt mature biofilms. Additionally, the pitavastatin/fluconazole combination reduced colony-forming units (CFUs) in *Caenorhabditis elegans* infected with *C. auris* by 14-92%. The synergy between pitavastatin and fluconazole was attributed to the inhibition of the fungal efflux machinery (Eldesouky et al., 2020b). These findings underscore the potential of aprepitant and pitavastatin as adjunctive treatments for *C. auris* infections, particularly when used in combination with established antifungals.

Synergistic interactions between ospemifene, a non-hormonal estrogen receptor modulator used to treat dyspareunia, and itraconazole against *C. auris* and other fungal species have been reported by Eldesouky et al. (2020c). The combination of ospemifene and itraconazole exhibited a synergistic effect against five *C. auris* isolates, with FICI values ranging from 0.14 to 0.27. Treatment of *Caenorhabditis elegans* infected with *C. auris* using this combination resulted in a dramatic 96% reduction in fungal burden. The synergistic action of the ospemifene/itraconazole combination is thought to involve interference with fungal efflux systems, which facilitates the enhanced intracellular accumulation of azoles, thus improving their antifungal efficacy (Eldesouky et al. 2020c). The potential of combining HIV PIs with azole antifungals against azole-resistant *C. auris* has also been explored in several studies. Synergistic effects were observed with combinations such as atazanavir with fluconazole or itraconazole (Elgammal et al. 2023a), saquinavir with itraconazole or voriconazole (Elgammal et al. 2023b), lopinavir with fluconazole, itraconazole, or voriconazole (Eldesouky et al. 2021; Salama et al. 2023), and ritonavir with fluconazole, itraconazole, or voriconazole (Salama et al., 2023), as well as atazanavir and saquinavir with posaconazole (Elgammal et al. 2024) against *C. auris*. Notably, the combination of itraconazole with atazanavir, saquinavir, or lopinavir, as well as posaconazole with atazanavir or saquinavir, demonstrated synergistic effects against all *C. auris* isolates tested. In *in vivo* studies, the best drug combinations significantly reduced fungal burden in murine kidneys by up to 95%, compared to vehicle controls, antifungals alone, or HIV PIs alone (Elgammal et al. 2023a; Elgammal et al. 2023b; Salama et al. 2023; Elgammal et al. 2024). Mechanistic studies indicated that HIV PIs compromised the efflux capacity of *C. auris*, resulting in enhanced susceptibility to azole drugs, thereby improving the overall antifungal efficacy of the combination therapy. These findings highlight the potential of combining HIV PIs with azole antifungals as a promising therapeutic strategy against resistant *C. auris* infections.

4. *Candida tropicalis*

Over the past two decades, there has been a significant shift in the epidemiology of invasive *Candida* species. In many regions, the proportion of

C. albicans infections has declined, with a corresponding rise in infections caused by non-*albicans Candida* species. *C. tropicalis*, which typically ranks as the second or third most common cause of candidemia after *C. albicans*, has emerged as one of the most clinically significant species. This species is associated with relatively high mortality rates, ranging from 40% to 60%. The virulence of *C. tropicalis* is attributed to several key factors, including strong adhesion to host tissues, the production of hydrolytic enzymes, and the formation of robust biofilms, all of which contribute to its pathogenicity. A growing concern is the increasing prevalence of antifungal-resistant *C. tropicalis* isolates, with resistance reported across multiple classes of antifungal agents. Despite the introduction of new antifungal drugs, *C. tropicalis* infections remain a serious challenge to public health. In response, clinical practices have evolved, with increased use of prophylactic and empirical antifungal therapies, particularly in high-risk patients, to mitigate the impact of these infections (Puig-Asensio et al., 2014; Spruijtenburg et al., 2023).

In a study conducted by Yousfi et al. (2019), the potential antifungal activity of ribavirin was assessed against several *Candida* species, including isolates of *C. tropicalis*. Ribavirin, a guanosine analogue known for its broad-spectrum antiviral activity against many RNA and DNA viruses, is commonly used in combination with interferon alpha for the treatment of hepatitis C virus (HCV). The researchers tested a collection of 96 *Candida* strains, which included 5 *C. tropicalis* isolates. Antifungal susceptibility was evaluated using nine antifungal agents from four therapeutic classes: polyenes (amphotericin B), azoles (fluconazole, posaconazole, voriconazole, and itraconazole), echinocandins (anidulafungin, caspofungin, and micafungin), and nucleoside analogues (5-flucytosine). The MIC values were determined for each drug. The *C. tropicalis* isolates exhibited the highest susceptibility to ribavirin, with MICs primarily ≤3.02 μM (≤0.75 μg/mL). Notably, ribavirin in combination with posaconazole showed the most significant synergistic effect, especially for the *C. tropicalis* strains. However, the precise mechanism of action of ribavirin against *Candida* species remains unclear, warranting further investigation.

In a study by Jothi et al. (2023), the antifungal activity of 3-bromopyruvate (3BP), an established anticancer drug, was evaluated against *Candida* species responsible for vulvovaginal candidiasis (VVC) and recurrent vulvovaginal candidiasis (RVVC), including *C. tropicalis*. The research aimed to explore the potential repurposing of 3BP as an antifungal treatment. The study revealed that the MIC of 3BP was 256 μg/mL for all

Candida species tested. Moreover, 3BP demonstrated significant anti-*Candida* efficacy, with complete killing (99.9%) of all *Candida* strains occurring within 20 minutes when used at 2×MIC. Further experiments targeting ergosterol and sorbitol indicated that 3BP's antifungal properties do not involve disruption of the cell wall or ergosterol. Instead, 3BP was found to induce a cascade of pre-apoptotic events, including phosphatidylserine externalization, nuclear condensation, and ROS accumulation, as evidenced by propidium iodide, DAPI, and DCFH-DA staining methods. Additionally, 3BP effectively eradicated mature *Candida* biofilms, achieving up to 90% biofilm clearance. The safety and toxicity of 3BP were evaluated through *in vitro* erythrocyte lysis and *in vivo Galleria mellonella* survival assays, both confirming that 3BP is non-toxic at concentrations up to 512 μg/mL.

El-Ganiny et al. (2022) investigated the phenotypic mechanisms of azole drug resistance in clinical *Candida* isolates, focusing on the expression of efflux pump genes in resistant strains using RT-PCR. The study also explored the potential synergistic effects of haloperidol and pantoprazole in combination with fluconazole against fluconazole-resistant *Candida* isolates. Antifungal susceptibility testing was conducted on 122 *Candida* isolates using three antifungal agents: fluconazole, voriconazole, and amphotericin B. Among the eight *C. tropicalis* isolates analyzed, only one exhibited resistance to both fluconazole and voriconazole, with no resistance observed against amphotericin B. The researchers assessed biofilm formation and efflux activity in a fluconazole and voriconazole-resistant *C. tropicalis* strain, finding strong biofilm formation and efflux pump activity. Efflux gene expression was evaluated in the azole-resistant isolates, with the *MDR1* gene of *C. tropicalis* showing only a modest up-regulation (1.16-fold increase). The resistant strain's MIC of fluconazole was then tested in the presence of subinhibitory concentrations (¼×MIC) of haloperidol and pantoprazole. Pantoprazole reduced the MIC of fluconazole by 5-fold, while haloperidol achieved a 4-fold reduction. Both haloperidol-fluconazole and pantoprazole-fluconazole combinations effectively lowered the MIC of fluconazole in the resistant *C. tropicalis* isolates. Overall, the study highlighted that azole-resistant *Candida* isolates have the ability to form biofilms and overexpress efflux pumps, underscoring the potential for combination therapies to combat azole resistance in *Candida* spp.

5. *Candida parapsilosis*

Candida parapsilosis is commonly found as a commensal yeast in the skin microbiota, gastrointestinal tract, and mucous membranes. However, it can also cause infections in immunocompromised and immunosuppressed individuals (Trofa et al. 2008; Branco et al. 2023). The increasing incidence of *C. parapsilosis* infections can be attributed to several factors, including its strong affinity for intravascular devices and prosthetic materials, colonization of the gastrointestinal tract, prolonged hospitalizations, and specific treatment regimens. Additionally, *C. parapsilosis* possesses virulence attributes, such as biofilm formation and adhesion, particularly on intravenous catheters and parenteral nutrition devices (Dotis et al. 2012; Ziccardi et al. 2015; Zuo et al. 2021). Studies have highlighted *C. parapsilosis* as the second or third most common cause of candidemia, with its prevalence varying by patient population and geographic location. For example, a 2017 study encompassing 1,122 candidemia cases across nine surveillance sites in the USA found that *C. parapsilosis* was responsible for 14% of cases. In Latin America, *C. parapsilosis* accounted for 26.5% of 672 candidemia cases reported between 2008 and 2010. In European countries such as Greece, Portugal, Spain, Italy, and Australia, it contributed to 20-25% of *Candida* episodes, while in Japan and China, it accounted for 20-23% of bloodstream infections. In India and other Asian regions, *C. parapsilosis* was identified as the third most prevalent *Candida* species. In Brazil, a five-year surveillance study reported that *C. parapsilosis* was responsible for 14.4% of total candidemia cases (Tóth et al. 2019; Tsay et al. 2020; Kajihara et al. 2022; Muzaheed et al. 2022).

The growing challenge of antifungal resistance in *C. parapsilosis* is another significant concern for global healthcare systems. Prolonged or inappropriate use of antifungal therapies has led to the emergence of resistant strains, particularly against azoles and echinocandins (Pristov and Ghannoum, 2019; Zeng et al. 2021; Daneshnia et al. 2022; Daneshnia et al. 2023), underscoring the urgent need for novel treatment strategies. In a study by Yousif et al. (2019), the antiviral drug ribavirin was tested for its antifungal activity against 100 *Candida* species, including six *C. parapsilosis* strains. The results demonstrated ribavirin's effectiveness, with MIC values of ≤3.12 μM (≤0.75 μg/mL) against the *C. parapsilosis* isolates. Fungicidal activity was observed at concentrations ≥2-fold the MIC, highlighting ribavirin's potential as a therapeutic option for *C. parapsilosis* infections.

Eldesouky and colleagues (2021) explored the interaction between lopinavir, an HIV PI, and azole antifungal drugs against *C. parapsilosis*. Their study revealed that the combination of lopinavir and itraconazole was significantly more effective than the use of either drug alone in susceptibility tests. The results demonstrated a broad-spectrum synergistic interaction between lopinavir and itraconazole, with a calculated FICI of 0.2, indicating a strong synergistic effect. This finding suggests that lopinavir may enhance the antifungal activity of itraconazole, offering a promising approach for treating *C. parapsilosis* infections, particularly those that are resistant to conventional antifungal therapies.

Selective serotonin reuptake inhibitors (SSRIs), commonly prescribed as antidepressants and first-line treatments for premenstrual syndrome, were serendipitously found to possess antifungal properties. This discovery arose when three patients with chronic vulvovaginal candidiasis, treated with sertraline for premenstrual syndrome, reported the resolution of their candidiasis symptoms (Tekintas et al. 2020). To further investigate this phenomenon, the antifungal activity of serotonin, fluoxetine, and paroxetine was tested against three clinical strains and one ATCC reference strain of *C. parapsilosis*, both individually and in combination with fluconazole. While the SSRIs alone showed no inhibitory effect on the growth of these isolates, variation was observed in their susceptibility to fluconazole. However, when each SSRI was combined with fluconazole, the results indicated a potential synergistic effect against *C. parapsilosis*. These findings suggest that SSRIs may serve as promising adjuvants to enhance the efficacy of fluconazole, potentially offering a novel therapeutic approach for treating *C. parapsilosis* infections.

Silva et al. (2017) explored the antifungal effects of fluoxetine, paroxetine, and sertraline against fluconazole-resistant strains of *C. albicans*, *C. tropicalis*, *C. parapsilosis*, and *C. glabrata*, aiming to better understand their mechanisms of action. Among the species tested, *C. parapsilosis* exhibited the greatest sensitivity to sertraline, with a MIC range of 1-2 mg/mL. The study demonstrated that sertraline not only inhibited fungal growth, as indicated by MIC determinations, but also reduced biofilm formation, although direct biofilm analysis was not conducted for sertraline. Fluoxetine, at a concentration of 4 mg/mL, significantly reduced biofilm activity, as evidenced by MTT assays. The researchers also assessed membrane damage by exposing the strains to different concentrations (MIC, 2×MIC, 4×MIC), revealing dose-dependent damage to the plasma membrane in treated samples compared to controls. In a subsequent study, Oliveira et al. (2018) further evaluated the efficacy of fluoxetine and sertraline in disrupting mature *Candida* biofilms. Both SSRIs exhibited biofilm activity by reducing metabolic activity and promoting the disaggregation of biofilm biomass in a dose-dependent manner. Testing concentrations ranged from 0.0015 mg/mL to 3.9 mg/mL. After 48 hours of exposure to *C. parapsilosis*, fluoxetine at 0.12 mg/mL led to a 38% reduction in biomass and a 60% reduction in metabolic activity. Sertraline at the same concentration resulted in a 50% reduction in biomass and a 70% reduction in metabolic activity. These findings underscore the potential of SSRIs, as non-classical antifungal agents, to inhibit *Candida* biofilms, impacting both biomass and metabolic functions *in vitro*.

Another innovative approach to utilizing unconventional drugs involves formulating them into alternative treatment modalities. Mucoadhesive gels, in particular, have gained attention for their potential in managing vaginitis. While topical azoles are commonly used for this purpose, many women report side effects such as hypersensitivity and burning sensations. These drawbacks, along with issues like limited penetration into the epithelium and the requirement for application before bedtime, have prompted the search for more effective alternatives. One such alternative being explored is the mucoadhesive gel formulation of disulfiram, a dithiocarbamate traditionally used in the treatment of alcoholism. Disulfiram induces discomfort in patients who consume alcohol, serving as a deterrent for alcohol intake. Studies have shown that disulfiram exhibits efficacy against a variety of conditions, including scabies, lice, bacterial and fungal infections, and even cancer (Das et al. 2019; Li et al. 2020; Lajarin-Reinares et al. 2022a; Lajarin-Reinares et al. 2022b). *In vitro* and *in vivo* tests have confirmed the susceptibility of *C. albicans*, *C. glabrata*, and *C. parapsilosis* to disulfiram, with MIC values of 2 μg/mL, 4 μg/mL, and 8 μg/mL, respectively, demonstrating significant antifungal activity. This efficacy is attributed to disulfiram's chelating properties, which allow it to sequester copper and disrupt various metabolic pathways within cells. Additionally, disulfiram targets thiol groups in cysteine residues of several cellular proteins, further contributing to its antifungal effect. The development of a mucoadhesive gel formulation of disulfiram can enhance the treatment of vaginal candidiasis by improving drug delivery. This formulation would ensure better bioavailability, stronger adhesive capacity, and more convenient application, potentially increasing patient adherence and leading to better clinical outcomes. As such, the combination of disulfiram in a mucoadhesive gel represents a promising strategy for the topical treatment of vaginal candidiasis (Lajarin-Reinares et al. 2023).

6. *Candida glabrata*

Candida glabrata has become an increasingly significant pathogen in the context of invasive candidiasis, accounting for approximately 15-25% of clinical cases. In the USA and Northwestern Europe, it is the second most commonly isolated *Candida* species, while in Asia, it ranks as the third or fourth most prevalent invasive pathogen (Hassan et al., 2021; Kumar et al., 2019). The rising incidence of *C. glabrata* infections is linked to higher morbidity and mortality rates, particularly in immunocompromised individuals. This increased risk is partially attributed to the species' ability to produce virulence attributes such as biofilms and hydrolytic enzymes, as well as its growing resistance to commonly used antifungal agents (Frías-De-León et al., 2021; Hassan et al., 2021). *C. glabrata* is intrinsically resistant to fluconazole, and there are also reports of isolates exhibiting resistance to amphotericin B and echinocandins.

A computational analysis using molecular docking was performed to predict the binding energy between receptor and ligand molecules, assessing their potential to form a stable complex. The study focused on the calcium channel encoded by the *Cch1* gene in *C. glabrata* and evaluated several FDA-approved drugs for their binding affinity with *Cch1*. These drugs included antihypertensives (amlodipine besylate, carvedilol), anti-inflammatory agents (diclofenac), hypolipidemics (lovastatin, mevastatin, simvastatin), an anti-malarial (artesunate), a neuroprotective drug (etazolate), and an antifungal (butenafine). Among these, amlodipine besylate, a ligand for voltage-gated calcium channels (VGCC) in mammals, which shares homology with *Cch1*, exhibited the lowest binding energy with the protein. Low binding energy suggests a stronger interaction between the drug and the *Cch1* target, positioning amlodipine besylate as a promising candidate for further in vitro and *in vivo* testing against *C. glabrata* (Ahmad et al., 2016).

The potential synergistic effect between fluconazole and pitavastatin, a drug commonly used to treat hypercholesterolemia and dyslipidemia, was observed in *C. glabrata* strains, with a FICI ranging from 0.38 to 0.50, indicating strong synergy. In addition, the anti-biofilm activity of the combination was evaluated using sub-inhibitory concentrations of fluconazole (2 μg/mL) in combination with ½×MIC of pitavastatin (32 μg/mL). This combination resulted in a ~70% reduction in *C. glabrata* biofilm formation, whereas fluconazole or pitavastatin alone only caused a ~10% reduction. Furthermore, an *in vivo* study using the *Caenorhabditis elegans* infection model demonstrated that the combination of pitavastatin (at ½×MIC for *C.*

glabrata, 32 μg/mL) and fluconazole (at 2, 8, or 32 μg/mL) significantly reduced fungal loads by approximately 84-93%, compared to individual treatments, which showed similar efficacy to the untreated control (Eldesouky et al., 2020).

Antidepressant drugs, particularly the widely used SSRIs fluoxetine and sertraline, have been investigated for their effects on *Candida* spp. biofilms, including their ability to reduce biofilm biomass and metabolic viability of sessile cells. Concentrations ranging from 3.9 to 0.015 mg/mL were examined. Among the strains tested with fluoxetine, *C. glabrata* exhibited the highest $SMIC_{50}$ (MIC for sessile cells) of 0.24 mg/mL, indicating greater resistance of sessile cells to the drug. However, at the highest concentration (3.9 mg/mL), fluoxetine caused a significant 60% reduction in yeast metabolic activity. In terms of biomass, *C. glabrata* showed greater susceptibility to fluoxetine, with biofilm disintegration observed across all concentrations, achieving an 82% reduction at 3.9 mg/mL. In contrast, sertraline had a more pronounced impact on both metabolic activity and biomass. At 3.9 mg/mL, sertraline reduced *C. glabrata* biofilm biomass by 88% and metabolic activity by approximately 90%. Lower concentrations (ranging from 0.12 to 0.9 mg/mL) led to biomass inhibition of 60% to 90% and metabolic activity suppression of 70% to 80%, demonstrating the compound's potent anti-biofilm effects (Oliveira et al., 2018).

The Pharmakon 1600, a library of 1600 known drugs from both the USA and international pharmacopeias, was screened for activity against *C. glabrata*. The results were evaluated using the BIC_{50}, which represents the compound concentration that inhibits biofilm formation by 50%. Among the drugs tested, three compounds—drospirenone (used in birth control and menopausal hormone therapy), perhexiline (a prophylactic antianginal agent), and toremifene (used in treating advanced breast cancer in postmenopausal women)—were selected for further investigation. The BIC_{50} values were >400 μM for drospirenone, 60 μM for perhexiline, and 28 μM for toremifene. To enhance the efficacy of conventional antifungal treatments, the combination effects of these compounds with classical antifungals were assessed. The FICI values indicated a synergistic effect with caspofungin (FICI ≤0.5) across all concentrations. For amphotericin B, only drospirenone, at three concentrations (100, 50, and 25 μM), achieved a FICI <0.5. These results support further exploration of these drugs for use in coating medical devices or implants, potentially enhancing their effectiveness in conventional therapy by reducing biofilm formation (Delattin et al. 2014).

Rosato et al. (2016) explored the potential of non-steroidal anti-inflammatory drugs (NSAIDs) such as aspirin, diclofenac, and ibuprofen in combination with the classic antifungal anidulafungin to inhibit *C. glabrata* growth and biofilm formation. NSAIDs target the synthesis of prostaglandins, lipid molecules involved in mammalian metabolism, which are also produced by fungi like *Cryptococcus* and *Candida* through arachidonic acid. The study found that aspirin inhibited *C. glabrata* growth at 0.20 mM, diclofenac at 100 mM, and ibuprofen at 0.20 mM. Anidulafungin alone inhibited yeast growth at 2 µg/mL. When combined with anidulafungin, the NSAIDs exhibited a synergistic effect, with FICI values ranging from 0.1 to 0.3, indicating synergism (FICI <0.5). The biofilm inhibition was also enhanced by the drug combinations: anidulafungin plus aspirin, diclofenac, or ibuprofen inhibited 49.78%, 65.96%, and 62.33% of biofilm formation, respectively, compared to the individual agents, which inhibited 37.42%, 32.02%, and 57.23%, respectively. These findings highlight the potential of combining NSAIDs with antifungal agents to improve treatment efficacy against *C. glabrata* biofilms.

An ENZO FDA-approved drug library, consisting of 640 drugs, along with the oncology collection from the Finnish Institute of Molecular Medicine (FIMM), was screened to identify compounds that could target both primary oncological diseases and secondary *Candida* infections commonly affecting immunocompromised patients. Among the promising results, methionine maleate demonstrated an MIC of 3.57 µg/mL against *C. glabrata* (Stylianou et al., 2014). This study provides valuable insight into the development of antineoplastic therapies with dual antifungal activity, offering potential therapeutic options for patients at risk of invasive fungal infections while undergoing cancer treatment.

The potential synergy between ribavirin, an antiviral agent, and various azoles was investigated against azole-resistant *C. glabrata* isolates using the microdilution method. The combination of ribavirin with fluconazole demonstrated synergy in one *C. glabrata* strain (FICI ≤ 0.5), while no synergistic effect was observed in the other four strains. In combinations with itraconazole, synergistic activity was observed in two itraconazole-resistant *C. glabrata* strains. Additionally, when ribavirin was combined with posaconazole, only one strain exhibited a synergistic response. These findings suggest that ribavirin can enhance the efficacy of fluconazole, itraconazole, and posaconazole against certain *C. glabrata* strains, highlighting its potential as a therapeutic adjunct in combating azole resistance (Yousif et al., 2019).

The study explored the potential synergistic effects of drugs from the Prestwick Library (PL) of FDA/EMA-approved compounds against azole-resistant *C. glabrata* clinical isolates. This evaluation was conducted both in the absence and presence of subinhibitory concentrations of a small panel of non-antifungal drugs, namely paromomycin, primaquine, and ibuprofen. Paromomycin, an aminoglycoside antibiotic, is used for treating acute and chronic intestinal amebiasis; primaquine is used for treating and preventing malaria and treating *Pneumocystis* pneumonia; and ibuprofen, a nonsteroidal anti-inflammatory drug, is commonly used to relieve pain, fever, and inflammation. The combination of paromomycin with β-escin, and primaquine with celecoxib, exhibited notable synergistic activity against *C. glabrata*, with FICI values of 0.5 and 0.13–0.16, respectively. β-escin is a natural mixture of triterpenoid saponins isolated from horse chestnut (*Aesculus hippocastanum*) seeds, known for its vasoprotective, anti-inflammatory, anti-edematous, and antinociceptive properties. Celecoxib is a nonsteroidal anti-inflammatory drug that specifically inhibits cyclooxygenase 2 (COX-2). Additionally, the study highlighted that the primaquine + celecoxib combination significantly reduced the MIC for both drugs—by 8-fold for primaquine and greater than 16-fold for celecoxib. These findings suggest that these drug combinations may offer therapeutic advantages in clinical settings, particularly for managing *C. glabrata* infections (Vallières et al., 2020).

7. *Candida krusei*

Candida krusei infections account for approximately 2.8% of all *Candida* infections worldwide (Pfaller et al., 2019; Jamiu et al., 2021), yet the species is notably prevalent in patients with hematological malignancies (Conen et al., 2008; Jansen et al., 2022; Otto and Babady, 2023). In fact, *C. krusei* infections are associated with significantly higher mortality rates (44% - 67%) compared to other common *Candida* species (WHO, 2022). One of the major challenges in managing these infections is the intrinsic resistance of *C. krusei* to fluconazole (Khalifa et al., 2022) and its reduced susceptibility to amphotericin B and flucytosine (Pfaller et al., 2008), complicating the selection of effective antifungal therapies. Auranofin, a thiol-containing compound typically used to treat rheumatoid arthritis, has demonstrated antifungal activity against *C. krusei*, with a MIC value of ≤1 μg/mL (Wiederhold et al., 2017). Another promising repurposed drug is ebselen, an organoselenium antioxidant and anti-inflammatory agent, which has been under investigation for various therapeutic applications, including in cancer (Bijian et al., 2012), Zika virus (Simanjuntak et al., 2018), and COVID-19 (Sies and Parnham, 2020). Ngo et al. (2016) reported significant antifungal activity of ebselen against *Candida* species, including *C. krusei*, with MIC values of ≤1.56 µg/mL. Both auranofin and ebselen share a common mechanism of action: inhibition of thioredoxin reductase (TrxR), an essential enzyme in the antioxidant defense system of fungi. This inhibition leads to increased intracellular oxidative stress, ultimately resulting in fungal cell death (May et al., 2018). These findings suggest that auranofin and ebselen could be valuable candidates for the treatment of *C. krusei* infections, particularly given the challenges posed by the species' resistance to standard antifungal agents.

Geldanamycin, an antitumor agent, is a well-known inhibitor of heat shock protein 90 (Hsp90), a chaperone protein critical for regulating drug resistance, cell signaling, virulence, and apoptosis (Dai et al., 2012; Hossain et al., 2020; Robbins and Cowen, 2023). Due to its ability to modulate these cellular processes, geldanamycin has been explored as a potential novel antifungal therapy. Zhang et al. (2013) demonstrated its antifungal efficacy against various *Candida* species, including both *C. albicans* and non-*albicans Candida* species, with *C. krusei* isolates exhibiting MIC values of 12.8 µg/mL. Geldanamycin's mechanism of action in fungi involves the inhibition of Hsp90, which disrupts critical cellular processes such as protein folding, cellular signaling, and stress responses. This interference impairs the ability of *Candida* species to adapt to hostile environments, thereby increasing their vulnerability to antifungal treatments and enhancing the host's immune response (Hossain et al., 2020). These findings suggest that geldanamycin, by targeting a crucial cellular chaperone, could be a promising adjunct to current antifungal therapies, particularly for drug-resistant *Candida* infections.

Metergoline, a serotonin antagonist, has been explored for its potential antifungal activity against various fungal species, including *C. krusei*. A study by Kang et al. (2010) demonstrated its antifungal effectiveness against clinical *C. krusei* isolates resistant to fluconazole. The researchers also noted that metergoline exhibited significant synergism when combined with amphotericin B at subinhibitory concentrations, with FICI values ranging from 0.375 to 0.5. Additionally, metergoline was found to inhibit the production of extracellular phospholipases in a dose-dependent manner, suggesting a potential mechanism of action that involves disrupting fungal virulence factors. However, the precise mechanism by which metergoline exerts its antifungal effects against *C. krusei* remains unclear and warrants further investigation.

8. *Cryptococcus* spp.

Species belonging to the *Cryptococcus* genus are commonly found in the environment, particularly in soil and bird droppings. These fungi are responsible for causing cryptococcosis and cryptococcal meningitis, which affect approximately 194,000 people globally each year. While the mortality rate for treated patients is around 60%, untreated cases often result in near 100% fatality (Denning, 2024). Cryptococcosis is a significant cause of death, accounting for roughly 20% of all AIDS-related fatalities (Zhao et al., 2023). *C. neoformans* primarily affects immunocompromised individuals, while *C. gattii* is more commonly associated with infections in immunocompetent individuals (Rajasingham et al., 2017; Zaragoza, 2019; Firacative et al., 2021; Baker et al., 2022; Zhao et al., 2023).

Currently, the treatment of *Cryptococcus* infections relies on three main antifungal agents with established *in vivo* efficacy: amphotericin B (administered intravenously), flucytosine (oral), and fluconazole (oral) (Ngan et al., 2022). The WHO recommends initial therapy with amphotericin B alone or in combination with flucytosine, followed by maintenance treatment with either fluconazole or flucytosine (Maziarz and Perfect, 2016; WHO, 2022; Huang et al., 2023; Sousa et al., 2023). However, these treatments are associated with significant toxicity, which can lead to additional complications during prolonged use. Moreover, *Cryptococcus* species have developed various mechanisms of resistance to both flucytosine and the azole class of antifungals. Furthermore, the *Cryptococcus* genus is intrinsically resistant to echinocandins (Vu et al., 2018; Papon and Goldman, 2021; Ngan et al., 2022; Oliveira et al., 2022).

Quetiapine and olanzapine, widely prescribed antipsychotic medications, are commonly used to treat psychiatric conditions such as bipolar disorder, schizophrenia, and manic episodes (Zubiaur et al., 2021; Kishi et al., 2022; Otis, 2022; Maan et al., 2023). In patients with cryptococcal meningitis, the increased intracranial pressure often leads to severe headaches, which can trigger psychotic episodes (Seelig et al., 2019). Medications that cross the blood-brain barrier and can address both psychosis and cryptococcal meningitis present a promising therapeutic option (Ogundeji et al., 2017). Both quetiapine and olanzapine have shown antifungal activity against *C.*

neoformans and *C. gattii* at concentrations lower than the recommended maximum blood dosage, with quetiapine demonstrating greater efficacy (Ogundeji et al., 2017; Skov et al., 2015; Lu et al., 2018). These antipsychotics reduce the overall diameter of *Cryptococcus* cells and cause rupture of their capsular structure, a key virulence factor (Ogundeji et al., 2017). The *Cryptococcus* capsule enables the pathogen to evade phagocytosis, induce macrophage apoptosis, resist degradation in the phagolysosome, and impede complement system activity (Yang et al., 2022). Quetiapine and olanzapine exert their antifungal effects by modulating the molecular flow at the plasma membrane of *Cryptococcus* cells (Ogundeji et al., 2017). Additionally, both drugs have immunomodulatory properties, promoting the production of pro-inflammatory cytokines such as IL-6 and IFN-γ by macrophages, which enhances their phagocytic activity (Ogundeji et al., 2017). Moreover, these antipsychotics have demonstrated synergistic effects when used in combination with traditional antifungal therapies like amphotericin B or fluconazole (Ogundeji et al., 2017).

Benzimidazole derivatives exhibit a broad spectrum of biological activities and have diverse clinical applications (Singh et al., 2012). In particular, mebendazole, a treatment for hydatid disease (Aydin et al., 2022), and fenbendazole, an anthelmintic agent widely approved for use in domestic animals and the food industry (European Medicines Agency), have shown promise as antifungal agents against *Cryptococcus* species (Joffe et al., 2017; de Oliveira et al., 2020). Both mebendazole and fenbendazole display low toxicity in macrophages while demonstrating potent activity against *C. neoformans* and *C. gattii*, with fenbendazole proving to be more effective. Treatment with these compounds induces significant changes in the cellular ultrastructure, including shortened and reduced capsular fibers, loss of the spherical shape, and the appearance of intracellular furrows (Joffe et al., 2017; de Oliveira et al., 2020). Additionally, mebendazole and fenbendazole reduce the intracellular proliferation of *Cryptococcus* in infected macrophages (Joffe et al., 2017), and fenbendazole inhibits vomocytosis, a process critical to *Cryptococcus* pathogenesis (de Oliveira et al., 2020). The mechanism of action of mebendazole involves interactions with Scramblase proteins and the nucleolar protein Nop16, leading to disruption of plasma membrane integrity (Joffe et al., 2017). On the other hand, fenbendazole inhibits the polymerization of β-tubulin, resulting in altered microtubule profiles in *Cryptococcus* cells (de Oliveira et al., 2020). It is well-known that benzimidazole derivatives undergo rapid hepatic metabolism, producing sulfone derivatives (Gottschall et al., 1990; Varlamova et al., 2021), which, *in vitro*, show no anti-*Cryptococcus* activity (de Oliveira et al., 2020). Furthermore, oral administration of fenbendazole or mebendazole showed no efficacy in murine models of cryptococcosis (de Oliveira et al., 2020). This limitation was overcome by intranasal administration of fenbendazole, which led to a remarkable 100% survival rate in infected mice after 27 days of treatment (de Oliveira et al., 2020). In combination with amphotericin B, fenbendazole exhibited synergistic effects, while mebendazole demonstrated an additive effect. However, translating these promising results to human clinical settings requires further investigation (Joffe et al., 2017; de Oliveira et al., 2020).

Tamoxifen, a widely known adjuvant in breast cancer treatment, functions as a selective estrogen receptor modulator (SERM). By competitively binding to estrogen receptors (ERs), it alters estrogenic responses in the body, inhibiting the stimulatory effects of estradiol (E2) in estrogen-sensitive tissues, particularly in the breast (Wibowo et al., 2011; Komm and Mirkin, 2014; Wibowo and Wassersug, 2014; Wibowo et al., 2016; Hai et al., 2019). In addition to its established role in cancer therapy, tamoxifen has shown promising *in vitro* antifungal activity against clinical isolates of *Cryptococcus* species, including *C. neoformans* and *C. gattii* (Hai et al., 2019).

Tamoxifen was found to exhibit synergistic effects when combined with amphotericin B, and an additive interaction was noted when paired with fluconazole (Hai et al., 2019). However, clinical findings from Ngan et al. (2021) did not entirely align with these *in vitro* observations. In their study, interactions between tamoxifen and both amphotericin B and fluconazole were less pronounced compared to the synergism reported by Hai et al. (2019). Moreover, the clinical trial revealed that tamoxifen treatment resulted in a 34% mortality rate among patients, with the drug failing to significantly improve yeast clearance from cerebrospinal fluid in patients with cryptococcal meningitis, whether they were HIV-infected or not (Ngan et al., 2021). These findings suggest that, while tamoxifen demonstrates potential antifungal properties *in vitro*, its clinical efficacy, particularly in the treatment of cryptococcal meningitis, remains limited. Further research is needed to explore its potential in combination therapies or as part of a broader treatment strategy.

Sertraline has been shown to exhibit anti-cryptococcal activity in several studies. It has demonstrated efficacy in eliminating both viable and quiescent *C. neoformans* cells, significantly reducing the fungal burden in the brains of mice with cryptococcosis (Zhai et al., 2012). *In vitro* research suggests that sertraline may exert its antifungal effects through modulation of lipid metabolism, as it induces the formation of intracellular lipid droplets within *C. neoformans* cells (Breuer et al., 2022). These findings highlight sertraline's potential as a therapeutic adjunct in treating cryptococcosis. Despite promising *in vitro* and *in vivo* results, clinical studies of sertraline's efficacy in conjunction with standard antifungal therapy have yielded mixed outcomes. In a clinical trial, the addition of sertraline to standard induction therapy significantly enhanced fungal clearance from cerebrospinal fluid by 34% to 48%, compared to patients receiving only standard treatment (Alhadab et al., 2019). However, sertraline did not reduce mortality in patients with HIV-associated cryptococcal meningitis. Furthermore, adverse events were reported in both the sertraline and placebo groups, with an incidence of 31% and 33%, respectively, over an 18-week period (Rhein et al., 2019). These clinical outcomes suggest that while sertraline may improve fungal clearance, its impact on patient survival and safety remains uncertain. Therefore, while sertraline shows promise in preclinical models, further investigation is needed to better understand its mechanisms of action, safety profile, and potential role in combination therapies for cryptococcosis. In contrast, rapamycin, an immunosuppressant drug approved for preventing organ transplant rejection, has also demonstrated significant anti-cryptococcal effects. At a concentration of 2.5 μM (2.285 μg/mL), rapamycin inhibited *C. neoformans* cell viability by 53%, increased cell size by 19.93%, and reduced the polysaccharide capsule by 72.14%. It also elevated lipid body formation by 20%, chitin content by 127%, and secretion of aspartic peptidase by 32.55%, while decreasing urease activity (Andrade et al., 2023). Rapamycin exerts its effects through FKBP12-dependent inhibition of Tor1, disrupting key cellular processes and impairing *C. neoformans* virulence in animal models (Cruz et al., 2001; Cruz et al., 1999). These findings support the potential of rapamycin as an adjunctive therapy for cryptococcosis, with effects on both cellular structure and virulence factors.

Freitas et al. (2023) explored the potential of repurposing the antimalarial drugs amodiaquine (AQ) and halofantrine (HAL) as treatments for *C. neoformans* and *C. gattii*, highlighting their intriguing potential against cryptococcosis. Given the established mechanisms of action of these antimalarials in inhibiting *Plasmodium* heme polymerase and inducing oxidative stress, their application to *Cryptococcus* infections is promising. This approach is supported by the fact that *Cryptococcus* species are particularly susceptible to ROS and heavily rely on iron metabolism, which may be disrupted by these antimalarials. Amodiaquine and halofantrine demonstrated significant inhibitory effects on the growth of *C. neoformans* and *C. gattii*, both when used alone (with MICs $\geq$ 512 µg/mL) and in combination with amphotericin B (MICs ranging from 0.24 µg/mL to 0.39 µg/mL). The drugs exhibited a range of effects on the fungal cells, including disruption of ergosterol synthesis, which impacts the fungal plasma membrane, as well as a reduction in melanin and capsular production. Additionally, these treatments induced oxidative and nitrosative stress in the fungal cells, contributing to their antifungal activity. *In vivo* experiments involving combinations of halofantrine and amphotericin B, as well as amodiaquine and amphotericin B, demonstrated a reduction in lethality in mice infected with *Cryptococcus* spp. These results suggest that the combination of antimalarial drugs with conventional antifungal agents, such as amphotericin B, could represent a promising therapeutic strategy for cryptococcosis, warranting further investigation for clinical application.

9. *Aspergillus* spp.

The rise in *Aspergillus*-related infections, particularly those caused by *A. fumigatus*, parallels the increasing number of immunosuppressed and immunocompromised individuals. These patients, whose immune systems are compromised, are more vulnerable to severe infections caused by filamentous fungi like *A. fumigatus*. Conditions such as cystic fibrosis further increase susceptibility to these infections (Latgé and Chamilos, 2019; Kanaujia et al., 2023). *Aspergillus* infections are primarily transmitted through the inhalation of airborne conidial, which are ubiquitous in the environment. Environmental factors influence conidia concentrations, which can vary significantly, ranging from 1 to 10^8 conidia per mm^3. Once inhaled, these conidia can lead to a spectrum of diseases, ranging from non-invasive allergic reactions to chronic and invasive pulmonary infections. Understanding the immunological status of patients is critical for predicting the severity and progression of *Aspergillus*-related diseases. This knowledge is essential for developing effective prevention strategies and for tailoring treatment approaches based on an individual's immune function. As the prevalence of immunosuppression continues to increase, managing *Aspergillus* infections becomes ever more crucial in clinical practice (Latgé and Chamilos, 2019; Kanaujia et al., 2023).

The treatment of aspergillosis typically involves the use of triazoles, polyenes, or echinocandins, with azoles being the most commonly prescribed antifungal agents (Kanaujia et al., 2023). However, their clinical effectiveness can be limited by significant drug-drug interactions, side effects, and the emergence of resistant strains (Kanaujia et al., 2023). The increasing prevalence of azole-resistant *Aspergillus* strains is strongly linked to the widespread use of azole fungicides in agriculture, given the ubiquitous presence of *Aspergillus* species in the environment (Berger et al., 2017). Mortality rates from invasive pulmonary aspergillosis remain high, exceeding 50-60% in severely immunosuppressed patients (Osherov and Kontoyiannis, 2016). This unacceptably high mortality rate is largely attributed to delayed diagnoses and the limited efficacy of current antifungal therapies, underscoring the urgent need for improved diagnostic methods and novel treatment strategies (Osherov and Kontoyiannis, 2016).

Auranofin, a gold(I) complex consisting of a gold center coordinated to a thiosugar and triethylphosphine, was originally introduced in 1985 for the treatment of rheumatoid arthritis. It is known to inhibit multiple inflammatory pathways (Wiederhold et al., 2017). Recent studies have highlighted its antifungal potential, particularly against *Aspergillus* species, with *A. fumigatus* exhibiting fungicidal effects at MICs ranging from 2-16 μg/mL (Wiederhold et al., 2017; Chen et al., 2023). Auranofin also demonstrates significant antibiofilm activity against *A. fumigatus* at 6 μg/mL, achieved by downregulating key transcriptional regulators, SomA and MedA, which are essential for biofilm formation and development (Chen et al., 2023). Notably, auranofin has shown synergistic effects when combined with itraconazole or amphotericin B, particularly against resistant *A. fumigatus* strains (Chen et al., 2023). Its antimicrobial activity is primarily mediated through the induction of oxidative stress, and in *Aspergillus* species, auranofin also downregulates the expression of thioredoxin reductase (AfTrxR), a critical component of the fungal antioxidant defense system (Wiederhold et al., 2017; Chen et al., 2023). These mechanisms suggest that auranofin holds promise as a novel adjunct in the treatment of *Aspergillus* infections, particularly in the context of drug-resistant strains.

Holbrook et al. (2017) synthesized a series of bromperidol derivatives with varying R-group sizes, which demonstrated synergistic activity when combined with posaconazole against *A. terreus*, as evidenced by the checkerboard assay. The phenothiazine compounds chlorpromazine and trifluoperazine are known to inhibit calmodulin, leading to membrane modifications, alterations in cyclic nucleotide metabolism, and DNA intercalation (Vitale et al., 2007). In tests against *A. fumigatus*, chlorpromazine exhibited a geometric mean MIC of 42.2 μg/mL, while trifluoperazine showed a geometric mean of 34.3 μg/mL (Vitale et al., 2007). Disulfiram, a compound historically used in alcohol aversion therapy since 1949, inhibits ATP-binding cassette (ABC) drug transporters, indicating its potential as an antifungal agent. Disulfiram has demonstrated fungicidal activity against a range of fungal species, including *A. fumigatus*, with MIC_{50} and MIC_{90} values of 2 μg/mL and 8 μg/mL, respectively (Khan et al., 2007). In combination with the polyene natamycin, disulfiram (at 20 μM) proved effective in an *in vivo* mouse model of *A. fumigatus* keratitis, reducing IL-1β secretion, attenuating macrophage and neutrophil recruitment, and lowering myeloperoxidase and nitric oxide synthase levels, which helped preserve corneal transparency (Yan et al., 2022). Additionally, the antidepressant sertraline has shown promising anti-*Aspergillus* activity. In *Galleria mellonella* and murine models of *A. fumigatus* infection, sertraline treatment (15 mg/kg) improved larval survival and health index scores. At a dose of 10 mg/kg, sertraline also reduced pulmonary fungal burden to levels comparable to control treatments of 3 mg/kg amphotericin B and 10 mg/kg voriconazole in the murine model (Treviño-Rangel et al., 2019). These findings suggest that repurposing existing drugs such as disulfiram and sertraline, in combination with established antifungals, may offer new therapeutic strategies against *Aspergillus* infections, particularly in the context of drug resistance.

The antimycobacterial drug clofazimine, at a concentration of 16 μg/mL, has been shown to enhance the activity of caspofungin (0.125 μg/mL) and posaconazole (0.0625 μg/mL) against *A. fumigatus*, resulting in reduced hyphal growth and significant alterations in hyphal morphology (Robbins et al., 2016). Minocycline, an antibiotic, also exhibits synergistic effects when combined with azoles, significantly increasing larval survival ($P < 0.001$) in an *in vivo* model of *A. fumigatus* infection in *Galleria mellonella* (Gao et al., 2020). In a screening using two compound libraries—the Pathogen Box and the National Institutes of Health Clinical Collection—Reis et al. (2021) identified miltefosine as a promising compound with fungicidal activity against *A. fumigatus* at 10 μM. Miltefosine exerts its antifungal effects by interfering with sphingolipid biosynthesis in *A. fumigatus* (Reis et al., 2021). Additionally, other compounds showing potential anti-*Aspergillus* activity include econazole nitrate (antifungal), fluvastatin (used to treat hypercholesterolemia), mesoridazine (neuroleptic), cisapride (gastroprokinetic agent), oxiconazole nitrate (antifungal), indinavir sulfate (antiretroviral protease inhibitor), enalaprilat (for hypertension), vincristine sulfate (chemotherapy agent), and iodoquinol (antiprotozoal) (Reis et al., 2021). These findings highlight the potential for repurposing existing drugs from diverse therapeutic classes to combat *Aspergillus* infections, particularly in light of emerging drug resistance. Such repurposing could offer novel avenues for treatment, especially when combined with traditional antifungal therapies.

The fungal histone deacetylase (HDAC)-heat shock protein 90 (Hsp90)-calcineurin axis plays a vital role in the fungal stress response and cell wall repair mechanisms during treatment with conventional antifungal agents. Inhibition of this axis could potentially enhance the efficacy of antifungal drugs by disrupting these critical protective pathways (Osherov and Kontoyiannis, 2016). Immunomodulators such as tacrolimus (FK506) and cyclosporine, which inhibit the calcineurin pathway, have been shown to boost the activity of antifungals like caspofungin and isavuconazole against various *Aspergillus* species, including *A. fumigatus*, *A. flavus*, *A. nidulans*, *A. niger*, and *A. terreus* (Kontoyiannis et al., 2002; Schwarz and Dannaoui, 2020). Furthermore, combining tacrolimus with amphotericin B or voriconazole has shown synergistic inhibitory effects against *Aspergillus* biofilms. This combination may also shift the activity of azoles from fungistatic to fungicidal, enhancing their antifungal potency (Gao and Sun, 2015). The antitumor Hsp90 inhibitor geldanamycin, at a concentration of 4 μg/mL, when combined with caspofungin (1 μg/mL) or tacrolimus (0.1 μg/mL), demonstrates fungicidal activity against azole-resistant strains of *A. fumigatus* (Lamoth et al., 2013). These findings suggest that targeting the HDAC-Hsp90-calcineurin axis, either through the use of immunomodulators or Hsp90 inhibitors, holds promise for enhancing the effectiveness of current antifungal therapies, particularly in overcoming resistance mechanisms in *Aspergillus* species.

10. *Scedosporium/Lomentospora* spp.

The genera *Scedosporium* and *Lomentospora* consist of filamentous fungi that are widely distributed in the environment, often in areas significantly influenced by human activities, such as agricultural settings, urban soils, polluted ponds, and sediments (Mello et al., 2019). These fungi are classified as opportunistic human pathogens, primarily affecting individuals who are immunocompromised or immunosuppressed. However, they can also infect immunocompetent patients, typically causing localized diseases like mycotic mycetoma (Ramirez-Garcia et al., 2017; Mello et al., 2019). Clinical manifestations of *Scedosporium* and *Lomentospora* infections can range from pulmonary colonization to invasive systemic disease, with more severe outcomes in patients with central nervous system (CNS) involvement, where mortality rates can exceed 80% (Ramirez-Garcia et al., 2017). These alarming mortality rates highlight the critical need for early detection and effective treatment strategies, particularly for vulnerable populations at heightened risk of severe infections.

The treatment of scedosporiosis presents significant challenges due to the high resistance of *Scedosporium* species to current antifungal therapies, including amphotericin B, 5-fluorocytosine, and first-generation triazoles (Ramirez-Garcia et al., 2017). This issue is particularly pronounced in *Lomentospora prolificans* (formerly *S. prolificans*), which exhibits pan-antifungal resistance, further complicating treatment options. The European Guidelines recommend voriconazole and surgical debridement, where feasible, as first-line treatments for scedosporiosis. However, even with these interventions, mortality rates remain alarmingly high, exceeding 65%, and reaching 100% in cases of disseminated infection (Ramirez-Garcia et al., 2017). In addition, infections caused by *Scedosporium* and *Lomentospora* are often characterized by biofilm formation, which presents another layer of difficulty in treatment. These biofilms are dense, robust structures composed of mycelial cells embedded in a polymeric extracellular matrix (Mello et al., 2019). *In vitro* studies have demonstrated that the resistance of *Scedosporium* and *Lomentospora* biofilms is due to a combination of mechanisms, including enhanced antioxidant responses, increased efflux pump activity, high cellular density, and the presence of an extracellular matrix that limits the penetration of antifungal agents to the fungal cells (Mello et al., 2022). This complex resistance profile underscores the need for novel therapeutic strategies and a better understanding of the biofilm-related mechanisms involved in *Scedosporium* and *Lomentospora* infections.

A screening of compounds from the Pandemic Response Box® library against *Scedosporium* species identified three promising candidates—alexidine, amorolfine, and olorofim—that inhibit fungal growth at concentrations ranging from 0.15 to 60 μM. Alexidine, a bis-biguanide antimicrobial, amorolfine, a morpholine derivative, and olorofim, an orotomide, also demonstrated the ability to inhibit biofilm formation in *S. boydii*, *S. dehoogii*, *S. apiospermum*, and *L. prolificans* (Rollin-Pinheiro et al., 2023). Both alexidine and amorolfine exhibited synergistic activity with voriconazole and caspofungin against *S. aurantiacum*, and all three compounds induced oxidative stress and morphological changes in *Scedosporium* species. A similar investigation using the Pathogen Box Library revealed that auranofin and iodoquinol also exhibit antifungal activity against *Scedosporium* and *Lomentospora* species. Auranofin, primarily used in rheumatoid arthritis treatment, displayed a MIC of 5 μM against *S.*

aurantiacum, *S. apiospermum*, *S. boydii*, *S. dehoogii*, and *L. prolificans*. Iodoquinol, commonly used to treat amoebiasis, inhibited the growth of these fungi at concentrations ranging from 0.625 to 5 μM (Rollin-Pinheiro et al., 2021). Both compounds significantly impaired biofilm formation, inhibited biomass production, and caused morphological alterations in the fungi. Additionally, iodoquinol showed synergistic activity with caspofungin against *S. aurantiacum* (Rollin-Pinheiro et al., 2021), and another study reported the additive effects of auranofin and voriconazole against specific isolates of *Scedosporium* and *Lomentospora* species (Yaakoub et al., 2021). Honokiol, a biphenolic compound derived from *Magnolia officinalis*, traditionally used in oriental medicine as an anxiolytic agent, has also been shown to exhibit antifungal activity against *Scedosporium* and *Lomentospora* species. The MIC of honokiol ranges from 4 to 16 μg/mL, and its antifungal action is attributed to mitochondrial dysfunction, oxidative stress, and induction of apoptosis (Yaakoub et al., 2021). Honokiol upregulates 17 genes related to oxidoreduction functions in these fungi, including those encoding superoxide dismutase, catalases, and components of the glutathione and thioredoxin systems. Given these mechanisms, a combination of honokiol with auranofin is suggested, as honokiol induces mitochondrial dysfunction and oxidative stress, while auranofin inhibits thioredoxin reductases. This combination has demonstrated additive effects, offering a potential synergistic strategy for treating *Scedosporium* and *Lomentospora* infections (Yaakoub et al., 2021). These findings highlight the potential for novel combination therapies targeting oxidative stress and biofilm formation to enhance antifungal efficacy against resistant *Scedosporium* and *Lomentospora* species.

Miltefosine also exhibits promising activity against *Scedosporium* and *Lomentospora* species. It effectively inhibits planktonic growth of *S. aurantiacum*, *S. boydii*, *S. apiospermum*, *S. dehoogii*, and *L. prolificans* at concentrations ranging from 2 to 4 μg/mL. Additionally, miltefosine reduces biofilm biomass by 50% to 60% at concentrations of 16 and 32 μg/mL, respectively. The antifungal effects of miltefosine are attributed to its induction of oxidative stress, mitochondrial dysfunction, alterations in membrane lipid rafts, and changes in the cell morphology of *S. aurantiacum* (Rollin-Pinheiro et al., 2021). Another compound evaluated for activity against *Scedosporium* and *Lomentospora* is the antidepressant sertraline. Sertraline demonstrates antifungal activity against *L. prolificans*, *S. boydii*, and *S. apiospermum* at concentrations ranging from 8 to 32 μg/mL. Furthermore, sertraline shows synergistic effects when combined with amphotericin B against *S. apiospermum* and *L. prolificans* (Villanueva-

Lozano et al., 2019). The antifungal mechanism of sertraline is thought to involve destabilization of the plasma membrane, interference with vesicle transport, and inhibition of protein synthesis, all of which contribute to its antifungal activity (Zhai et al., 2012; Villanueva-Lozano et al., 2019). Together, these findings suggest that miltefosine and sertraline, through distinct mechanisms, may offer novel therapeutic options or synergistic combinations for combating infections caused by *Scedosporium* and *Lomentospora* species.

Conclusion

In conclusion, the limited availability of effective antifungal agents and the increasing prevalence of drug resistance in fungal infections underscore the urgent need for innovative treatment strategies. As conventional antifungal therapies face growing challenges in terms of efficacy, drug repositioning has emerged as a promising therapeutic approach. Repurposing existing medications for novel uses not only expedites research but also conserves valuable time and resources. This strategy provides a unique opportunity to identify compounds that may exhibit antifungal activity against resistant strains, including those from high-priority pathogens listed by the WHO, such as *Candida* spp., *Cryptococcus* spp., *Aspergillus* spp., and *Scedosporium/Lomentospora* spp.. The studies discussed in this chapter collectively emphasize the substantial potential of drug repositioning in uncovering new therapeutic options, offering hope for more effective treatments for resistant fungal infections.

Disclaimer

No conflicts of interest to declare.

References

Ahmad, H., Gupta, P., Tufchi, N., Pant, K., Kumar, N. (2016). *In silico* modeling and docking of cch1 protein of *Candida glabrata* with FDA-approved drugs: A drug repurposing approach. *Asian Journal of Pharmaceutical and Clinical Research, 9*, 113-118.

Ahmad, S., & Alfouzan, W. (2021). *Candida auris*: epidemiology, diagnosis, pathogenesis, antifungal susceptibility, and infection control measures to combat the spread of infections in healthcare facilities. *Microorganisms*, *9*(4), 807. https://doi.org/10.3390/microorganisms9040807.

Alanís-Ríos, S. A., González, G. M., Andrade, A., Becerril-García, M. A., Bonifaz, A., Robledo-Leal, E. R., Montoya, A. M., & Treviño-Rangel, R. J. (2022). Evaluation of the synergistic antifungal activity of micafungin and voriconazole plus sertraline against *Candida auris*. *Brazilian Journal of Microbiology*, *53*(4), 2003–2008. https://doi.org/10.1007/s42770-022-00817-y.

Alhadab, A. A., Rhein, J., Tugume, L., Musubire, A., Williams, D. A., Abassi, M., Nicol, M. R., Meya, D. B., Boulware, D. R., Brundage, R. C., & ASTRO-CM Study Team (2019). Pharmacokinetics-pharmacodynamics of sertraline as an antifungal in HIV-infected Ugandans with cryptococcal meningitis. *Journal of Pharmacokinetics and Pharmacodynamics*, *46*(6), 565–576. https://doi.org/10.1007/s10928-019-09657-0.

Al-Majed, A. R., Assiri, E., Khalil, N. Y., & Abdel-Aziz, H. A. (2015). Losartan: comprehensive profile. *Profiles of Drug Substances, Excipients, and Related Methodology*, *40*, 159–194. https://doi.org/10.1016/bs.podrm.2015.02.003.

Anagha, K., Shihabudheen, P., & Uvais, N. A. (2021). Side effect profiles of selective serotonin reuptake inhibitors: a cross-sectional study in a naturalistic setting. *The Primary Care Companion for CNS Disorders*, *23*(4), 20m02747. https://doi.org/10.4088/PCC.20m02747.

Andrade, I. B., Alves, V., Pereira, L., Miranda, B., Corrêa-Junior, D., Galdino Figueiredo-Carvalho, M. H., Santos, M. V., Almeida-Paes, R., & Frases, S. (2023). Effect of rapamycin on *Cryptococcus neoformans*: cellular organization, biophysics and virulence factors. *Future Microbiology*, *18*, 1061–1075. https://doi.org/10.2217/fmb-2023-0097.

Arendrup, M. C., & Patterson, T. F. (2017). Multidrug-resistant *Candida*: epidemiology, molecular mechanisms, and treatment. *The Journal of Infectious Diseases*, *216*(suppl_3), S445–S451. https://doi.org/10.1093/infdis/jix131.

Ashburn, T. T., & Thor, K. B. (2004). Drug repositioning: identifying and developing new uses for existing drugs. *Nature Reviews: Drug Discovery*, *3*(8), 673–683. https://doi.org/10.1038/nrd1468.

Aydin, Y., Ulas, A. B., Ahmed, A. G., & Eroglu, A. (2022). Pulmonary hydatid cyst in children and adults: diagnosis and management. *The Eurasian Journal of Medicine*, *54*(Suppl1), 133–140. https://doi.org/10.5152/eurasianjmed.2022.22289.

Baker, R. P., Chrissian, C., Stark, R. E., & Casadevall, A. (2022). *Cryptococcus neoformans* melanization incorporates multiple catecholamines to produce polytypic melanin. *The Journal of Biological Chemistry*, *298*(1), 101519. https://doi.org/10.1016/j.jbc.2021.101519.

Barnes P. J. (2013). Theophylline. *American Journal of Respiratory and Critical Care Medicine*, *188*(8), 901–906. https://doi.org/10.1164/rccm.201302-0388PP.

Barratt, M. J., Frail, D. E. (2012). The business case and current strategies to repurpose shelved candidates and marketed drugs. *In* Drug repositioning (bringing new life to shelved assets and existing drugs) drug repositioning, 7–32. https://doi.org/10.1002/9781118274408.ch1.

Barreto, T. L., Rossato, L., de Freitas, A. L. D., Meis, J. F., Lopes, L. B., Colombo, A. L., & Ishida, K. (2020). Miltefosine as an alternative strategy in the treatment of the emerging fungus *Candida auris*. *International Journal of Antimicrobial Agents*, *56*(2), 106049. https://doi.org/10.1016/j.ijantimicag.2020.106049.

Bazzi, T., De Melo, S. M. P., Fihera, R. A., Kommers, G. D. (2016). Clinical-epidemiological, histomorphological, and histochemical characteristics of sporotrichosis. *Brazilian Journal of Veterinary Medicine*., v.36, n.4, p.303-311, 2016.

Berger, S., El Chazli, Y., Babu, A. F., & Coste, A. T. (2017). Azole resistance in *Aspergillus fumigatus*: A consequence of antifungal use in agriculture?. *Frontiers in Microbiology*, *8*, 1024. https://doi.org/10.3389/fmicb.2017.01024.

Bijian, K., Zhang, Z., Xu, B., Jie, S., Chen, B., Wan, S., Wu, J., Jiang, T., & Alaoui-Jamali, M. A. (2012). Synthesis and biological activity of novel organoselenium derivatives targeting multiple kinases and capable of inhibiting cancer progression to metastases. *European Journal of Medicinal Chemistry*, *48*, 143–152. https://doi.org/10.1016/j.ejmech.2011.12.006.

Bowyer, P., & Denning, D. W. (2014). Environmental fungicides and triazole resistance in *Aspergillus*. *Pest Management Science*, *70*(2), 173–178. https://doi.org/10.1002/ps.3567.

Branco, J., Miranda, I. M., & Rodrigues, A. G. (2023). *Candida parapsilosis* virulence and antifungal resistance mechanisms: a comprehensive review of key determinants. *Journal of Fungi*, *9*(1), 80. https://doi.org/10.3390/jof9010080.

Breuer, M. R., Dasgupta, A., Vasselli, J. G., Lin, X., Shaw, B. D., & Sachs, M. S. (2022). The antidepressant sertraline induces the formation of supersized lipid droplets in the human pathogen *Cryptococcus neoformans*. *Journal of Fungi*, *8*(6), 642. https://doi.org/10.3390/jof8060642.

Brownsey, R. W., Boone, A. N., Elliott, J. E., Kulpa, J. E., & Lee, W. M. (2006). Regulation of acetyl-CoA carboxylase. *Biochemical Society Transactions*, *34*(Pt 2), 223–227. https://doi.org/10.1042/BST20060223.

Bustamante, C., Ochoa, R., Asela, C., & Muskus, C. (2019). Repurposing of known drugs for leishmaniasis treatment using bioinformatic predictions, *in vitro* validations and pharmacokinetic simulations. *Journal of Computer-aided Molecular Design*, *33*(9), 845–854. https://doi.org/10.1007/s10822-019-00230-y.

Calgin, M. K., & Cetinkol, Y. (2018). Distribution and antifungal susceptibility patterns of *Candida* species at a university hospital in Northern Turkey. *Journal of Infection in Developing Countries*, *12*(2), 97–101. https://doi.org/10.3855/jidc.9698.

Chen, M., Xu, Y., Hong, N., Yang, Y., Lei, W., Du, L., Zhao, J., Lei, X., Xiong, L., Cai, L., Xu, H., Pan, W., & Liao, W. (2018). Epidemiology of fungal infections in China. *Frontiers of Medicine*, *12*(1), 58–75. https://doi.org/10.1007/s11684-017-0601-0.

Chen, P., Yang, J., Jin, Y., Lu, C., Feng, Z., Gao, F., Chen, Y., Wang, F., Shang, Z., & Lin, W. (2023). *In vitro* antifungal and antibiofilm activities of auranofin against itraconazole-resistant *Aspergillus fumigatus*. *Journal de Mycologie Medicale*, *33*(2), 101381. https://doi.org/10.1016/j.mycmed.2023.101381.

Chen, T., Yao, J., Quan, K., Xu, J., Hang, X., Tong, Q., Liu, G., Luo, P., Zeng, L., Feng, G., & Bi, H. (2024). Repurposing a human acetyl-CoA carboxylase inhibitor firsocostat to treat fungal candidiasis alone and in combination. *Antimicrobial Agents and Chemotherapy*, *68*(1), e0113123. https://doi.org/10.1128/aac.01131-23.

Cheng, Y. S., Roma, J. S., Shen, M., Mota Fernandes, C., Tsang, P. S., Forbes, H. E., Boshoff, H., Lazzarini, C., Del Poeta, M., Zheng, W., & Williamson, P. R. (2021). Identification of antifungal compounds against multidrug-resistant *Candida auris* utilizing a high-throughput drug-repurposing screen. *Antimicrobial Agents and Chemotherapy*, *65*(4), e01305-20. https://doi.org/10.1128/AAC.01305-20.

Chow, E. W., Clancey, S. A., Billmyre, R. B., Averette, A. F., Granek, J. A., Mieczkowski, P., Cardenas, M. E., & Heitman, J. (2017). Elucidation of the calcineurin-Crz1 stress response transcriptional network in the human fungal pathogen *Cryptococcus neoformans*. *PLoS Genetics*, *13*(4), e1006667. https://doi.org/10.1371/journal.pgen.1006667.

Conen, A., Weisser, M., Jörg, C. O., Battegay, M., Trampuz, A., Frei, R., Christen, S., Heim, D., Gratwohl, A., & Flückiger, U. (2008). *Candida krusei* - a serious complication in patients with hematological malignancies: successful treatment with caspofungin. *Transplant Infectious Disease: an Official Journal of the Transplantation Society*, *10*(1), 66–70. https://doi.org/10.1111/j.1399-3062.2007.00238.x.

Cowen, L. E., Sanglard, D., Howard, S. J., Rogers, P. D., & Perlin, D. S. (2014). Mechanisms of antifungal drug resistance. *Cold Spring Harbor Perspectives in Medicine*, *5*(7), a019752. https://doi.org/10.1101/cshperspect.a019752.

Cristina, M. L., Spagnolo, A. M., Sartini, M., Carbone, A., Oliva, M., Schinca, E., Boni, S., & Pontali, E. (2023). An overview on *Candida auris* in healthcare settings. *Journal of Fungi*, *9*(9), 913. https://doi.org/10.3390/jof9090913.

Cruz, M. C., Cavallo, L. M., Görlach, J. M., Cox, G., Perfect, J. R., Cardenas, M. E., & Heitman, J. (1999). Rapamycin antifungal action is mediated via conserved complexes with FKBP12 and TOR kinase homologs in *Cryptococcus neoformans*. *Molecular and Cellular Biology*, *19*(6), 4101–4112. https://doi.org/10.1128/MCB.19.6.4101.

Cruz, M. C., Goldstein, A. L., Blankenship, J., Del Poeta, M., Perfect, J. R., McCusker, J. H., Bennani, Y. L., Cardenas, M. E., & Heitman, J. (2001). Rapamycin and less immunosuppressive analogs are toxic to *Candida albicans* and *Cryptococcus neoformans* via FKBP12-dependent inhibition of TOR. *Antimicrobial Agents and Chemotherapy*, *45*(11), 3162–3170. https://doi.org/10.1128/AAC.45.11.3162-3170.2001.

Dadar, M., Tiwari, R., Karthik, K., Chakraborty, S., Shahali, Y., & Dhama, K. (2018). *Candida albicans* - biology, molecular characterization, pathogenicity, and advances in diagnosis and control - An update. *Microbial Pathogenesis*, *117*, 128–138. https://doi.org/10.1016/j.micpath.2018.02.028.

Dai, B., Wang, Y., Li, D., Xu, Y., Liang, R., Zhao, L., Cao, Y., Jia, J., & Jiang, Y. (2012). Hsp90 is involved in apoptosis of *Candida albicans* by regulating the calcineurin-caspase apoptotic pathway. *PloS One*, *7*(9), e45109. https://doi.org/10.1371/journal.pone.0045109.

Daneshnia, F., Arastehfar, A., Lombardi, L., Binder, U., Scheler, J., Vahedi Shahandashti, R., Hagen, F., Lass-Flörl, C., Mansour, M. K., Butler, G., & Perlin, D. S. (2023). *Candida parapsilosis* isolates carrying mutations outside *FKS1* hotspot regions confer high echinocandin tolerance and facilitate the development of echinocandin resistance. *International Journal of Antimicrobial Agents*, *62*(1), 106831. https://doi.org/10.1016/j.ijantimicag.2023.106831.

Daneshnia, F., de Almeida Júnior, J. N., Arastehfar, A., Lombardi, L., Shor, E., Moreno, L., Verena Mendes, A., Goreth Barberino, M., Thomaz Yamamoto, D., Butler, G., Perlin, D. S., & Colombo, A. L. (2022). Determinants of fluconazole resistance and echinocandin tolerance in *C. parapsilosis* isolates causing a large clonal candidemia outbreak among COVID-19 patients in a Brazilian ICU. *Emerging Microbes & Infections*, *11*(1), 2264–2274. https://doi.org/10.1080/22221751.2022.2117093.

Das, S., Garg, T., Chopra, S., & Dasgupta, A. (2019). Repurposing disulfiram to target infections caused by non-tuberculous mycobacteria. *The Journal of Antimicrobial Chemotherapy*, *74*(5), 1317–1322. https://doi.org/10.1093/jac/dkz018.

Delattin, N., De Brucker, K., Vandamme, K., Meert, E., Marchand, A., Chaltin, P., Cammue, B. P., & Thevissen, K. (2014). Repurposing as a means to increase the activity of amphotericin B and caspofungin against *Candida albicans* biofilms. *The Journal of Antimicrobial Chemotherapy*, *69*(4), 1035–1044. https://doi.org/10.1093/jac/dkt449.

Denning D. W. (2024). Global incidence and mortality of severe fungal disease. *The Lancet. Infectious diseases*, S1473-3099(23)00692-8. Advance online publication. https://doi.org/10.1016/S1473-3099(23)00692-8.

Donlin, M. J., & Meyers, M. J. (2022). Repurposing and optimization of drugs for discovery of novel antifungals. *Drug Discovery Today*, *27*(7), 2008–2014. https://doi.org/10.1016/j.drudis.2022.04.021.

Dos Reis, T. F., Horta, M. A. C., Colabardini, A. C., Fernandes, C. M., Silva, L. P., Bastos, R. W., Fonseca, M. V. L., Wang, F., Martins, C., Rodrigues, M. L., Silva Pereira, C., Del Poeta, M., Wong, K. H., & Goldman, G. H. (2021). Screening of chemical libraries for new antifungal drugs against *Aspergillus fumigatus* reveals sphingolipids are involved in the mechanism of action of miltefosine. *mBio*, *12*(4), e0145821. https://doi.org/10.1128/mBio.01458-21.

Dotis, J., Prasad, P. A., Zaoutis, T., & Roilides, E. (2012). Epidemiology, risk factors and outcome of *Candida parapsilosis* bloodstream infection in children. *The Pediatric Infectious Disease Journal*, *31*(6), 557–560. https://doi.org/10.1097/INF.0b013e31824da7fe.

Du, H., Bing, J., Hu, T., Ennis, C. L., Nobile, C. J., & Huang, G. (2020). *Candida auris*: Epidemiology, biology, antifungal resistance, and virulence. *PLoS Pathogens*, *16*(10), e1008921. https://doi.org/10.1371/journal.ppat.1008921.

Eldesouky, H. E., Lanman, N. A., Hazbun, T. R., & Seleem, M. N. (2020). Aprepitant, an antiemetic agent, interferes with metal ion homeostasis of *Candida auris* and displays potent synergistic interactions with azole drugs. *Virulence*, *11*(1), 1466–1481. https://doi.org/10.1080/21505594.2020.1838741.

Eldesouky, H. E., Salama, E. A., Hazbun, T. R., Mayhoub, A. S., & Seleem, M. N. (2020). Ospemifene displays broad-spectrum synergistic interactions with itraconazole through potent interference with fungal efflux activities. *Scientific Reports*, *10*(1), 6089. https://doi.org/10.1038/s41598-020-62976-y.

Eldesouky, H. E., Salama, E. A., Lanman, N. A., Hazbun, T. R., & Seleem, M. N. (2020). Potent synergistic interactions between lopinavir and azole antifungal drugs against emerging multidrug-resistant *Candida auris*. *Antimicrobial Agents and Chemotherapy, 65*(1), e00684-20. https://doi.org/10.1128/AAC.00684-20.

Eldesouky, H. E., Salama, E. A., Li, X., Hazbun, T. R., Mayhoub, A. S., & Seleem, M. N. (2020). Repurposing approach identifies pitavastatin as a potent azole chemosensitizing agent effective against azole-resistant *Candida* species. *Scientific Reports, 10*(1), 7525. https://doi.org/10.1038/s41598-020-64571.

Elgammal, Y., Salama, E. A., & Seleem, M. N. (2023). Atazanavir resensitizes *Candida* auris to azoles. *Antimicrobial Agents and Chemotherapy, 67*(5), e0163122. https://doi.org/10.1128/aac.01631-22.

Elgammal, Y., Salama, E. A., & Seleem, M. N. (2023). Saquinavir potentiates itraconazole's antifungal activity against multidrug-resistant *Candida auris in vitro* and *in vivo*. *Medical Mycology, 61*(9), myad081. https://doi.org/10.1093/mmy/myad081.

Elgammal, Y., Salama, E. A., & Seleem, M. N. (2024). Enhanced antifungal activity of posaconazole against *Candida auris* by HIV protease inhibitors, atazanavir and saquinavir. *Scientific Reports, 14*(1), 1571. https://doi.org/10.1038/s41598-024-52012-8.

El-Ganiny, A. M., Kamel, H. A., Yossef, N. E., Mansour, B., & El-Baz, A. M. (2022). Repurposing pantoprazole and haloperidol as efflux pump inhibitors in azole resistant clinical *Candida albicans* and non-*albicans* isolates. *Saudi pharmaceutical journal: the Official Publication of the Saudi Pharmaceutical Society, 30*(3), 245–255. https://doi.org/10.1016/j.jsps.2022.01.011.

European medicines agency. Science medicines health. https://www.ema.europa.eu/en/medicines/veterinary/EPAR/panacur-aquasol Accessed in: 02/march/2024 - 15:30.

Firacative, C., Trilles, L., & Meyer, W. (2021). Recent advances in *Cryptococcus* and cryptococcosis. *Microorganisms, 10*(1), 13. https://doi.org/10.3390/microorganisms10010013.

Frías-De-León, M. G., Hernández-Castro, R., Conde-Cuevas, E., García-Coronel, I. H., Vázquez-Aceituno, V. A., Soriano-Ursúa, M. A., Farfán-García, E. D., Ocharán-Hernández, E., Rodríguez-Cerdeira, C., Arenas, R., Robledo-Cayetano, M., Ramírez-Lozada, T., Meza-Meneses, P., Pinto-Almazán, R., & Martínez-Herrera, E. (2021). *Candida glabrata* antifungal resistance and virulence factors, a perfect pathogenic combination. *Pharmaceutics, 13*(10), 1529. https://doi.org/10.3390/pharmaceutics13101529.

Gao, L., & Sun, Y. (2015). *In vitro* interactions of antifungal agents and tacrolimus against *Aspergillus* biofilms. *Antimicrobial Agents and Chemotherapy, 59*(11), 7097–7099. https://doi.org/10.1128/AAC.01510-15.

Gao, L., Sun, Y., Yuan, M., Li, M., & Zeng, T. (2020). *In vitro* and *in vivo* study on the synergistic effect of minocycline and azoles against pathogenic fungi. *Antimicrobial Agents and Chemotherapy, 64*(6), e00290-20. https://doi.org/10.1128/AAC.00290-20.

Garcia-Solache, M. A., & Casadevall, A. (2010). Global warming will bring new fungal diseases for mammals. *mBio, 1*(1), e00061-10. https://doi.org/10.1128/mBio.00061-10.

Geddes-McAlister, J., & Shapiro, R. S. (2019). New pathogens, new tricks: emerging, drug-resistant fungal pathogens and future prospects for antifungal therapeutics. *Annals of the New York Academy of Sciences*, *1435*(1), 57–78. https://doi.org/10.1111/nyas.13739.

Gowri, M., Jayashree, B., Jeyakanthan, J., & Girija, E. K. (2020). Sertraline as a promising antifungal agent: inhibition of growth and biofilm of *Candida auris* with special focus on the mechanism of action *in vitro*. *Journal of Applied Microbiology*, *128*(2), 426–437. https://doi.org/10.1111/jam.14490.

Granato, M. Q., Mello, T. P., Nascimento, R. S., Pereira, M. D., Rosa, T. L. S. A., Pessolani, M. C. V., McCann, M., Devereux, M., Branquinha, M. H., Santos, A. L. S., & Kneipp, L. F. (2021). Silver(I) and copper(II) complexes of 1,10-phenanthroline-5,6-dione against *Phialophora verrucosa*: a focus on the interaction with human macrophages and *Galleria mellonella* larvae. *Frontiers in Microbiology*, *12*, 641258. https://doi.org/10.3389/fmicb.2021.641258.

Hai, T. P., Van, A. D., Ngan, N. T. T., Nhat, L. T. H., Lan, N. P. H., Vinh Chau, N. V., Thwaites, G. E., Krysan, D., & Day, J. N. (2019). The combination of tamoxifen with amphotericin B, but not with fluconazole, has synergistic activity against the majority of clinical isolates of *Cryptococcus neoformans*. *Mycoses*, *62*(9), 818–825.

Hamdy, R. F., Zaoutis, T. E., & Seo, S. K. (2017). Antifungal stewardship considerations for adults and pediatrics. *Virulence*, *8*(6), 658–672. https://doi.org/10.1080/21505594.2016.1226721.

Hassan, Y., Chew, S. Y., & Than, L. T. L. (2021). *Candida glabrata*: Pathogenicity and resistance mechanisms for adaptation and survival. *Journal of fungi*, *7*(8), 667. https://doi.org/10.3390/jof7080667.

Holbrook, S. Y. L., Garzan, A., Dennis, E. K., Shrestha, S. K., & Garneau-Tsodikova, S. (2017). Repurposing antipsychotic drugs into antifungal agents: Synergistic combinations of azoles and bromperidol derivatives in the treatment of various fungal infections. *European Journal of Medicinal Chemistry*, *139*, 12–21. https://doi.org/10.1016/j.ejmech.2017.07.030.

Hossain, S., Veri, A. O., & Cowen, L. E. (2020). The proteasome governs fungal morphogenesis via functional connections with hsp90 and cAMP-protein kinase A signaling. *mBio*, *11*(2), e00290-20. https://doi.org/10.1128/mBio.00290-20.

Huang, W., Liao, G., Baker, G. M., Wang, Y., Lau, R., Paderu, P., Perlin, D. S., & Xue, C. (2016). Lipid flippase subunit cdc50 mediates drug resistance and virulence in *Cryptococcus neoformans*. *mBio*, *7*(3), e00478-16. https://doi.org/10.1128/mBio.00478-16.

Jamiu, A. T., Albertyn, J., Sebolai, O. M., & Pohl, C. H. (2021). Update on *Candida krusei*, a potential multidrug-resistant pathogen. *Medical Mycology*, *59*(1), 14–30. https://doi.org/10.1093/mmy/myaa031.

Jansen, J. C., Verhasselt, H. L., & Rath, P. M. (2022). P357 Update on risk factors for *Candida krusei*-fungemia. *Medical Mycology*, *60*(Suppl 1), myac072P357. https://doi.org/10.1093/mmy/myac072.P357.

Ji, C., Liu, N., Tu, J., Li, Z., Han, G., Li, J., & Sheng, C. (2020). Drug repurposing of haloperidol: discovery of new benzocyclane derivatives as potent antifungal agents against cryptococcosis and candidiasis. *ACS Infectious Diseases*, *6*(5), 768–786. https://doi.org/10.1021/acsinfecdis.9b00197.

Joffe, L. S., Schneider, R., Lopes, W., Azevedo, R., Staats, C. C., Kmetzsch, L., Schrank, A., Del Poeta, M., Vainstein, M. H., & Rodrigues, M. L. (2017). The Anti-helminthic compound mebendazole has multiple antifungal effects against *Cryptococcus neoformans*. *Frontiers in Microbiology*, *8*, 535. https://doi.org/10.3389/fmicb.2017.00535.

Jothi, R., Hong, S. T., Enkhtsatsral, M., Pandian, S. K., & Gowrishankar, S. (2023). ROS mediated anticandidal efficacy of 3-bromopyruvate prevents vulvovaginal candidiasis in mice model. *PloS One*, *18*(12), e0295922. https://doi.org/10.1371/journal.pone.0295922.

Kajuhara T., Yahara K., Nagi M., Kitamura N., Hirabayashi A., Hosaka Y., Abe M., Miyazaki Y., Sugai M. (2022). Distribution, trends, and antifungal susceptibility of *Candida* species causing candidemia in Japan, 2010-2019: A retrospective observational study based on national surveillance data. *Medical Mycology*. 60, myac071. https://doi.org/10.1093/mmy/myac071.

Kanafani, Z. A., & Perfect, J. R. (2008). Antimicrobial resistance: resistance to antifungal agents: mechanisms and clinical impact. *Clinical Infectious Diseases: an Official Publication of the Infectious Diseases Society of America*, *46*(1), 120–128. https://doi.org/10.1086/524071.

Kanaujia, R., Singh, S., & Rudramurthy, S. M. (2023). Aspergillosis: an update on clinical spectrum, diagnostic schemes, and management. *Current Fungal Infection Reports*, 1–12. Advance online publication. https://doi.org/10.1007/s12281-023-00461-5.

Kang, K., Wong, K. S., Seneviratne, C. J., Samaranayake, L. P., Fong, W. P., & Tsang, P. W. (2010). *In vitro* synergistic effects of metergoline and antifungal agents against *Candida krusei*. *Mycoses*, *53*(6), 495–499. https://doi.org/10.1111/j.1439-0507.2009.01747.x.

Khalifa, H. O., Hubka, V., Watanabe, A., Nagi, M., Miyazaki, Y., Yaguchi, T., & Kamei, K. (2022). Prevalence of antifungal resistance, genetic basis of acquired azole and echinocandin resistance, and genotyping of *Candida krusei* recovered from an international collection. *Antimicrobial Agents and Chemotherapy*, *66*(2), e0185621. https://doi.org/10.1128/AAC.01856-21.

Khan, S., Singhal, S., Mathur, T., Upadhyay, D. J., & Rattan, A. (2007). Antifungal potential of disulfiram. *Japanese Journal of Medical Mycology*, *48*(3), 109–113. https://doi.org/10.3314/jjmm.48.109.

Kishi, T., Ikuta, T., Matsuda, Y., Sakuma, K., Okuya, M., Nomura, I., Hatano, M., & Iwata, N. (2022). Pharmacological treatment for bipolar mania: a systematic review and network meta-analysis of double-blind randomized controlled trials. *Molecular Psychiatry*, *27*(2), 1136–1144. https://doi.org/10.1038/s41380-021-01334-4.

Kollef, M. H., Torres, A., Shorr, A. F., Martin-Loeches, I., & Micek, S. T. (2021). Nosocomial infection. *Critical Care Medicine*, *49*(2), 169–187. https://doi.org/10.1097/CCM.0000000000004783.

Komm, B. S., & Mirkin, S. (2014). An overview of current and emerging SERMs. *The Journal of Steroid Biochemistry and Molecular Biology*, *143*, 207–222. https://doi.org/10.1016/j.jsbmb.2014.03.003.

Kontoyiannis, D. P., Lewis, R. E., Osherov, N., Albert, N. D., & May, G. S. (2003). Combination of caspofungin with inhibitors of the calcineurin pathway attenuates growth *in vitro* in *Aspergillus* species. *The Journal of Antimicrobial Chemotherapy*, *51*(2), 313–316. https://doi.org/10.1093/jac/dkg090.

Kumar, K., Askari, F., Sahu, M. S., & Kaur, R. (2019). *Candida glabrata*: a lot more than meets the eye. *Microorganisms*, *7*(2), 39. https://doi.org/10.3390/microorganisms 7020039.

Lajarin-Reinares, M., Martinez-Esteve, E., Pena-Rodríguez, E., Cañellas-Santos, M., Bulut, S., Karabelas, K., Clauss, A., Nieto, C., Mallandrich, M., & Fernandez-Campos, F. (2022). The efficacy and biopharmaceutical properties of a fixed-dose combination of disulfiram and benzyl benzoate. *International Journal of Molecular Sciences*, *23*(18), 10969. https://doi.org/10.3390/ijms231810969.

Lajarin-Reinares, M., Naveira-Souto, I., Mallandrich, M., Suñer-Carbó, J., Llagostera Casas, M., Calvo, M. A., & Fernandez-Campos, F. (2023). Repurposing disulfiram as an antifungal agent: development of a new disulfiram vaginal mucoadhesive gel. *Pharmaceutics*, *15*(5), 1436. https://doi.org/10.3390/pharmaceutics15051436.

Lajarin-Reinares, M., Pena-Rodríguez, E., Cañellas-Santos, M., Rosell-Vives, E., Cortés, P., Casas, M. L., Calvo, M. À., & Fernandez-Campos, F. (2022). Repurposing disulfiram as an antimicrobial agent in topical infections. *Antibiotics*, *11*(12), 1752. https://doi.org/10.3390/antibiotics11121752.

Lamoth, F., Juvvadi, P. R., Gehrke, C., & Steinbach, W. J. (2013). *In vitro* activity of calcineurin and heat shock protein 90 Inhibitors against *Aspergillus fumigatus* azole- and echinocandin-resistant strains. *Antimicrobial Agents and Chemotherapy*, *57*(2), 1035–1039. https://doi.org/10.1128/AAC.01857-12.

Lara, V. S., Silva, R. A. D., Ferrari, T. P., Santos, C. F. D., & Oliveira, S. H. P. (2023). Losartan plays a fungistatic and fungicidal activity against *Candida albicans* biofilms: drug repurposing for localized candidosis. *Assay and Drug Development Technologies*, *21*(4), 157–165. https://doi.org/10.1089/adt.2023.013.

Latgé, J. P., & Chamilos, G. (2019). *Aspergillus fumigatus* and aspergillosis in 2019. *Clinical Microbiology Reviews*, *33*(1), e00140-18. https://doi.org/10.1128/CMR. 00140-18.

Lee, Y., Puumala, E., Robbins, N., & Cowen, L. E. (2021). Antifungal drug resistance: molecular mechanisms in *Candida albicans* and beyond. *Chemical Reviews*, *121*(6), 3390–3411. https://doi.org/10.1021/acs.chemrev.0c00199.

Li, H., Wang, J., Wu, C., Wang, L., Chen, Z. S., & Cui, W. (2020). The combination of disulfiram and copper for cancer treatment. *Drug Discovery Today*, *25*(6), 1099–1108. https://doi.org/10.1016/j.drudis.2020.04.003.

Liao, K., & Sun, L. (2018). Roles of the Hsp90-calcineurin pathway in the antifungal activity of Honokiol. *Journal of Microbiology and Biotechnology*, *28*(7), 1086–1093. https://doi.org/10.4014/jmb.1801.01024.

Maan, J. S., Ershadi, M., Khan, I., & Saadabadi, A. (2023). Quetiapine. In *StatPearls*. StatPearls Publishing.

May, H. C., Yu, J. J., Guentzel, M. N., Chambers, J. P., Cap, A. P., & Arulanandam, B. P. (2018). Repurposing Auranofin, Ebselen, and PX-12 as antimicrobial agents targeting the thioredoxin system. *Frontiers in Microbiology*, *9*, 336. https://doi.org/10.3389/fmicb.2018.00336.

Maziarz, E. K., & Perfect, J. R. (2016). Cryptococcosis. *Infectious Disease Clinics of North America*, *30*(1), 179–206. https://doi.org/10.1016/j.idc.2015.10.006.

Mello, T. P., Aor, A. C., Branquinha, M. H., & Santos, A. L. S. (2020). Insights into the interaction of *Scedosporium apiospermum, Scedosporium aurantiacum, Scedosporium minutisporum, and Lomentospora prolificans* with lung epithelial cells. *Brazilian Journal of Microbiology*, *51*(2), 427–436. https://doi.org/10.1007/s42770-019-00183-2.

Mello, T. P., Oliveira, S. S. C., Branquinha, M. H., & Santos, A. L. S. (2022). Decoding the antifungal resistance mechanisms in biofilms of emerging, ubiquitous and multidrug-resistant species belonging to the *Scedosporium/Lomentospora* genera. *Medical Mycology*, *60*(6), myac036. https://doi.org/10.1093/mmy/myac036.

Miró-Canturri, A., Ayerbe-Algaba, R., & Smani, Y. (2019). Drug repurposing for the treatment of bacterial and fungal infections. *Frontiers in Microbiology*, *10*, 41. https://doi.org/10.3389/fmicb.2019.00041.

Muzaheed. Alshehri B. A., Rabaan A. A., El-Masry O. S., Acharya S., Alzahrani F. M., Multair A. A., Alhumaid S., Al-Tawfiq J. A., Muhammad J., Khan A., Dhama K., Al-Omari A. (2022). A 20-year retrospective clinical analysis of *Candida* infections in tertiary center: Single-center experience. *Journal of Infection and Public Health.* 15:69-74. https://doi.org/10.1096/j.jiph.2021.11.014.

Nett, J. E., & Andes, D. R. (2016). Antifungal agents: spectrum of activity, pharmacology, and clinical indications. *Infectious Disease Clinics of North America*, *30*(1), 51–83. https://doi.org/10.1016/j.idc.2015.10.012.

Ngan, N. T. T., Flower, B., & Day, J. N. (2022). Treatment of cryptococcal meningitis: how have we got here and where are we going?. *Drugs*, *82*(12), 1237–1249. https://doi.org/10.1007/s40265-022-01757-5.

Ngan, N. T. T., Thanh Hoang Le, N., Vi Vi, N. N., Van, N. T. T., Mai, N. T. H., Van Anh, D., Trieu, P. H., Lan, N. P. H., Phu, N. H., Chau, N. V. V., Lalloo, D. G., Hope, W., Beardsley, J., White, N. J., Geskus, R., Thwaites, G. E., Krysan, D., Tai, L. T. H., Kestelyn, E., Binh, T. Q., Day, J. N. (2021). An open label randomized controlled trial of tamoxifen combined with amphotericin B and fluconazole for cryptococcal meningitis. *eLife*, *10*, e68929. https://doi.org/10.7554/eLife.68929.

Ngo, H. X., Shrestha, S. K., & Garneau-Tsodikova, S. (2016). Identification of ebsulfur analogues with broad-spectrum antifungal activity. *ChemMedChem*, *11*(14), 1507–1516. https://doi.org/10.1002/cmdc.201600236.

Nosengo N. (2016). Can you teach old drugs new tricks?. *Nature*, *534*(7607), 314–316. https://doi.org/10.1038/534314a.

Ogundeji, A. O., Pohl, C. H., & Sebolai, O. M. (2017). The repurposing of anti-psychotic drugs, quetiapine and olanzapine, as anti-*Cryptococcus* drugs. *Frontiers in Microbiology*, *8*, 815. https://doi.org/10.3389/fmicb.2017.00815.

Oliveira, A. S., Martinez-de-Oliveira, J., Donders, G. G. G., Palmeira-de-Oliveira, R., & Palmeira-de-Oliveira, A. (2018). Anti-*Candida* activity of antidepressants sertraline and fluoxetine: effect upon pre-formed biofilms. *Medical Microbiology and Immunology*, *207*(3-4), 195–200. https://doi.org/10.1007/s00430-018-0539-0.

Oliveira, H. C., Joffe, L. S., Simon, K. S., Castelli, R. F., Reis, F. C. G., Bryan, A. M., Borges, B. S., Medeiros, L. C. S., Bocca, A. L., Del Poeta, M., & Rodrigues, M. L. (2020). Fenbendazole controls *in vitro* growth, virulence potential, and animal infection in the *Cryptococcus* model. *Antimicrobial Agents and Chemotherapy*, *64*(6), e00286-20. https://doi.org/10.1128/AAC.00286-20.

Oliveira, N. K., Bhattacharya, S., Gambhir, R., Joshi, M., & Fries, B. C. (2022). Novel ABC transporter associated with fluconazole resistance in aging of *Cryptococcus neoformans*. *Journal of Fungi, 8*(7), 677. https://doi.org/10.3390/jof8070677.

Osherov, N., & Kontoyiannis, D. P. (2017). The anti-*Aspergillus* drug pipeline: Is the glass half full or empty?. *Medical Mycology*, *55*(1), 118–124. https://doi.org/10.1093/mmy/myw060.

OTIS (Organization of Teratology Information Specialists). (2022). Quetiapina (Seroquel®). In *Mother to Baby Fact Sheets*.

Otto, C., & Babady, N. E. (2023). Epidemiology and outcomes of non-*albicans Candida* bloodstream infections in transplant recipients and cancer patients. *Mycopathologia*, *188*(6), 863–871. https://doi.org/10.1007/s11046-023-00765-7.

Padder, S. A., Ramzan, A., Tahir, I., Rehman, R. U., & Shah, A. H. (2022). Metabolic flexibility and extensive adaptability governing multiple drug resistance and enhanced virulence in *Candida albicans*. *Critical Reviews in Microbiology*, *48*(1), 1–20. https://doi.org/10.1080/1040841X.2021.1935447.

Papon, N., & Goldman, G. H. (2021). Unraveling caspofungin resistance in *Cryptococcus neoformans*. *mBio*, *12*(2), e00156-21. https://doi.org/10.1128/mBio.00156-21.

Patton, L. L., McKaig, R., Strauss, R., Rogers, D., & Eron, J. J., Jr (2000). Changing prevalence of oral manifestations of human immuno-deficiency virus in the era of protease inhibitor therapy. *Oral Surgery, Oral Medicine, Oral Pathology, Oral Radiology, and Endodontics*, *89*(3), 299–304. https://doi.org/10.1016/s1079-2104(00)70092-8.

Pereirinha P. R. R, Figueira A. R. R. 2016. Desenvolvimento de sistemas mucoadesivos para a administração controlada de fármacos. *Boletim informativo Geum.* 7(2):63-81, abr-jun. ISSN 2237-7387.

Perfect J. R. (2017). The antifungal pipeline: a reality check. *Nature Reviews. Drug discovery*, *16*(9), 603–616. https://doi.org/10.1038/nrd.2017.46.

Perlroth, J., Choi, B., & Spellberg, B. (2007). Nosocomial fungal infections: epidemiology, diagnosis, and treatment. *Medical Mycology*, *45*(4), 321–346. https://doi.org/10.1080/13693780701218689.

Peyclit, L., Yousfi, H., Rolain, J. M., & Bittar, F. (2021). Drug repurposing in medical mycology: identification of compounds as potential antifungals to overcome the emergence of multidrug-resistant fungi. *Pharmaceuticals*, *14*(5), 488. https://doi.org/10.3390/ph14050488.

Pfaller, M. A., Diekema, D. J., Gibbs, D. L., Newell, V. A., Nagy, E., Dobiasova, S., Rinaldi, M., Barton, R., Veselov, A., & Global Antifungal Surveillance Group (2008). *Candida krusei*, a multidrug-resistant opportunistic fungal pathogen: geographic and temporal trends from the ARTEMIS DISK Antifungal Surveillance Program, 2001 to 2005. *Journal of Clinical Microbiology*, *46*(2), 515–521. https://doi.org/10.1128/JCM.01915-07.

Pfaller, M. A., Diekema, D. J., Turnidge, J. D., Castanheira, M., & Jones, R. N. (2019). Twenty years of the SENTRY antifungal surveillance program: results for *Candida* species from 1997-2016. *Open Forum Infectious Diseases*, *6*(Suppl 1), S79–S94. https://doi.org/10.1093/ofid/ofy358.

Prasad, R., Nair, R., & Banerjee, A. (2019). Emerging mechanisms of drug resistance in *Candida albicans*. *Progress in Molecular and Subcellular Biology*, *58*, 135–153. https://doi.org/10.1007/978-3-030-13035-0_6.

Pristov, K. E., & Ghannoum, M. A. (2019). Resistance of *Candida* to azoles and echinocandins worldwide. *Clinical Microbiology and Infection*, *25*(7), 792–798. https://doi.org/10.1016/j.cmi.2019.03.028.

Puig-Asensio, M., Padilla, B., Garnacho-Montero, J., Zaragoza, O., Aguado, J. M., Zaragoza, R., Montejo, M., Muñoz, P., Ruiz-Camps, I., Cuenca-Estrella, M., Almirante, B., CANDIPOP Project, GEIH-GEMICOMED (SEIMC), & REIPI (2014). Epidemiology and predictive factors for early and late mortality in *Candida* bloodstream infections: a population-based surveillance in Spain. *Clinical Microbiology and Infection*, *20*(4), O245–O254. https://doi.org/10.1111/1469-0691.12380.

Rajasingham, R., Smith, R. M., Park, B. J., Jarvis, J. N., Govender, N. P., Chiller, T. M., Denning, D. W., Loyse, A., & Boulware, D. R. (2017). Global burden of disease of HIV-associated cryptococcal meningitis: an updated analysis. *The Lancet. Infectious Diseases*, *17*(8), 873–881. https://doi.org/10.1016/S1473-3099(17)30243-8.

Ramos, L. S., Silva, L. N., de Mello, T. P., Frota, H. F., Branquinha, M. H., & Santos, A. L. S. (2022). Prospective medicines against the widespread, emergent, and multidrug-resistant opportunistic fungal pathogen *Candida auris*: a breath of hope. *Current Topics in Medicinal Chemistry*, *22*(16), 1297–1305. https://doi.org/10.2174/1568026622666220520153748.

Rhein, J., Huppler Hullsiek, K., Tugume, L., Nuwagira, E., Mpoza, E., Evans, E. E., Kiggundu, R., Pastick, K. A., Ssebambulidde, K., Akampurira, A., Williams, D. A., Bangdiwala, A. S., Abassi, M., Musubire, A. K., Nicol, M. R., Muzoora, C., Meya, D. B., Boulware, D. R., & ASTRO-CM team (2019). Adjunctive sertraline for HIV-associated cryptococcal meningitis: a randomised, placebo-controlled, double-blind phase 3 trial. *The Lancet. Infectious Diseases*, *19*(8), 843–851. https://doi.org/10.1016/S1473-3099(19)30127-6.

Robbins, N., & Cowen, L. E. (2023). Roles of Hsp90 in *Candida albicans* morphogenesis and virulence. *Current Opinion in Microbiology*, *75*, 102351. https://doi.org/10.1016/j.mib.2023.102351.

Robbins, N., Spitzer, M., Yu, T., Cerone, R. P., Averette, A. K., Bahn, Y. S., Heitman, J., Sheppard, D. C., Tyers, M., & Wright, G. D. (2015). An antifungal combination matrix identifies a rich pool of adjuvant molecules that enhance drug activity against diverse fungal pathogens. *Cell Reports*, *13*(7), 1481–1492. https://doi.org/10.1016/j.celrep.2015.10.018.

Rollin-Pinheiro, R., Almeida, Y. C., Rochetti, V. P., Xisto, M. I. D. D. S., Borba-Santos, L. P., Rozental, S., & Barreto-Bergter, E. (2021). Miltefosine against *Scedosporium* and *Lomentospora* species: antifungal activity and its effects on fungal cells. *Frontiers in Cellular and Infection Microbiology*, *11*, 698662. https://doi.org/10.3389/fcimb.2021.69866.

Rollin-Pinheiro, R., Borba-Santos, L. P., da Silva Xisto, M. I. D., de Castro-Almeida, Y., Rochetti, V. P., Rozental, S., & Barreto-Bergter, E. (2021). Identification of promising antifungal drugs against *Scedosporium* and *Lomentospora* species after screening of Pathogen Box Library. *Journal of Fungi*, *7*(10), 803. https://doi.org/10.3390/jof7100803.

Rollin-Pinheiro, R., Xisto, M. I. D. D. S., de Castro-Almeida, Y., Rochetti, V. P., Borba-Santos, L. P., Fontes, Y. D. S., Ferreira-Pereira, A., Rozental, S., & Barreto-Bergter, E. (2023). Pandemic Response Box® library as a source of antifungal drugs against *Scedosporium* and *Lomentospora* species. *PloS One*, *18*(2), e0280964. https://doi.org/10.1371/journal.pone.0280964.

Rosato, A., Catalano, A., Carocci, A., Carrieri, A., Carone, A., Caggiano, G., Franchini, C., Corbo, F., & Montagna, M. T. (2016). *In vitro* interactions between anidulafungin and nonsteroidal anti-inflammatory drugs on biofilms of *Candida* spp. *Bioorganic & Medicinal Chemistry*, *24*(5), 1002–1005. https://doi.org/10.1016/j.bmc.2016.01.026.

Rossato, L., Camargo Dos Santos, M., Vitale, R. G., de Hoog, S., & Ishida, K. (2021). Alternative treatment of fungal infections: synergy with non-antifungal agents. *Mycoses*, *64*(3), 232–244. https://doi.org/10.1111/myc.13203.

Salahuddin., Shaharyar, M., & Mazumder, A. (2017). Benzimidazoles: A biologically active compounds. *Arabian Journal of Chemistry*, 10, S157-S173.

Salama, E. A., Eldesouky, H. E., Elgammal, Y., Abutaleb, N. S., & Seleem, M. N. (2023). Lopinavir and ritonavir act synergistically with azoles against *Candida auris in vitro* and in a mouse model of disseminated candidiasis. *International Journal of Antimicrobial Agents*, *62*(3), 106906. https://doi.org/10.1016/j.ijantimicag.2023.106906.

Samaranayake, Y. H., & Samaranayake, L. P. (1994). *Candida krusei*: biology, epidemiology, pathogenicity and clinical manifestations of an emerging pathogen. *Journal of Medical Microbiology*, *41*(5), 295–310. https://doi.org/10.1099/00222615-41-5-295.

Sangenito, L. S., d'Avila-Levy, C. M., Branquinha, M. H., & Santos, A. L. S. (2022). Repositioning drug strategy against *Trypanosoma cruzi*: lessons learned from HIV aspartyl peptidase inhibitors. *Memorias do Instituto Oswaldo Cruz*, *117*, e210386. https://doi.org/10.1590/0074-02760210386.

Satoh, K., Makimura, K., Hasumi, Y., Nishiyama, Y., Uchida, K., & Yamaguchi, H. (2009). *Candida auris* sp. nov., a novel ascomycetous yeast isolated from the external ear canal of an inpatient in a Japanese hospital. *Microbiology and Immunology*, *53*(1), 41–44. https://doi.org/10.1111/j.1348-0421.2008.00083.x.

Sauna, Z. E., Shukla, S., & Ambudkar, S. V. (2005). Disulfiram, an old drug with new potential therapeutic uses for human cancers and fungal infections. *Molecular bioSystems*, *1*(2), 127–134. https://doi.org/10.1039/b504392a.

Schwarz, P., & Dannaoui, E. (2020). *In vitro* interaction between isavuconazole and tacrolimus, cyclosporin a, or sirolimus against *Aspergillus* species. *Journal of Fungi*, *6*(3), 103. https://doi.org/10.3390/jof6030103.

Scorzoni, L., de Paula E Silva, A. C., Marcos, C. M., Assato, P. A., de Melo, W. C., de Oliveira, H. C., Costa-Orlandi, C. B., Mendes-Giannini, M. J., & Fusco-Almeida, A. M. (2017). Antifungal therapy: new advances in the understanding and treatment of mycosis. *Frontiers in Microbiology*, *8*, 36. https://doi.org/10.3389/fmicb.2017.00036.

Seelig, S., Ryus, C. R., Harrison, R. F., Wilson, M. P., & Wong, A. H. (2019). Cryptococcal meningoencephalitis presenting as a psychiatric emergency. *The Journal of Emergency Medicine*, *57*(2), 203–206. https://doi.org/10.1016/j.jemermed.2019.03.016.

Shi, H., Zhang, Y., Zhang, M., Chang, W., & Lou, H. (2021). Molecular mechanisms of azole resistance in four clinical *Candida albicans* isolates. *Microbial Drug Resistance*, *27*(12), 1641–1651. https://doi.org/10.1089/mdr.2020.0413.

Sies, H., & Parnham, M. J. (2020). Potential therapeutic use of ebselen for COVID-19 and other respiratory viral infections. *Free Radical Biology & Medicine*, *156*, 107–112. https://doi.org/10.1016/j.freeradbiomed.2020.06.032.

Silva, R. A. C., da Silva, C. R., de Andrade Neto, J. B., da Silva, A. R., Campos, R. S., Sampaio, L. S., do Nascimento, F. B. S. A., da Silva Gaspar, B., da Cruz Fonseca, S. G., Josino, M. A. A., Grangeiro, T. B., Gaspar, D. M., de Lucena, D. F., de Moraes, M. O., Cavalcanti, B. C., & Nobre Júnior, H. V. (2017). *In vitro* anti-*Candida* activity of selective serotonin reuptake inhibitors against fluconazole-resistant strains and their activity against biofilm-forming isolates. *Microbial Pathogenesis*, *107*, 341–348. https://doi.org/10.1016/j.micpath.2017.04.008.

Simanjuntak, Y., Liang, J. J., Chen, S. Y., Li, J. K., Lee, Y. L., Wu, H. C., & Lin, Y. L. (2018). Ebselen alleviates testicular pathology in mice with Zika virus infection and prevents its sexual transmission. *PLoS pathogens*, *14*(2), e1006854. https://doi.org/10.1371/journal.ppat.1006854.

Singh, H. K., & Saadabadi, A. (2023). Sertraline. In *StatPearls*. StatPearls Publishing.

Singh, S., Fatima, Z., Ahmad, K., & Hameed, S. (2020). Repurposing of respiratory drug theophylline against *Candida albicans*: mechanistic insights unveil alterations in membrane properties and metabolic fitness. *Journal of Applied Microbiology*, *129*(4), 860–875. https://doi.org/10.1111/jam.14669.

Skov, L., Johansen, S. S., & Linnet, K. (2015). Postmortem femoral blood reference concentrations of aripiprazole, chlorprothixene, and quetiapine. *Journal of Analytical Toxicology*, *39*(1), 41–44. https://doi.org/10.1093/jat/bku121.

Sousa, N. S. O., Almeida, J. D. R., Frickmann, H., Lacerda, M. V. G., & Souza, J. V. B. (2023). Searching for new antifungals for the treatment of cryptococcosis. *Revista da Sociedade Brasileira de Medicina Tropical*, *56*, e01212023. https://doi.org/10.1590/0037-8682-0121-2023.

Spruijtenburg, B., Badali, H., Abastabar, M., Mirhendi, H., Khodavaisy, S., Sharifisooraki, J., Taghizadeh Armaki, M., de Groot, T., & Meis, J. F. (2022). Confirmation of fifth *Candida auris* clade by whole genome sequencing. *Emerging Microbes & Infections*, *11*(1), 2405–2411. https://doi.org/10.1080/22221751.2022.2125349.

Spruijtenburg, B., Meijer, E. F. J., Xiao, M., Shawky, S. M., Meis, J. F., de Groot, T., & El-Kholy, M. A. (2023). Genotyping and susceptibility testing uncovers large azole-resistant *Candida tropicalis* clade in Alexandria, Egypt. *Journal of Global Antimicrobial Resistance*, *34*, 99–105. https://doi.org/10.1016/j.jgar.2023.06.012.

Strollo, S., Lionakis, M. S., Adjemian, J., Steiner, C. A., & Prevots, D. R. (2016). Epidemiology of hospitalizations associated with invasive candidiasis, United States, 2002-2012. *Emerging Infectious Diseases*, *23*(1), 7–13. https://doi.org/10.3201/eid2301.161198.

Stylianou, M., Kulesskiy, E., Lopes, J. P., Granlund, M., Wennerberg, K., & Urban, C. F. (2014). Antifungal application of nonantifungal drugs. *Antimicrobial Agents and Chemotherapy*, *58*(2), 1055–1062. https://doi.org/10.1128/AAC.01087-1.

Tekintaş, Y., Temel, A., Ateş, A., Eraç, B., Metin, D. Y., Hilmioğlu Polat, S., & Hoşgör Limoncu, M. (2020). Antifungal and antibiofilm activities of selective serotonin reuptake inhibitors alone and in combination with fluconazole. *Turkish Journal of Pharmaceutical Sciences*, *17*(6), 667–672. https://doi.org/10.4274/tjps.galenos.2019.65481.

Tóth R., Nosek J., Mora-Montes H. M., Gabaldon T., Bliss J. M., Nosanchuck J. D., Turner S. A., Butler G., Vágvölgyi C., Gáscer A. (2019). *Candida parapsilosis* from genes to the besides. *Clinical Microbiology Reviews*. 32e00111-18 https://doi.org/10.1128./CRM/00111-18.

Treviño-Rangel, R. J., Villanueva-Lozano, H., Méndez-Galomo, K. S., Solís-Villegas, E. M., Becerril-García, M. A., Montoya, A. M., Robledo-Leal, E. R., & González, G. M. (2019). *In vivo* evaluation of the antifungal activity of sertraline against *Aspergillus fumigatus*. *The Journal of Antimicrobial Chemotherapy*, *74*(3), 663–666. https://doi.org/10.1093/jac/dky455.

Trofa, D., Gácser, A., & Nosanchuk, J. D. (2008). *Candida parapsilosis*, an emerging fungal pathogen. *Clinical Microbiology Reviews*, *21*(4), 606–625. https://doi.org/10.1128/CMR.00013-08 surveillance data. *Medical Mycology*. 60, myac071. https://doi.org/10.1093/mmy/myac071.

Tsay S. V., Mu Y., Williams S., Epson E., Nadle J., Bamberg W. M., Barter D. M., Johnston H. L., Farley M. M., Harb S., Thomas S., Bonner L. A., Harrison L. H., Hollick R., Marceaux K., Mody R. K., Pattee B., Davis S. S., Phipps E. C., Tesini B. L., Gellert A. B., Zhang A. Y., Schaffaner W., Hillis S., Ndi D., Graber C. R., Jackson B. R., Chiller T., Magill S., Vallabhananeni S. (2020). Burden of candidemia in the United States, 2017. *Clinical Infections Diseases*. 71(9):e449-453. https://doi.10.1093.cid/ciaa.193.

Tsirilakis, K., Kim, C., Vicencio, A. G., Andrade, C., Casadevall, A., & Goldman, D. L. (2012). Methylxanthine inhibit fungal chitinases and exhibit antifungal activity. *Mycopathologia*, *173*(2-3), 83–91. https://doi.org/10.1007/s11046-011-9483-x.

Vallières, C., Singh, N., Alexander, C., & Avery, S. V. (2020). Repurposing nonantifungal approved drugs for synergistic targeting of fungal pathogens. *ACS Infectious Diseases*, *6*(11), 2950–2958. https://doi.org/10.1021/acsinfecdis.0c00405.

Vanzolini, T., & Magnani, M. (2024). Old and new strategies in therapy and diagnosis against fungal infections. *Applied Microbiology and Biotechnology*, *108*(1), 147. https://doi.org/10.1007/s00253-023-12884-8.

Verma, R., Pradhan, D., Nayek, A., Singh, H., Jain, A. K., & Khan, L. A. (2022). Target-based drug repurposing against *Candida albicans*-A computational modeling, docking, and molecular dynamic simulations study. *Journal of Cellular Biochemistry*, *123*(2), 289–305. https://doi.org/10.1002/jcb.30163.

Vitale, R. G., Afeltra, J., Meis, J. F., & Verweij, P. E. (2007). Activity and post antifungal effect of chlorpromazine and trifluopherazine against *Aspergillus*, *Scedosporium* and zygomycetes. *Mycoses*, *50*(4), 270–276. https://doi.org/10.1111/j.1439-0507.2007.01371.x.

von Lilienfeld-Toal, M., Wagener, J., Einsele, H., Cornely, O. A., & Kurzai, O. (2019). Invasive fungal infection. *Deutsches Arzteblatt International*, *116*(16), 271–278. https://doi.org/10.3238/arztebl.2019.0271.

Vu, K., Thompson, G. R., 3rd, Roe, C. C., Sykes, J. E., Dreibe, E. M., Lockhart, S. R., Meyer, W., Engelthaler, D. M., & Gelli, A. (2018). Flucytosine resistance in *Cryptococcus gattii* is indirectly mediated by the FCY2-FCY1-FUR1 pathway. *Medical Mycology*, *56*(7), 857–867. https://doi.org/10.1093/mmy/myx135.

Walkup, J. T., Albano, A. M., Piacentini, J., Birmaher, B., Compton, S. N., Sherrill, J. T., Ginsburg, G. S., Rynn, M. A., McCracken, J., Waslick, B., Iyengar, S., March, J. S., & Kendall, P. C. (2008). Cognitive behavioral therapy, sertraline, or a combination in childhood anxiety. *The New England Journal of Medicine*, *359*(26), 2753–2766. https://doi.org/10.1056/NEJMoa0804633.

Wibowo, E., & Wassersug, R. (2014). Estrogen in men. *American Scientist*, 102(6), 452-460.

Wibowo, E., Pollock, P. A., Hollis, N., & Wassersug, R. J. (2016). Tamoxifen in men: a review of adverse events. *Andrology*, *4*(5), 776–788. https://doi.org/10.1111/andr.12197.

Wibowo, E., Schellhammer, P., & Wassersug, R. J. (2011). Role of estrogen in normal male function: clinical implications for patients with prostate cancer on androgen deprivation therapy. *The Journal of Urology*, *185*(1), 17–23. https://doi.org/10.1016/j.juro.2010.08.094.

Wiederhold N. P. (2021). Emerging Fungal infections: new species, new names, and antifungal resistance. *Clinical Chemistry*, *68*(1), 83–90. https://doi.org/10.1093/clinchem/hvab217.

Wiederhold, N. P., Patterson, T. F., Srinivasan, A., Chaturvedi, A. K., Fothergill, A. W., Wormley, F. L., Ramasubramanian, A. K., & Lopez-Ribot, J. L. (2017). Repurposing auranofin as an antifungal: *In vitro* activity against a variety of medically important fungi. *Virulence*, *8*(2), 138–142. https://doi.org/10.1080/21505594.2016.1196301.

Wiederhold, N. P., Patterson, T. F., Srinivasan, A., Chaturvedi, A. K., Fothergill, A. W., Wormley, F. L., Ramasubramanian, A. K., & Lopez-Ribot, J. L. (2017). Repurposing auranofin as an antifungal: *In vitro* activity against a variety of medically important fungi. *Virulence*, *8*(2), 138–142. https://doi.org/10.1080/21505594.2016.1196301.

World Health Organization. (2022). WHO fungal priority pathogens list to guide research, development and public health action. Geneva: *World Health Organization.* License: CC BY-NC-SA 3.0 IGO.

Wouters, O. J., McKee, M., & Luyten, J. (2020). Estimated research and development investment needed to bring a new medicine to market, 2009-2018. *JAMA*, *323*(9), 844–853. https://doi.org/10.1001/jama.2020.1166.

Wu, Y., Totten, M., Memon, W., Ying, C., & Zhang, S. X. (2020). *In vitro* antifungal susceptibility of the emerging multidrug-resistant pathogen *Candida auris* to miltefosine alone and in combination with amphotericin B. *Antimicrobial Agents and Chemotherapy*, *64*(2), e02063-19. https://doi.org/10.1128/AAC.02063-19.

Yaakoub, H., Staerck, C., Mina, S., Godon, C., Fleury, M., Bouchara, J. P., & Calenda, A. (2021). Repurposing of auranofin and honokiol as antifungals against *Scedosporium* species and the related fungus *Lomentospora prolificans*. *Virulence*, *12*(1), 1076–1090. https://doi.org/10.1080/21505594.2021.1909266.

Yan, H., Yang, H., Wang, L., Sun, X., Han, L., Cong, P., Chen, X., Lu, D., & Che, C. (2022). Disulfiram inhibits IL-1β secretion and inflammatory cells recruitment in *Aspergillus* fumigatus keratitis. *International Immunopharmacology*, *102*, 108401. https://doi.org/10.1016/j.intimp.2021.108401.

Yang, C., Huang, Y., Zhou, Y., Zang, X., Deng, H., Liu, Y., Shen, D., & Xue, X. (2022). *Cryptococcus* escapes host immunity: What do we know?. *Frontiers in Cellular and Infection Microbiology*, *12*, 1041036. https://doi.org/10.3389/fcimb.2022.1041036.

Yousfi, H., Cassagne, C., Ranque, S., Rolain, J. M., & Bittar, F. (2019). Repurposing of ribavirin as an adjunct therapy against invasive *Candida* strains in an *in vitro* study. *Antimicrobial Agents and Chemotherapy*, *63*(10), e00263-19. https://doi.org/10.1128/AAC.00263-19.

Zaragoza O. (2019). Basic principles of the virulence of *Cryptococcus*. *Virulence*, *10*(1), 490–501. https://doi.org/10.1080/21505594.2019.1614383.

Zhai, B., Wu, C., Wang, L., Sachs, M. S., & Lin, X. (2012). The antidepressant sertraline provides a promising therapeutic option for neurotropic cryptococcal infections. *Antimicrobial Agents and Chemotherapy*, *56*(7), 3758–3766. https://doi.org/10.1128/AAC.00212-12.

Zhang, J., Liu, W., Tan, J., Sun, Y., Wan, Z., & Li, R. (2013). Antifungal activity of geldanamycin alone or in combination with fluconazole against *Candida* species. *Mycopathologia*, *175*(3-4), 273–279. https://doi.org/10.1007/s11046-012-9612-1.

Zhao, Y., Ye, L., Zhao, F., Zhang, L., Lu, Z., Chu, T., Wang, S., Liu, Z., Sun, Y., Chen, M., Liao, G., Ding, C., Xu, Y., Liao, W., & Wang, L. (2023). *Cryptococcus neoformans*, a global threat to human health. *Infectious Diseases of Poverty*, *12*(1), 20. https://doi.org/10.1186/s40249-023-01073-4.

Zheng, Y. J., Xie, T., Wu, L., Liu, X. Y., Zhu, L., Chen, Y., Mao, E. Q., Han, L. Z., Chen, E. Z., & Yang, Z. T. (2021). Epidemiology, species distribution, and outcome of nosocomial *Candida* spp. bloodstream infection in Shanghai: an 11-year retrospective analysis in a tertiary care hospital. *Annals of Clinical Microbiology and Antimicrobials*, *20*(1), 34. https://doi.org/10.1186/s12941-021-00441-y.

Zhou, J., Li, J., Cheong, I., Liu, N. N., & Wang, H. (2021). Evaluation of artemisinin derivative artemether as a fluconazole potentiator through inhibition of Pdr5. *Bioorganic & Medicinal Chemistry*, *44*, 116293. https://doi.org/10.1016/j.bmc.2021.116293.

Zhu, C., Liao, B., Ye, X., Zhou, Y., Chen, X., Liao, M., Cheng, L., Zhou, X., & Ren, B. (2021). Artemisinin elevates ergosterol levels of *Candida albicans* to synergise with amphotericin B against oral candidiasis. *International Journal of Antimicrobial Agents*, *58*(3), 106394. https://doi.org/10.1016/j.ijantimicag.2021.106394.

Ziccardi, M., Souza, L. O., Gandra, R. M., Galdino, A. C., Baptista, A. R., Nunes, A. P., Ribeiro, M. A., Branquinha, M. H., & Santos, A. L. S. (2015). *Candida parapsilosis (sensu lato)* isolated from hospitals located in the Southeast of Brazil: Species distribution, antifungal susceptibility and virulence attributes. *International Journal of Medical Microbiology: IJMM, 305*(8), 848–859. https://doi.org/10.1016/j.ijmm.2015.08.003.

Zubiaur, P., Soria-Chacartegui, P., Villapalos-García, G., Gordillo-Perdomo, J. J., & Abad-Santos, F. (2021). The pharmacogenetics of treatment with olanzapine. *Pharmacogenomics, 22*(14), 939–958. https://doi.org/10.2217/pgs-2021-0051.

Zuo, X. S., Liu, Y., & Hu, K. (2021). Epidemiology and risk factors of candidemia due to *Candida parapsilosis* in an intensive care unit. *Revista do Instituto de Medicina Tropical de São Paulo, 63*, e20. https://doi.org/10.1590/S1678-9946202163020.

Chapter 2

Drug Repurposing for Effective Alzheimer's and Parkinson's Medicines: Potential Approaches for Drug Discovery

**Md. Jawaid Akhtar*
and Shah Alam Khan**
Department of Pharmaceutical Chemistry, National University of Science and Technology, Bousher, Muscat, Sultanate of Oman

Abstract

Neurological diseases such as Alzheimer's disease (AD) and Parkinson's disease (PD) currently lack effective treatments. The currently used drugs are often associated with challenges such as drug resistance and toxicity. Drug repurposing is a process of identifying new clinical indications from existing drugs. It offers a time-efficient and cost-effective alternative to traditional drug discovery. The known safety profiles, toxicity data, and targets of existing drugs significantly speeds up the drug development process. In this chapter, we discuss the importance of drug repurposing and explore various approaches employed in this field. Techniques such as high throughput screening, the utilization of vast transcriptome datasets from gene expression databases, and the analysis of structural information enable the discovery of new indications for approved drugs. These methods facilitate the generation of drug-target connectivity maps, linking small molecules to disease associated gene expression signatures, and represent groundbreaking advances in drug repurposing. The chapter also sheds light on drugs found to be effective against AD and PD from the existing approved drugs. It covers natural

* Corresponding Author's Email: mjawaid@nu.edu.om

In: Drug Repurposing
Editors: Shah Alam Khan and Md. Jawaid Akhtar
ISBN: 979-8-89530-422-8

products, antihypertensive, anticancer, anti-inflammatory, antidiabetics, antibiotics, etc., found to be promising candidates against AD and PD. The chapter aims to support the scientific community and pharmaceutical industries by providing them with updated information that could help them refine and direct the development of novel repurposed drugs as anti-Alzheimer's and anti-Parkinsonian agents.

Keywords: Alzheimer's disease, drug discovery, drug repurposing, machine learning, Parkinson's disease

1. Introduction

Life expectancy is affected by the prevalence of neurodegenerative diseases including Alzheimer's and Parkinson's diseases. These diseases currently affect more than 50 million of the world's population. It involves the progressive death of the neurons in the brain (Arbo et al., 2022). Alzheimer's disease (AD) is characterized by memory impairment, decreased cognitive ability, and personality changes and involves accumulation of the phosphorylated tau (*p*-tau in neurofibrillary tangles) and amyloid-β accumulation in walls of blood vessels and extracellular spaces (Das et al., 2021). It is one of the major causes of dementia (Arbo et al., 2022; Atiya et al., 2023). Acetylcholine (ACh) decrease is linked with cognitive impairment in AD patients and is explained by the cholinergic hypothesis. The acetylcholinesterase (AChE) enzyme in the cerebral cortex and synaptic clefts and butyrylcholinesterase (BuChE) in the glial cells are responsible for the ACh metabolism and therefore, inhibitors of these enzymes are used as treatment strategy (Akkaya et al., 2024). So far, three cholinesterase inhibitors namely galantamine, rivastigmine, and donepezil have been approved by the FDA for the treatment of AD which acts by increasing the brain levels of ACh. The other treatment for AD involves N-methyl-D-aspartate (NMDA) receptor antagonists and monoclonal antibodies. The two recently FDA-approved monoclonal antibodies include lecanemab and donanemab. The main therapeutic purpose of these drugs is to slow down the progression of AD and thus be indicated for treatment of mild AD. The drug aducanumab was controversially approved from 2003 to 2022 but never allowed for medical use by a European medical agency (Grabowska et al., 2023). Mild to severe AD is treated by FDA-approved memantine that reduces the glutamate in the brain. Parkinson's disease (PD) is associated with the loss of the dopaminergic neurons in the substantia nigra and is associated with muscular rigidity, bradykinesia, postural and gait impairment. The treatment involves levodopa/carbidopa (precursor of dopamine), inhibitors of enzymes for dopamine metabolism such as monoamine oxidase-B (MAO-B), catechol-*O*-methyltransferase (COMT), and agonist of dopamine receptors (Arbo et al., 2022). The drugs approved by FDA are safinamide and istradefylline as an add-on treatment with PD therapy.

Neurodegenerative diseases have seen little progress in the development of newer therapeutics due to complex disease heterogeneity. Furthermore, molecules targeting these diseases fail at the different stages of the clinical trials due to safety concerns creating a big monetary burden on the pharmaceutical industries and thus requiring a novel drug discovery process. Drug repurposing or positioning is one of the effective strategies for finding new clinical indications for existing drugs (Atiya et al., 2023). It involves identifying the new therapeutic use of the existing drugs utilizing the classical drug discovery i.e., knowing its pharmacological effects or binding receptors and drug activity (Advani et al., 2020). Drug repurposing is an interesting strategy for the search for new therapeutics for treating neurological disorders at a lower cost and in much less time than developing the drugs from the beginning (Arbo et al., 2022). Drug repurposing increases the drug discovery from the known toxicity and safety profile of any drugs through a fast-track process (Shalaby et al., 2023). Many success stories highlight the role of drug repurposing in successfully treating various diseases and securing approval. Sildenafil approved for treating erectile dysfunction was developed as an antihypertensive drug, and thalidomide was approved for treating multiple myeloma in 2009 which was introduced as an antiemetic drug in the 1950s. The drug minoxidil approved for androgenetic alopecia was developed for hypertension whereas zidovudine approved for HIV/AIDS was developed for treating cancers. One of the drugs approved for AD treatment is galantamine which was marketed for treating paralytic and neuropathic patients (Das et al., 2024). This book chapter discusses the treatment strategy for AD and PD by repurposing the approved drugs indicated for some other disease conditions. It also describes different approaches/methods of drug repurposing.

2. Different Methods of Drug Repurposing

Traditional drug development includes target identification and then lead optimization followed by preclinical and clinical phases. In the drug repurposing process, existing drugs are tested against known targets preclinically and then moved to phase II clinical trials (Kulkarni et al., 2023). Drug repurposing is a cost-effective, safe, and low-risk approach for finding new indications of novel therapeutics in AD. It helps in identifying new therapeutic applications from the existing approved drug-related data. The challenges of the gathered data are heterogeneity, limited availability of drug-disease interaction, and diversity. The drug repurposing methods include knowledge-based where the known drugs and their targets are identified, the underlying mechanism and biomarkers then high throughput screening of thousands of compounds to test the compounds with activity. The dysregulated pathways can be identified in steps through single-cell genomic data and then those pathways can be targeted for identifying new drugs. The signature-based method identifies the gene expression profile (elevated or downregulated), and pattern associated with the disease, and then small molecules, or drugs used to treat play an important role in identifying the biological pathways affected by the help of computational tools and then drug targeting the dysregulated pathway are identified as drug repositioning. The computational-based approach includes utilisation of multi-omics data related to the disease and the drugs are selected based on the interaction between the drug and the disease. The creation of an algorithm to check the interaction between drugs and diseases validates the process. Other computational analysis includes chemical structure matching, pathways mapping, adverse effect profile matching, literature mining, machine learning, analysis of the data mining, and genome-wide association studies with the established library of drugs. The phenotypes are based on discovering the genetic traits linked with human disease and are based on patient electronic health reports (EHR) that include medical history, diagnosis, side effects, and outcomes. The computer-based identified drug is then followed by the validation steps including cell-based assay, comparing prediction with known drugs, sensitivity, and specificity-based metrics, and then clinical trials (Rani et al., 2024). Some of the recent examples include end-to-end knowledge graph-based approach (EKGDR) computer-based drug repurposing. The EKGDR can generate the causes behind drug-disease interactions and complex interactions between drugs, targets, side effects, and other relevant entities then combine these messages to generate disease and drug nodes in an end-to-end manner. The methods generated the top 20 candidates for AD and PD. The top 20 drugs predicted for AD are flecainide, tiagabine, riluzole, pyridostigmine, ambenonium, aprindine, felbamate, vandetanib, vigabatrin, neostigmine, guanfacine, mexiletine, rufinamide, regorafenib, betaine, dasatinib, atomoxetine, buspirone, dronedarone, and stiripentol. Among them, most showed positive BBB except one dasatinib with negative permeability and two drugs betaine and stiripentol had unknown values. Similarly, the top 20 drugs predicted for PD are 9 with positive BBB (pergolide, rimantadine, metixene, mephentermine, orphenadrine, amrinone, milrinone, and phenelzine), 9 with negative BBB (dobutamine, epinephrine, ritodrine, methoxamine, orciprenaline, methyldopa, ergoloidmesylate, and salbutamol) and two with unknown values (nesiritide, vilanterol) (Tayebi & BabaAli, 2024). Another example of drug repurposing includes the exploration of the transcriptome data stored in the databases (GEO and ArrayExpress) of the gene expression. The transcriptome data help to analyse the drug repurposing and provide the detailed gene expression profile during disease onset and progression. It analyses the specific states and correlation of the data between the patient and healthy individual, drug-disease connection of the patient, predicts drug combinations, and gene analysis before and after drug intervention (He et al., 2023). Other techniques of drug repurposing include computational drug repurposing which became more important after COVID-19 when a fast solution to cope with the disease was needed. Computer drug purposing includes docking, pharmacophore identification, similarity searching, and machine learning. This all happens in several steps like preparation of the data set, building and validation, of the model, scoring and prioritization and, further validation by prediction of *in vitro* and *in vivo* methods (Urbina et al., 2021). Another approach is designing the hybrid drug by identifying the pharmacophore from the failed drug or the natural products to target AD (Padhi & Govindaraju, 2022). A total of 970 genes related to PD were collected from the databases and the potential drugs were identified by the proximity scores and Inverted Gene Set Enrichment Analysis of the drug target. Ten drugs were selected based on a false discovery rate < 0.1. The drugs were carfilzomib, afatinib, prednisolone acetate, isosorbide, arsenic trioxide, vorinostat, tucatinib, tazemetostat, avapritinib, and tixocortol (Quan et al., 2021). Single-cell RNA sequencing data of the brain tissue of the patients with AD was used to build the molecular network with genome-wide association study and multiple gene annotation resources. The computational approach (data-driven high throughput screening; GWAS and differential gene expression analysis of single-cell RNA seq) identified 54 drugs targeting MAPK and IGF1R signaling pathways. The drugs targeting MAPK and related pathways includes afatinib, lapatinib, nintedanib, and sorafenib. MEK inhibitors such as selumetinib and trametinib and Raf inhibitor dabrafenib have also been suggested for drug repurposing for AD (Parolo et al., 2023). The novel prediction framework on deep learning was used to identify the repurposed drugs for AD by Pan et al. (Pan et al., 2023). The steps were the formation of drug-target pair (DTP) network based on drug features and the targets and then association of DTP nodes as the edges and AD disease network followed by relationship generation between drug-drug, target-target, drug-target within and outside of drug-target pairs as quartet to develop finally AI based drug discovery network. A total 61 drugs out of 461 had been tested in clinical trials. The EHR of the University of Pennsylvania Health System taking individuals of mild cognitive impairment with hypertension between 2012 to 2022 were investigated. It was observed that five classes of the drug used were β-blockers, renin-angiotensin-system, calcium channel blockers, diuretics, and antihypertensives. The results showed that two classes of the drug reduced dementia i.e., diuretics and β-blockers. The drug under diuretics and β-blockers were furosemide and propranolol which showed protective effects (Li et al., 2023).

3. Drug Repurposing in Alzheimer's Disease (AD) and Parkinson's Disease (PD)

3.1. Anti-Inflammatory Drugs

The drugs acting against neuroinflammation act as a common point against neurodegenerative diseases and have been explored for drug options. The treatment with NSAIDs for long-term decreased AD and PD development but is still to be researched well as studies are showing no association with NSAIDs treatment and less risk of development of AD and PD. TNF-α, a proinflammatory cytokines, is found to be elevated in cerebrospinal fluid of the aged patients related to the cognitive impairment. The use of etanercept, an anti-TNF molecule, showed improvement in the long-term memory deficit preclinically. There is no confirmed favouring data for the change in cognition or behaviour improvement with the use of etanercept and its rare and serious adverse-effect risk is also very high (Arbo et al., 2022). Neuroinflammation is the immune response against the injury and infections. This results in the secretion of the proinflammatory mediators from the changes in the morphology of glial cells that increase the blood-brain barrier (BBB) permeability for the immune cells for the removal of misfolded proteins, pathogens, and cell debris and causes tissue repair. Continuous neuroinflammation is the cause of neurological diseases such as AD and PD (Yang & Zhou, 2019). The fenamate tolfenamic acid (TA) exhibiting anticancer effects has potential to cross the BBB and slow down the progression of AD. The two other analogues of TA; 2-(2-cyanophenylthio) benzoic acid (CBA) and niflumic acid (NA) target the neurodegenerative pathway are having better safety and efficacy. The CBA and NA at the dose of 100 mg/kg were compared with TA for the differential expression of the genes and proteomic analyses in C57BL/6 mice. The differential expressed genes were shown to be linked with the clearance of neurotoxic Aβ peptides, tau phosphorylation, and oxidative stress regulation. The bioinformatics analysis showed that the drugs CBA and NA were better in targeting neurodegenerative pathways i.e., autophagy, neuronal trafficking, and inflammation, and have a similar mode of action as that of TA (Shalaby et al., 2023). The targeting of the nuclear factor-erythroid factor 2-related factor 2 (NRF2) plays a critical role in the inflammation and provides cytoprotection in several diseases. The proteins may act as an attractive therapeutic target for AD, which led Bourdakou et al., to search for NRF2 inhibitors as new anti-Alzheimer's agents through an *in-silico* drug repurposing approach (Drug Gene Budger computational tools). The structural similarity from the ongoing NRF2 targeting drugs in clinical trials was compared and evaluated against NRF2 disease and then the comparison of the shortlisted drugs was done with the ongoing clinical trials drugs against targeted pathways followed by comparison with the existing drugs of AD. The five repurposed drugs (curcumin, trichostatin-A (TSA), panobinostat, parthenolide and entinostat) in the cell culture increased the NRF2 protein levels and transcriptional signature by downstream expression of the downstream heme oxygenase 1. The repurposed drugs TSA, panobinostat and entinostat act as histone deacetylase (HDAC) inhibitors and neuroprotectants. Among them, TSA was found to be the top among the repurposed drugs (Bourdakou et al., 2023). The HDAC class III sirtuins targets the histone and non-histone proteins and thus emerged as promising therapy against neurological disorders such as AD and PD. The levels of sirtuins affect brain health are involved in ageing and are responsible for neurological diseases (Yeong et al., 2020).

Table 1. The repurposing of anti-inflammatory drugs to treat neurodegenerative diseases

S. No	Drug name	Role as anti-inflammatory	Role in neurodegenerative disease	References
1	Acetylsalicylic acid and meloxicam	Analgesic	Reduce the loss of nigral neurons and maintain dopamine level 1-methyl-4-phenyl-1,2,3,6-tetrahydropyridine (MPTP) exposed mice.	(Arbo et al., 2022)
2	Ibuprofen	Analgesic	Reduce the loss of nigral neurons and maintain dopamine level 1-methyl-4-phenyl-1,2,3,6-tetrahydropyridine (MPTP) exposed mice.	(Arbo et al., 2022)
3	Celecoxib	NSAID and analgesic	Translocation of nuclear factor erythroid 2-related factor 2 (Nrf2) to nucleus neutralizing ROS and decreasing lipid peroxidation in SH-SY5Y cells incubated with Aβ. Rat treated with Aβ showed decreased proinflammatory cytokines, induced synaptic plasticity and prevented loss of long-term memory impairment.	(Arbo et al., 2022)
4	Indomethacin	NSAID	Preclinically beneficial effect against PD and AD. It showed anti-amyloidogenic effect and Aβ protection *in vitro*. It showed neuroprotective effect in mice exposed to MPTP. It reduces the amyloid pathology blocking the nuclear factor kappa B (NF-kB).	(Arbo et al., 2022)

The proteins involved in AD and PD cause IL-1β release (a pro-inflammatory cytokine). The drug canakinumab neutralizes the human IL-1β and suppresses inflammation. Similarly, inhibition of the mitogen-activated protein kinase p38 alfa (p38α) expressed in inflammation, synaptic dysfunction, and Aβ deposition showed favourable results. The drug neflamapimod showed improved activity in the Morris water maze test and reduced the IL-1β level in the aged rat hippocampus (Arbo et al., 2022). *In silico* study followed by an *in vitro* investigation (Thioflavin T-amyloid binding assay and protease digestion assay) confirmed naproxen to have high free binding energy and it was also found to cross BBB. It has the potential to treat AD by altering the aggregation and conformation of Aβ peptides (Sharma et al., 2021). The drug tarenflurbil is the active enantiomer of flurbiprofen and was being developed for the treatment of AD. It inhibits the activity of γ-secretase that generates Aβ peptide from APP. Although it failed in clinical trials due to poor BBB penetration (Séguy et al., 2022). A list of repurposed NSAIDs tested for the treatment of neurodegenerative diseases is presented in Table 1.

3.2. Anticancer and Immunosuppressant Drugs

The common background between cancer and neurodegenerative disorders led to the emergence of anticancer compounds as repurposed drugs with notable neuroprotective effects. There are common mutated genes found in both cancer and neurodegenerative disorders (Plun-Favreau et al., 2010). The signalling pathways BRCA1, Wnt, PIN1, p53, and common risk factors for the inflammation shared the common connection between cancer and AD. The genomic integrity between the two diseases depends upon the regulation of genes related to apoptosis, proliferation, and DNA repair proteins. AD and cancer have the commonality of DNA damage deposition in the neurons/glial cells, genomic instability, disruption of the ubiquitin-proteasome system (UPS), and autophagy. The aberrant cleavage of the APP and tau proteins is linked with the generation of Aβ peptides and neurofibrillary tangles (NFT) likewise certain cancer such as luminal breast cancer and neuroblastoma is associated with APP cleavage and tau deposition (Das et al., 2024). However, antineoplastic and immunosuppressant drugs are less favourable due to their adverse effects than the drugs acting on the nervous system (Grabowska et al., 2023).

3.2.1. Mechanism of the Kinase Inhibitors for Neurodegenerative Disorders

The complexity of CNS drug development and the lack of a valid target made it difficult to develop kinase inhibitors against neurological diseases. The identification and characterization of new kinases may be a potential target for the development of CNS drugs (Krahn et al., 2020).

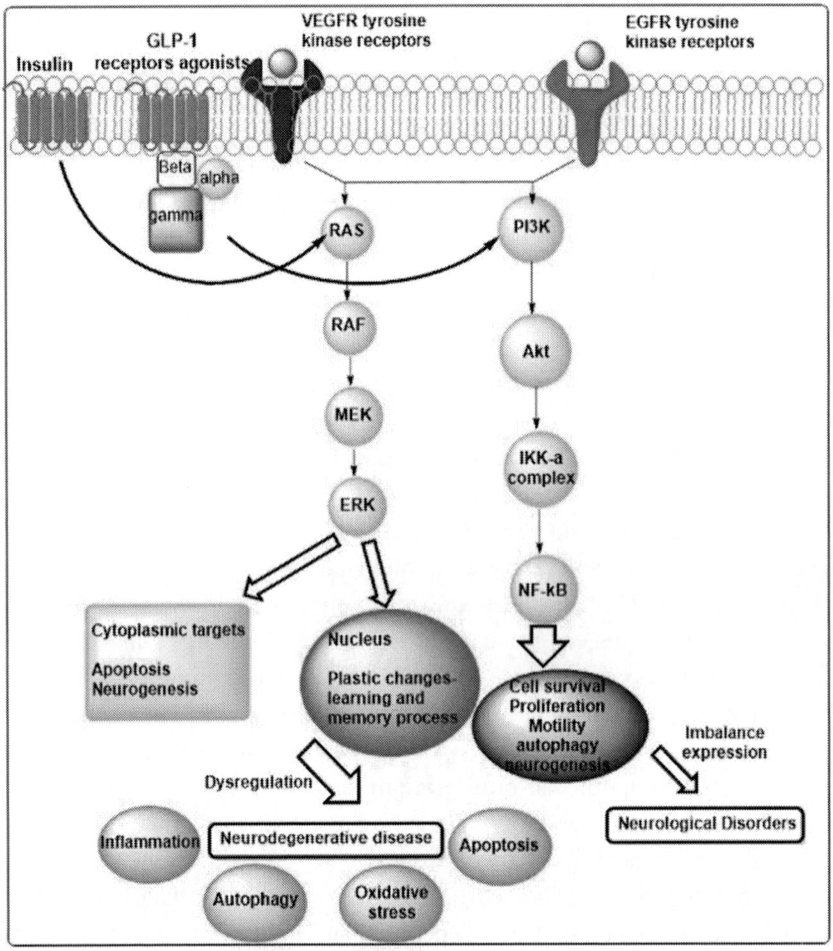

Figure 1. Ras/Raf/MAPK and P13K-AKT pathways for the development of neurodegenerative diseases.

The MAPK/ERK pathway serves the transmission of the cellular signals and plastic changes. After the phosphorylation of the ERK cytoplasmic or nuclear changes through sets of activating genes for differentiation, migration, and plastic changes are induced. The cytoplasmic target of the pERK involves apoptosis and neurogenesis which after the diffusion to the nucleus activates transcription factors. The sustained ERK activation with several immediate early genes c-Fos, Arc, and CREB led to the plastic changes and are related to learning and memory processes (Albert-Gascó et al., 2020). The dysregulated signalling pathways are causes of neurodegenerative disorders. The factors that influence neurodegeneration are inflammation, autophagy, oxidative stress, and apoptosis. Thus, controlling the dysregulated mechanism with kinase inhibitors plays an important role in the management of neurological disorders (Gravandi et al., 2023). Similarly, P13K-Akt pathways are involved in cell survival, proliferation, cell cycle, protein synthesis, neuronal morphology, and plasticity however imbalanced expression in the nervous system led to the progression of neurological diseases and are still under exploration (Long et al., 2021). Figure 1 illustrates the role of Ras/Raf/MAPK and P13K-AKT pathways in the development of neurodegenerative diseases.

3.2.2. Repurposed Anticancer Drugs against PD and AD

Advani et al., in the year 2021 studied the possibility of the repurposed drugs for AD from the approved anticancer drugs *via in silico* method based on omics data. AD-related genes were searched to find approved anticancer drugs related to AD targets. The list of 2914 AD-related genes and 49 potential druggable targets were identified. Furthermore, 15 AD-risk protein-protein interactions were found to be related to 30 approved anticancer drugs. The validation through drug-target interaction, functional analysis, and molecular mechanism resulted in identification of four repurposed drugs among which 3 are epidermal growth factor (EGFR) inhibitors. The three EGFR inhibitors include vandetanib, erlotinib, and gefitinib. It was identified that miRNA 200a shares common targets with three repurposed drugs. It was further suggested that the mechanisms involved in targeting AD were tau phosphorylation, autophagy, and neuroinflammation. The pathways targeted were PI3-Akt signalling, NF-kappa B pathway, and Ca^{2+} signalling (Advani & Kumar, 2021). It was observed that there is a higher occurrence of breast cancer (BRCA) in patients with PD. The integrated omics approach was used to identify the common genes associated with PD and BRCA. The dysregulated process, common regulatory mechanism, and transcription factors, miRNAs, and disease-specific gene expressions were compared for identification of repurposed drugs by LINCS L1000 analysis. The protein with both the increased neuronal α-syn uptake and inclusion body in PD as well as BRCA progression, invasion, and migration is Caveolin-1 (CAV1). The protein encoded by EIF4G1 components of the eIF4f complex for mRNA translation in PD and inflammatory breast cancer development. The protein sequestosome-1 (SQSTM1), WNT3, and UBTF are involved in both PD and BRCA. Among the drugs, lapatinib, mitoxantrone, neratinib, and palbociclib were the top interacting drugs with BRCA and have relations with PD. Tamoxifen and raloxifene are approved for positive metastatic breast cancer and invasive breast cancer and have beneficial effects in PD validated by CoDReS tools (Advani & Kumar, 2022). The drug lenalidomide inhibiting TNF-α, IL-6, IL-12, and IL-1 showed reduced microgliosis, expression of pro-inflammatory cytokines, and NF-κB activation preclinically (Arbo et al., 2022). The immunosuppressant drugs cannot be prescribed for the long term when administered systematically. The immunosuppressant rapamycin was found to be effective in the treatment of AD (Séguy et al., 2022). Its activity of abelson tyrosine kinase 1 (ABL1) was found to increase in stratum and substantia nigra thus it was postulated that ABL1 may be a target for PD. Therefore, nilotinib as an ABL inhibitor prescribed for leukaemia was postulated to prove beneficial in PD. The results of phase II study showed that the drug was safe and tolerable but lacked efficacy and deteriorated motor function and thus was not carried well to phase III trials (Fletcher et al., 2021). Repurposing of nilotinib, imatinib as c-Abl tyrosine kinase inhibitor (cAbl-TKI) as treatment of neurodegenerative diseases is considered as c-Abl is the regulator of the cell stress response. The TKI also promotes autophagy for the removal of the toxic tau and amyloid protein aggregates. The work also suggests c-Abl working as a promoter of the AD pathology *via* autophagy inhibition, tau phosphorylation, and p53 regulation. Figure 2 shows the role of p53 signalling against cancer and neurodegenerative diseases. The results showed that autophagy increased by 40% with the use of TKI. Inhibition of tau and amyloid aggregates were more *in vitro* than *in vivo* with respect to tau transgenic mice (Karelina, 2023). The olfactory dysfunction occurs in more than 90% of the cases of AD patients. The severity of AD is reflected by the hyperphosphorylated tau and Aβ deposition in the olfactory bulb and olfactory tract region. The early detection of the protein expression profile during progression by high-resolution mass spectrometry (HRMS) is an attractive approach for drug repurposing to reverse AD-related olfactory omics signatures (OMS). It was found by using a connectivity map-based drug repurposing approach that EGFR, aurora kinase, glycogen synthase kinase,

CDK inhibitors, and PKC ought to be top classes capable of reversing OMS. The insulin-like growth factor 1 (IGF-1), microtubules, and Polo-like kinase (PLK), PI3K inhibitors have potential to induce olfactory AD-associated gene expression changes. The results of an *in vitro* assay in pretreated SH-SY5Y human neuron-like cells with AG-147 (EGFR inhibitor) showed a neuroprotective effect in H_2O_2-induced damage whereas aurora kinase inhibitor, reversine decreased Aβ-induced neurotoxicity (Cartas-Cejudo et al., 2024). Rodriguez et al., used the DRIAD (drug repurposing in AD) a ML approach, and quantified the pathology of AD severity with the genes involved. The genes from the differentiated human neural cell cultures arise from the 80 FDA-approved drugs and clinically tested ones ranked the repurposed drugs. From the earlier work, it was found that top-scoring compounds such as TAE684, dovitinib, ruxolitinib, and fedratinib bind to the target genes RPS6KA1. Similarly, other genes NEK3, NEK6 and NEK9 showed a positive interaction with the top-scoring compounds and these genes are associated with microtubule function and tau phosphorylation. The results showed that the selected compounds for *in vitro* are majorly kinase inhibitors with anticancer activity. Among them, the top performers are the JAK inhibitors whereas other targets including JAK, ULK, and NEK are the most consistent targets. The top 15 FDA drugs according to the harmonic p values selected were ruxolitinib, vorinostat, regorafenib, tofacitinib, nilotinib, lapatinib, palbociclib, ruxolitinib, cabozantinib, baricitinib, nintedanib, tofacitinib, gefitinib, resveratrol, dasatinib and many drugs such as ruxolitinib inhibit more than one member of the Janus kinase (JAK1, JAK2 and JAK3 and TYK2). The drug raltitrexed was repurposed as anti-Alzheimer's and G9a/EHMT2 inhibitors. The work of G9a is to demethylate the H3K9 of chromatin and to repress the genes epigenetically leading to many diseases. It was also shown that raltitrexed targets G9a directly and lowers the total H3K9me2/H3K9 levels in AD transgenic *C. elegans* CL2006 worms. It also reduces Aβ aggregates and enhances the locomotive function of the worm highlighting its potential to act as an AD treatment (Jana et al., 2023). Table 2 shows the repurposing of the anticancer drugs against AD and PD.

Table 2. The repurposing of antineoplastic drugs to develop neurological disease therapeutics

S. No	Name of the drug	Role in cancer	Role in neurodegenerative diseases	References
1	Lenalidomide	Treating myeloma, lymphoma.	It showed a neuroprotective effect in the Amyotrophic Lateral Sclerosis (ALS) mouse model. A phase II trial in mild to moderate AD patients showed no cognition improvement.	(Das et al., 2024)
2	Vorinostat	Histone deacetylase inhibitors for T-cell lymphoma.	Potential drug candidates for Niemann-Pick Type C characterized by progressive neurodegeneration with ataxia, seizures, dementia, and Aβ and NFT deposition like AD. Phase I/II showed serious adverse effects such as vomiting, infections, gastrointestinal, and nervous system disorders.	(Das et al., 2024)
3	Daratumumab	Anti-CD38 antibody approved for treating multiple myeloma.	The drug is in phase II for treating mild to moderate AD.	(Das et al., 2024)
4	Dasatinib and quercetin	Dasatinib is highly effective against imatinib-resistant CML patients.	Reduce inflammation and senescent cells in ageing mice by altering the gut microbiota signature which is a potential risk factor in AD and PD.	(Das et al., 2024)
5	Bosutinib	Treating CML and Philadelphia chromosome-positive acute lymphoblastic leukaemia.	It facilitates the removal of amyloidogenic proteins in neurodegeneration and improves cognitive functions.	(Das et al., 2024)
6	Lapatinib	It is HER2 and EGFR inhibitor approved for advanced and metastasized breast cancer.	It has the potential to improve cognitive impairment by decreasing Aβ aggregation and modulating PI3K/Akt/GSK-3β signalling pathways. It also reduces oxidative stress and inflammation with reduced $A\beta_{1-42}$ and *p*-tau accumulation.	(Das et al., 2024)
7	Leuprolide	Treating advanced prostate cancer. It decreases luteinising hormone (LH) and secretion of FSH.	It decreases LH level and improves cognitive skills. It lowers amyloid and tau deposition in animal models.	(Das et al., 2024)
8	Nilotinib and masitinib	Nilotinib is a brain-penetrant oral Bcr-Abl TKI used to treat CML. Masitinib is c-kit inhibitor.	Nilotinib as Bcr-Abl inhibitor exhibited neuroprotective activity. Masitinib also acts as potential therapy for AD. In phase IIb/III masitinib reduces the cognitive decline and dementia progression.	(Das et al., 2024)

S. No	Name of the drug	Role in cancer	Role in neurodegenerative diseases	References
7	Sulforaphane	Preclinically the drug sulforaphane shows anticancer activities in multiple cancer models.	It also reduces Aβ burden and neurobehavioral deficit in AD mouse model and HDAC. It is also an inhibitor of NF-kB signalling.	(Das et al., 2024)
8	Tamibarotene	To treat acute promyelocytic leukaemia in Japan.	In preclinical models, it showed anti-AD by clearance of Aβ and APP targeting.	(Das et al., 2024; Parvez, 2023)
9	Nabilone	It is used to treat chemotherapy-induced nausea. It is a partial agonist for cannabinoid CB1 and CB2 receptors.	Cannabinoids decrease local and peripheral inflammation but there are no significant benefits in dementia and in neurological patients.	(Das et al., 2024)
10	Temsirolimus	It is an anticancer drug which is used to treat renal cell carcinoma.	It shows the clearance of Aβ in HEK293-APP695 cells by autophagy. It potentiates cellular apoptosis in APP/PS1 mice and improves memory and spatial learning.	(Das et al., 2024)
11	Dabrafenib	It is an anticancer drug against melanoma.	Th drug showed neuronal loss against the MPTP mouse model of PD.	(Fletcher et al., 2021)
12	Bexarotene	Retinoid X receptor (RXRs) agonist approved by FDA against cutaneous T-cell lymphoma.	The neuroprotective effect effectively eliminates Aβ plaques in mice models. It also induces ApoE4 overexpression and is responsible for the reduction of Aβ plaques.	(Parvez, 2023)
13	Resveratrol	Anticancer natural product.	It activates calcium/calmodulin-dependent protein kinase-β (CaMKKβ) by increasing intracellular calcium to promote autophagy.	(Liu et al., 2022)

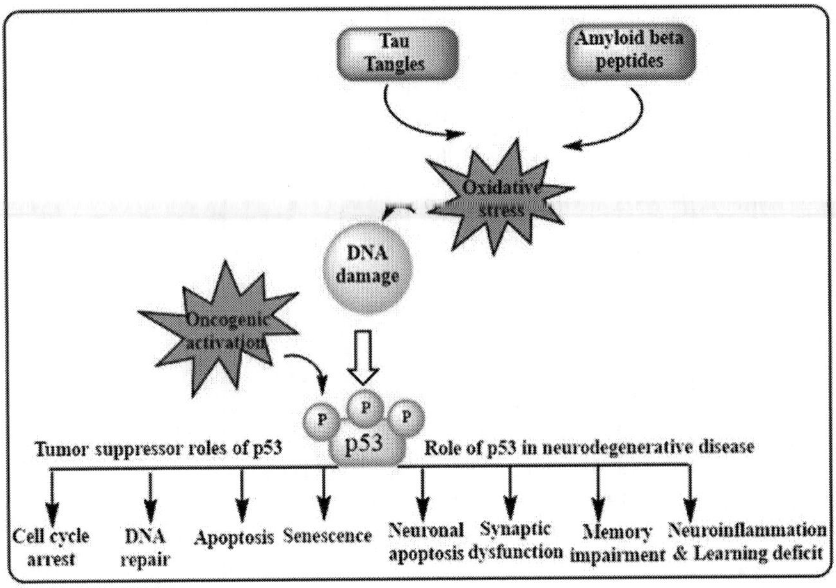

Figure 2. Role of p53 signalling in cancer and neurodegenerative disease.

3.3. Cholinesterase Inhibitors

The serious side effects of cholinesterase inhibitors against the use of neurodegenerative disorder AD led to the search drugs effective in other diseases. Among the drugs trimebutine, theophylline, and levamisole showed the most potent AChE inhibitory activity with an IC_{50} value of 68.70, 53.25, and 44.03% at a dose of 1000 µM. The molecular modelling confirmed the good fitness of trimebutine in the active site of AChE and is considered as a drug of the highest potential against AD (Akkaya et al., 2024). The structure-based scaffold repurposing approach was utilised by Dighe et al., in 2020. They developed a total of 6 cholinesterase inhibitors from the library of DNA gyrase inhibitors. The two compounds were identified as the most potent AChE and BuChE inhibitors. Compound **1** showed the most potent inhibition with an IC_{50} value of 6.10 µM whereas compound **2** was more active against BuChE with an IC_{50} value of 5.50 µM. *In-silico* studies showed that compound **1** demonstrates π-π stacking and H bond interactions within the active site of the AChE and compound **2** showed the binding in the choline and the catalytic binding sites of BuChE (Dighe et al., 2020). The drug repurposing screen for 11353 compounds with 3D shape similarity and pharmacophore modelling showed initial hit and eight molecules to be similar in structure with pyridostigmine. The structure-based screening using the pharmacophore-guided rescoring method identified 3 compounds to be potent AChE inhibitors. The 6-(2-(pyrrolidin-1-yl)pyrimidin-4-yl)-thiazolo[3,2-*a*]pyrimidine derivatives **3**, **4**, and **5** showed the most promising anticholinesterase activity with an IC_{50} values of 13.10, 16.02, and 6.22 μM. Compound **5** showed a similar activity to the FDA-approved drug galantamine. The compounds also showed hydrogen bonding and hydrophobic interaction within the active sites of the AChE. Figure 3 shows the structure of repurposed synthesized derivatives as potential AD therapeutics. The multi-targeted approach of inhibiting MAO-B and AChE from the FDA-approved drugs was repurposed against AD. The molecules after the virtual assessment and docking were subjected to induced fit (simulation) studies and binding energy calculation by MMGBSA. The results showed the 4 drugs (Dolutegravir, Rebamipide, Loracarbef, and Diflunisal) to be active as dual MAO-B and AChE inhibitors however in simulation with MAO-B two of the drugs showed false results i.e., Diflunisal and Rebamipide. The antiviral drug dolutegravir showed the most potent inhibition of 41% and 61% against MAO-B and AChE at 10μM dose. There were several stable hydrogen bond interactions within the active sites of the MAO-B and AChE receptors (Mateev et al., 2023).

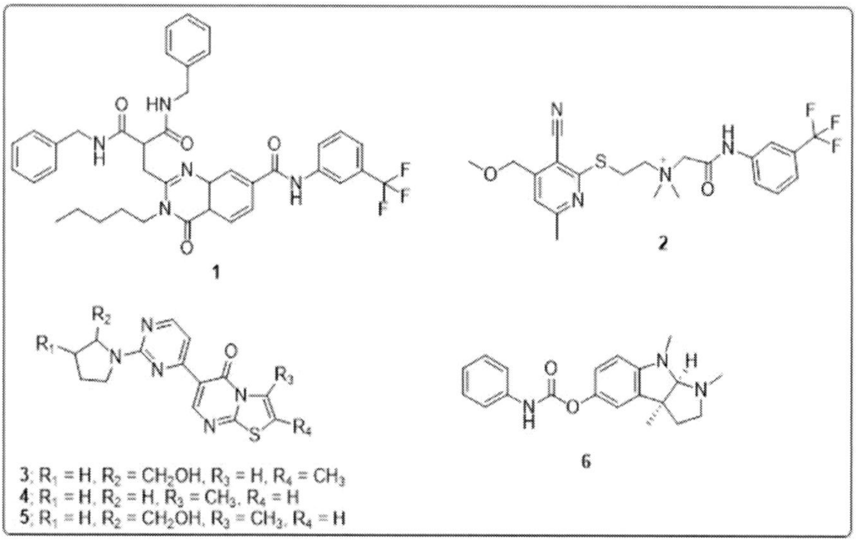

Figure 3. Chemical structures of some repurposed cholinesterase inhibitors.

Ballard et al., in 2020 through a systematic literature review and meta-analysis identified phenserine (**6**), a cholinesterase inhibitor, initially developed to replace physostigmine drug. It was found to be effective in ameliorating cognitive functions in AD by both cholinergic and non-cholinergic pathways but previously failed in phase III of AD-aimed clinical trials (Ballard et al., 2020).

3.4. CNS Drugs

The genetic targets for the drug discovery and repurposing were identified by the summary-data-based Mendelian randomization. This resulted in the identification of 116 AD, 3 ALS, 5 Lewy body dementia, 46 PD, and 9 progressive supranuclear palsy target genes by passing multiple test corrections. This led to the identification of 41 novel targets, 3 known targets, and 115 difficult targets and provided a new opportunity for the drug discovery, development, and repurposing approach. The drug riluzole was identified as a potential drug for AD by observing the drug-gene targets (Alvarado et al., 2023). The protein SORT1 is involved in various neurological disorders. The targeting of the SORT1 is challenging which led to the identification of two potential compounds, ergotamine and digitoxin by drug repurposing approach. The combined approach involved molecular docking, virtual screening, and molecular dynamics simulations of 3648-FDA approved drugs. The two hit molecules were analysed for the interaction with SORT1 and stability assessment *via* MD simulations for 300 ns that showed complex binding interactions. The structure-guided drug repurposing approach showed the significance of these molecules against SORT1 receptors (Almoyad et al., 2024). The drug fluoxetine was repurposed as a potent anti-AD drug-like compound and has the potential to inhibit transferrin. Transferrin acts as an iron transporter and is involved in the deposition and accumulation of iron and causes neurodegenerative and neuroferrinopathies. MD simulations showed the stable binding of drug fluoxetine with transferrin for 100 ns . Fluoxetine binds with the transferrin with a binding constant of $5.5 \times 10^5\ M^{-1}$ (Khan et al., 2022). Among the 573 repurposed drugs over the last 10 years, the second-generation antipsychotic clozapine emerged out to be the most promising repurposing candidate. The *in vivo* validation from the animal model was considered as most popular validation (Grabowska et al., 2023). The affinity of a drug target by a single model approach alone does not provide precise predictions for the success of any drug. Therefore, multifaceted features such as taking consideration of sequence, 3D structure, and the interaction between drugs and proteins using the drug target approach (MNNEL-DTA) were introduced by Zhao et al. The protein-protein target interactions were done on three AD-causing targets viz., spleen tyrosine kinase (STK), sigma 1 receptor (S1R), and medin screened for the corresponding dataset of ligands. The MNNEL-DTA includes various neural networks GNN, LSTM, and CNN integrating multiview features of the drug and protein and then identifying the potential therapeutic drug among a list of compounds. The approach identified memantine as a joint inhibitor of the S1R and STK whereas glutethimide (earlier prescribed for insomnia) was inhibitor of target medin. The drug memantine is already an FDA-approved drug for AD and the potential of glutethimide for AD treatment thus underscores the success in identifying the potent dual drug (Zhao et al., 2024). Several trials for the use of risperidone against AD are reported as it mitigates the psychological and behavioural symptoms associated with AD but still, there is no evidence of efficacy in the treatment of AD (Grabowska et al., 2023). The anticonvulsant drug zonisamide was found to be effective against levodopa-induced dyskinesia. It blocks the motor function in PD and is repurposed against dementia as it acts by blocking T-type calcium channels (Sahu et al., 2024). It also inhibits MAO-B, thereby decreasing the degradation of dopamine in neural and glial cells to improve the outcome of anti-PD treatment (Riederer & Müller, 2018). Paroxetine can be another good candidate for AD patients with depression (Savva et al., 2022). The fluorescence aggregation kinetics in the presence and absence of SSRIs and evaluating the change in the morphology of Aβ by transmission electron microscopy (TEM) and computational modelling using the software Discovery Studio Structure-Based-Design showed paroxetine was able to prevent the aggregation of both Aβ40 and Aβ42 peptides (Rao et al., 2020). The action of the drug pomalidomide as neuroprotective was tested for chronic 2-month administration after the infusion of human α-synuclein (H-αSynOs) for 1 month. The drug pomalidomide rescued the impairment in motor and coordination performance. The loss of 40-45% cells of the bilateral substantia nigra was abolished by the drug. The microglia showed proinflammatory production with increased anti-inflammatory cytokines TNF-α by the infusion of H-αSynOs whereas it is inhibited by the drug with an increase in the production of anti-inflammatory cytokine IL-10. The finding in the rodents' model of PD supports the repurposing of the drug pomalidomide as potential against PD (Palmas et al., 2022). Halliday et al., conducted *in vitro* and *in vivo* studies to study the neuroprotective effects of trazodone, an antidepressant drug. Trazodone was found to reverse eIF2α-P translational repression which is overactivated in AD patients and thus showed promising restoration of memory and prevented neurodegeneration in mice (Halliday et al., 2017). Grabowska et al., carried out a systematic review and included 124 relevant studies on AD published from 2012 to 2022 for a meta-analysis which resulted in the identification of 573 drug candidates for repurposing in AD. Amongst all the drug candidates, clozapine anti-psychotic drug appeared in six studies was identified as a promising drug molecule for further testing. Besides, adenosine, risperidone, tamoxifen, and verapamil were also identified in five studies (Grabowska et al., 2023). Lithium, an antimanic drug, which is used in the treatment of bipolar disorder has also been suggested to be a promising anti-AD drug repurposing candidate (Yu et al., 2021). Lithium by its ability to inhibit glycogen synthase kinase 3β (GSK3β), decreases tau hyperphosphorylation and thus can improve the cognitive functions in AD patients. Many ongoing clinical trials are evaluating its potential in AD (SA, 2024). Many studies have shown that methylphenidate, a CNS stimulant, can effectively reduce gait disorders of PD, as well as non-motor symptoms because of its ability to inhibit dopamine and noradrenaline reuptake, in the striatum and the prefrontal cortex (Devos et al., 2013).

3.5. Antidiabetics Drugs

It has been demonstrated that there is a link between neurodegenerative disorders such as PD, AD, and diabetes mellitus (DM). The risk factor of the DM with the neurodegenerative disorders is not fully understood but both share the common pathways leading to inflammation, oxidative stress, and mitochondrial dysfunctions. The brain cells are also unresponsive to insulin and cause insulin resistance thus leading to neurodegeneration. The change is related to the development of reduced neuroplasticity and memory that causes neurodegeneration. The commonality between neurodegeneration and DM led to exploration of the potential of antidiabetic drugs as drug repurposing for neurodegenerative diseases. Antidiabetic drugs are found to improve insulin resistance by reducing blood sugar levels and act as antioxidants, anti-inflammatory, and antiapoptotic (Arbo et al., 2022). The treatment strategy that reduces the pathologies of DM type 2 may be successful in treating neuroinflammation and neurodegenerative diseases (Kopp et al., 2022). The sulfonylurea antidiabetics such as tolbutamide, chlorpropamide, and glyburide were found to exhibit strong inhibitory activity against AChE. Among them glyburide was the most potent against AChE with an IC_{50} value of 0.74 μM which is comparable with the standard FDA-approved drug donepezil. Chlorpropamide and tolbutamide inhibited AChE with an IC_{50} value of 5.72 and 28.9 μM. The drug glyburide showed strong binding inside the catalytic sites and occupied the full gorge of AChE. The MD simulation studies of glyburide showed competitive inhibition against AChE (Baruah et al., 2021). The microtubule affinity regulating kinase 4 (MARK4) is targeted for various diseases and is linked to both DM and AD. The structure-based virtual screening of the FDA-approved drugs against MARK4 identified two hits' drugs linagliptin and empagliflozin interacting with the pocket of the protein critical residues. MD simulation showed good binding within the MARK4. The kinase assay using the malachite green showed these drugs to be potent inhibitors of MARK4 with an IC_{50} values of 7.63 and 7.10 μM for linagliptin and empagliflozin, respectively (Atiya et al., 2023). The results of the clinical as well as preclinical studies prove the beneficial effects of glucagon-like peptide-1 (GLP-1) agonists against neurodegenerative diseases. The results showed the positive effects for the PD patients treated with exenatide, a synthetic GLP-1. Therefore, a larger, multi-center Phase III study involving exenatide has been planned. Similarly, the results of semaglutide on the AD patients are promising in clinical trials (Chen et al., 2023). Liraglutide increased the glucose transport that is commonly found with AD. In another clinical study with liraglutide, there were no marked primary changes in the cortical glucose metabolic rates however there were marked secondary changes i.e., various markers of AD, tau phosphorylation, inflammatory markers, and Aβ levels. Also, the liraglutide treatment is safe and well-tolerated. Exenatide was also found to be well tolerated which reduced Aβ-42 concentration. In the moderate stage of PD, exenatide was found to improve mood and reduce depression. Semaglutide is also found to have potential for the treatment of AD and PD (Kopp et al., 2022). Harding utilised the mouse model of prion disease that covers many aspects of neurodegenerative diseases includeing endoplasmic reticulum stress, multiple dementias, synaptic loss, and cell death to repurpose the antidiabetics drugs. The preliminary investigation showed that metformin reduced motor disruption (Harding, 2023). Mendelian randomization was used to investigate 3000 genes codes for the druggable proteins for drug targets of PD.

Table 3. The repurposing of antidiabetics for the neurological diseases' therapeutics

S. No	Drug name	Role as anti-diabetic	Role in neurodegenerative disease	Reference
1	Biguanides (Metformin)	It inhibits glucose production and increases glucose utilisation.	It reduces ROS and proinflammatory cytokines IL-6, IL-1β, and TNF-α. Blocks the NF-kB and reduces the pathological symptoms of AD such as spatial memory and learning impairments, reduces Aβ plaque, and potentiate phagocytosis of Aβ and tau proteins. It decreases deposition of Aβ plaque, oxidative stress, inflammation and inflammatory cytokine levels and inhibits microglia-induced neuroinflammation in APP/PS1 mice model. It improves learning and memory ability and protects against MPTP neurotoxicity. It reduces the α-synuclein levels by 47.3% thus reducing the development of AD.	(Arbo et al., 2022; Birajdar et al., 2023; Khatri et al., 2021)
2	Thiazolidinediones (rosiglitazone and pioglitazone)	Peroxisome proliferator-activated receptor-γ (PPAR-γ) agonist.	PPAR-γ agonists increase the uptake of fatty acids and inhibit the proinflammatory genes expression and block NF-kB dependent genes expression, inhibit cytokines, chemokines, COX-2, iNOS in the brain tissue, inhibit pyrin domain-containing protein 3 (NLRP3) inflammasome and MAPK phosphatase 1 and caveolin-1-dependent pathways. These reduce dopaminergic neurodegeneration and maintains dopamine levels, improving motor and cognitive dysfunctions.	(Arbo et al., 2022; Birajdar et al., 2023)
			It was reported to cause a 28% reduction in the PD cases in type 2 DM patients under the prescription of glitazones than sulfonylureas or metformin whereas in another study there was no difference in newly diagnosed PD in T2DM patients compared to the control non-diabetes group.	(Fletcher et al., 2021)
3	Glucagon-like Peptide-1 (GLP-1) agonists	Promote the insulin secretion by pancreatic β cells and reduce glucagon secretion of pancreatic α cells.	These drugs reduce spatial learning and memory impairment, and reduce amyloid plaque and soluble Aβ in Aβ induced AD model of rodents.	(Arbo et al., 2022; Birajdar et al., 2023; Chen et al., 2023)

S. No	Drug name	Role as anti-diabetic	Role in neurodegenerative disease	Reference
			GLP-1 agonists reduce Aβ burden and tau hyperphosphorylation. They improve cognitive impairment and reduce brain atrophy.	
			Enrich neurogenesis, reduce apoptosis, relieve the neurons from oxidative stress, and reduce neuroinflammation.	
			Treatment with GLP-1 receptor agonists such as exenatide, liraglutide, and lixisenatide showed reduced incidence of PD. Exenatide showed neuroprotective action in rats lesioned with 6-hydroxydopamine (6-OHDA) and mouse model of 1-methyl-4-phenyl-1,2,3,6-tetrahydropyridine (MPTP). The action is lost in the presence of a GLP-1 antagonist confirming the action of exenatide in T2DM and PD.	(Fletcher et al., 2021)
4	DPP4 inhibitors (gliptins)	DPP4 inhibitors slow the degradation of incretin enhancing and prolonging GLP-1 signalling.	These drugs decrease Aβ load, tau phosphorylation, neuronal apoptosis, and oxidative stress markers. It reverses cognitive and memory impairment in AD experimental model and modulate synaptic plasticity. The alogliptin led to insulin sensitivity and decreased $A\beta_{1-42}$ mediated neurodegeneration.	(Arbo et al., 2022)
5	Sulfonylurea	Stimulate insulin secretion by closure of ATP-sensitive potassium channel (KATP) in pancreatic -cell led to the membrane depolarization, influx of calcium ions, and release of insulin from storage vesicles.	Sulfonylureas improve memory, maintain synaptic plasticity and inhibit AChE. These drugs decrease proinflammatory cytokines, microglia activation, TNF-α, and IL-6, in 1the hippocampus of rodents.	(Arbo et al., 2022)
6	Sodium-glucose co-transporter-2 (SGLT-2) inhibitors a) Dapagliflozin	SGLT-2 inhibitors decrease renal reabsorption of glucose and increase urinary excretion of glucose.	These drugs decrease oxidative stress, TNF-α, and NF-kB pathways and reduce neuroinflammation in rotenone-induced PD rat model.	(Birajdar et al., 2023)
	b) Empagliflozin		It showed anti-inflammatory, and autophagic actions. It decreased neuronal death, haemorrhage, and microglia loads. It showed improvement in cognitive impairment.	(Birajdar et al., 2023)

One of the genes encodes for subunit for metformin and it correlates with several studies on rodents that confirm the potential of metformin as neuroprotective agent. The retrospective cohort studies on 6000 patients with type 2 DM revealed that the use of metformin for more than 4 years reduces the PD incidence (Storm et al., 2021). The increase in the production and secretion of Aβ, and *p* tau is associated with insulin resistance-mediated dysregulation. It was evident that repaglinide targets the ATP binding cassette and increases neuronal survival by upregulation of the ATF-6 gene and blocking DREAM. This *in vitro* study showed that repaglinide increases cell viability by 80-85% and increases cell viability by 1.2-folds in SHSY-5Y cells with neurodegenerative conditions by treating with streptozotocin. An *in vivo* study against DM-linked AD Wistar rats showed that repaglinide significantly decreases Aβ, tau protein, TNF-α, IL-6, MDA, NO, and increment of GSH, SOD, and GSH levels. There is also an increased in retention and spatial memory in behavioural studies. A decrease in Bax, caspase-3 expression, and a significant increase in Bcl-2 and ATF-6 expression were also observed (Wadhwa & Taliyan, 2023). The drug pioglitazone acts on multiple targets, but enough amount of the drug is not able to reach brain tissues. The peptide saxagliptin is soluble in water, and hence it is not able to cross BBB. However, it has the potential to decrease Aβ1-42 peptide and tau levels and thereby could be effective against AD (Séguy et al., 2022). Table 3 lists the antidiabetic repurposed drugs as potential anti-Alzheimer's and anti-Parkinsonian agents.

3.6. Steroids

Neuroactive steroids are naturally synthesized by neurons and the glial cells in the peripheral nervous system (PNS), adrenal glands, and gonads. The steroids perform several functions regulating inflammatory conditions, neural plasticity and neural development. There is a decline in neurological steroids with age that might be linked to the development of AD and PD (Arbo et al., 2022). The aldosterone antagonist eplerinone was investigated for the treatment of AD using the STZ-induced AD Albino mice model. The anti-Alzheimer effects were assessed by the Morris water maze, elevated plus maze, open field, passive avoidance, and oxidative stress index measured by the levels of the SOD, GSH and MDA, CAT, and AChE in the brain tissue.

Table 4. Repurposing of the steroidal hormones against AD and PD

S. No	Steroid hormone	Status of hormones in AD and PD	Role in neurodegenerative disease	References
1	Estrogen	The brain levels of estrogen decrease in women. Decreased levels in estrogen in ovariectomy or aromatase inhibition caused activation of NF-kB activation, memory impairment, and Aβ deposition	Estradiol administration showed an anti-inflammatory effect, decreased Aβ deposition, and cognitive impairment. It also decreased NF-kB activation, cleared Aβ deposition, and reduced inflammatory genes expression such as TNF-α, COX-2, iNOS, and IL-1β.	(Arbo et al., 2022)
2	Progesterone	Decreased levels increased the risk of AD.	Progesterone administration decreased TNF-α and IL-1β and increased the neuronal survival and performance in the hippocampus of the rat.	(Arbo et al., 2022)
3	Androgen	Androgen decreases with ageing in AD in males.	Testosterone decreases Aβ deposition, improves cognitive performance and decreases AD-like pathology. Clinical trials showed improved cognitive function in hypogonadal male patients with testosterone whereas testosterone is not able to prevent dopamine depletion in MPTP or methamphetamine-induced PD.	(Arbo et al., 2022)
4	Dehydroepiandrosterone (DHEA)/ Dehydroepiandrosterone-sulphate (DHEA-S)	DHEA and DHEA-S levels also decline during ageing and are linked with AD.	DHEA and DHEA-S showed antioxidant and antiapoptotic effects in Aβ-induced cytotoxicity. They also decrease memory impairment and improve cognitive performance in mice models against age-related deposition of APP and Aβ.	(Arbo et al., 2022)

The brain tissues were also analysed for dopamine, serotonin, Aβ1−40, Aβ1−42, NF-κB, and TNF. The *in-silico* studies confirmed that the compounds interacted well within the receptors of AChE. The *in vivo* studies confirm the effectiveness of the drug eplerinone in treating and improving memory in AD and dementia (Hira et al., 2020). Table 4 presents the steroidal hormones repurposed for AD and PD.

3.7. Metal Chelators

The aggregation of the amyloid protein into toxic aggregate causes protein misfolding and oxidative stress that leads to the development of AD and PD. The divalent metals such as Fe^{2+}, Cu^{2+}, and Zn^{2+} play an important role in the progression of these diseases as these seemed to interact with the amyloid proteins such as Aβ and α-synuclein in AD and PD. Altered zinc expression leads to impaired DNA repair and synaptic loss whereas iron is directly involved in the Aβ, tau accumulation, and oxidative stress. The copper essential micronutrient deficiency affects overall health, and elevated levels are associated with worse survival outcomes in thyroid and hepatocellular carcinomas due to its interaction with Aβ and its deposition (Das et al., 2024). The chelating agents were investigated as a promising therapy in inhibiting the redox-active metals. The chelators include 8-hydroxyquinoline (8HQ) derivatives (VK-28, HLA-20, and MA-30) and SOD mimic compounds as chelating agents (1-BYT and 1,4-BYT). The chelating agents with MAO inhibitory activity include M-30 and HLA-20. The glutathione is three amino acids containing peptides Cys, Gly, and Glu having pleiotropic action against neurodegenerative diseases (Ben-Shushan & Miller, 2021). The iron chelators such as desferal, HQ, apomorphine, and deferiprone showed promising therapeutic effects against PD. The randomized double-blinded placebo-controlled clinical trials confirmed that deferiprone has the potential to reduce the iron content in the brain and is also very safe (Khatri et al., 2021). The results of clinical studies have provided evidence for repurposing of deferiprone in PD, but it is yet to be seen whether it will emerge as the iron chelator of choice. Table 5 enlists the peptides repurposed for AD and PD.

Table 5. Repurposing of the peptides against PD and AD

S. No	Neuropeptides families	Peptides	Neuroprotective action	References
1	Tachykinin	Neurokinin A (NKA)	Protects neuronal cells by associating with inflammatory cytokines.	(Ben-Shushan & Miller, 2021)
		Neurokinin B (NKB)	Acts against Cu^{2+}-induced calcium channel opening and synapse homeostasis.	
		Substance P (SP)	Decreases cognitive impairment.	
		Neuropeptide K (NPK)	Relaxes the contracted cerebral arteries.	
		Neuropeptide gamma (NPG)	Regulates function related to the reproduction associated with inflammatory cytokines.	
2	Not part of the subfamily of tachykinin	Neuropeptide W: NPW (Trp residues located in the N-terminal and C-terminal of the peptides).	Messenger in the brain network and activates the hypothalamus and pituitary adrenal.	(Ben-Shushan & Miller, 2021)
3	Not part of the subfamily of tachykinin	Neuropeptide Y; NPY (multiple tyrosine residues)	Reduces Aβ-induced cell death, decreases neurodegenerative disorders, decreases memory deficits and depressive-like behaviours, and ER-stress-induced cell death.	(Ben-Shushan & Miller, 2021)

3.8. Hypolipidemic Agents

Statins are new attention in the field of repurposed drugs for neurological disorders. The statins having anti-inflammatory and antioxidant activity exhibit neuroprotective activities. They are also found to decrease protein aggregation and amyloid plaque formation in AD and PD. Statins decrease the cholesterol mediated Aβ and α-synuclein aggregation in AD and PD. Statins protect from oxidative stress and decrease inflammatory cytokines such as s IL-1β, IL-6, and TNF-α. There are preclinical and clinical studies data that support the role of statins in AD and PD. Lovastatin decreases the Aβ deposition in humans whereas simvastatin led to a 71% decrease in the development of AD and was found to decrease Apo-lipoprotein E-4 (APOE4) with senile plaques. The cholesterol levels are worsened by the over-expression of the APOE4 genes. Several *in vitro* studies showed that lovastatin decreases senile plaques production whereas pravastatin is an effective HMGCOA reductase (catalyses the synthesis of cholesterol) inhibitor and simvastatin acts as a better neuroprotectant. Some preclinical studies showed that statins have a protective role against PD. Simvastatin suppresses NF-kB and preserves dopaminergic neurons and suppresses microglial activation in MPTP-induced models of rats. Similarly, lovastatin also reduced NF-kB, iNOS, TNF, IL-6, and IL-1β expression in inflammatory rodents' models (Bhat et al., 2021).

3.9. Phytochemicals

Phytochemicals may be repurposed as anti-AD drugs since they are found to be potential GSK-3β inhibitors which is one of the therapeutic drug targets for AD. Among the tested phytochemicals such as cuminaldehyde, alpha terpinyl acetate, eugenol, piperine, and cinnamaldehyde, the cuminaldehyde showed the best binding affinity *in silico* with GSK-3β receptor (PDB ID 1Q3D) with the highest docking score of -6.3. Also, it showed an important hydrogen bond interaction with amino acid Val135 inside the binding pocket (Chowdhury, 2021). ApoE is an attractive target for lipid metabolism and neurodegenerative diseases. Two compounds, ergotamine and dihydroergocristine showed competitive binding with ApoE receptors. The ergotamine and dihydroergocristine were found to be stable in MD simulation. The binding of ergotamine and dihydroergocristine showed a more compact protein structure. Further, the RMSF analysis showed that the binding provides stability by prohibiting the residual fluctuations in the protein structure (Furkan et al., 2024). Sanz et al., performed the HTS assay to identify the molecule that have the potential to improve the locomotor in the D*J*-1*β* mutant flies and oxidative stress in the D*J*-1-deficient SH-SY5Y human cells. One of the compounds identified was vincamine from the leaves of *Vinca minor*. The compound was able to suppress PD-related symptoms and it also reduced oxidative stress in the PD model of flies and humans. Vincamine reduces the oxidative stress in D*J*-1-deficient cells in human cells and decreases apoptosis induced by oxidative stress, improving mitochondrial viability partially *via* the inhibition of voltage-gated sodium channels. This potential of vincamine to treat the PD might be because of the blockade of the voltage-gated sodium channel (Sanz et al., 2023).

3.10. Antiviral Drugs

The antiviral drug amantadine is an example of one of the successfully repurposed drugs for PD. The drug was repurposed taking evidence in relief of motor symptoms in patients taking it for influenza. Amantadine has now become the standard PD treatment especially against early motor symptoms and against levodopa-induced dyskinesia (Fletcher et al., 2021). Artificial intelligence (AI) and ML *in silico* techniques were used to identify the best potential drugs from the FDA-approved drugs based on the predicted expression of the PINK1. PD involves the disruption of the mitochondrial membrane by the α-synuclein protein. The damaged mitochondria are further degraded by mitophagy by the activation of the PINK1 pathways or increased expression of the PINK1 genes. Among the repurposed drugs from the approved FDA list, nitazoxanide showed promising activity against PD by the 10-gradient boosted tree ensemble models built on hash-based circular fingerprints (Haneczok et al., 2023). The use of HTS, docking, and ML identified tilorone as an effective inhibitor of *eel*AChE in the nanomolar range. Tilorone inhibits *eel*AChE and *h*AChE with an IC_{50} value of 14.4 and 64.4 nM and BuChE > 50 μM. Molecular docking studies also showed specific interaction within the hydrophobic active sites in the anionic site. The drug does not inhibit any of the kinases and only inhibits AChE among all the 44 toxicology targets of proteins (Vignaux et al., 2021).

3.11. Antihypertensives

There is a reported relationship between the increased risk of AD with hypertension (HTN). Thus, targeting HTN reduces the risk of AD. The calcium channel blockers (CCBs) and the ARBs mainly demonstrate the neuroprotective effects. The ARBs directly act on the brain and show neuroprotection. Valsartan is one of the potential drugs involved in inhibition of the oligomerization of Aβ peptides into high molecular oligomeric peptides. Valsartan also showed a 3.7-folds reduction in Aβ plaques formation in the mouse model as compared to the group receiving only vehicle. The other ARBs such as losartan also exists as a potential drug and acts as a neuroprotective agent in AD. The other category of CCBs easily cross the BBB induce cerebral vasodilation and improve the blood flow to the brain. Drugs such as verapamil, isradipine, nimodipine, and diltiazem showed neuroprotective effects *in vitro* in human neuroblastoma/MC65 cell lines induced neurotoxicity by Aβ protein precursor C-terminal fragment. Among them, isradipine was found to be the most potent. The *in vivo* studies of the drugs nitrendipine and nilvadipine showed a reduction in the brain levels of Aβ in the transgenic mouse model of AD. Nilvadipine was more potent than nitrendipine (Khatri et al., 2021). The antihypertensive drug isradipine showed a 19% reduction in PD than non-users. Preclinically, it reduces cell death in PD rodents' models but showed no benefits in slowing down the progression of PD clinically in patients under 3 years of treatment with isradipine (Fletcher et al., 2021). Autophagy is the process of clearance of the proteins to decrease their toxicities in the body. Felodipine an L-type CCB shows autophagy and inhibits aggregation of neurodegenerative-associated proteins. It also clears α-synuclein in the mouse brain and thus promising in the use of neurodegeneration (Siddiqi et al., 2019). The computational drug repurposing screening was carried out for more than 1300 drugs for the treatment of apoE4 generelated to AD. Among them, bumetanide emerged as the most potent one. The drug tested in apoE4 mice with and without Aβ accumulation decreased the electrophysiological, pathological, and cognitive deficit. The single-nucleus RNA sequencing showed a decrease in AD signatures in these mice. Drug exposure in humans showed lower AD prevalence over 65 years of age (Taubes et al., 2021). Telmisartan blocks the angiotensin I (AT1 type 1) receptor and has been assessed for repurposing to treat AD. It is shown to decrease neuroinflammation and tau phosphorylation and improves Aβ1–42 clearance. It also shows a neuroprotective effect against the ischemia mice model and improves memory and cognitive impairment (Parvez, 2023). Fasudil, a selective inhibitor of rho kinase 1 and 2, is a vasodilator that has been found to be effective in pulmonary HTN. Clinically, it has been prescribed for cerebral vasospasm. Additionally, it has been shown to directly modulate aggregation of α-synuclein, hallmark of PD, both *in vitro* and in mice to improve memory and cognitive functions. It is currently being investigated under a phase II clinical trial for treatment of neurodegenerative diseases (Ballard et al., 2020).

3.12. Antibiotics

Mitochondrial function dysfunction is one of the causative factors of AD. The antibiotic minocycline shows efficacy against oxidative and mitochondrial stress and acts as a neuroprotective agent. The flies were overexpressed for Aβ42 protein in the brain of Drosophila using UAS/GAL-4 scheme. It was found that minocycline ceased the Aβ42 induced symptoms including behavioural changes and altered the membrane potential. It abolished AD-like manifestation *via* significant decrease in the expression of JNK/caspase 3-mediated pathway. This suggests that minocycline reduces neuronal loss, attenuates apoptosis, and improves mitochondrial function (Khatoon et al., 2022). The antibiotic ceftriaxone showed positive results in phase I and II in the ALS patient but failed to increase the lifespan of the ALS patients (Fletcher et al., 2021). Other antibiotics rifamycin, and rapamycin have shown reduction in Aβ levels, inflammatory cytokines, and activation of microglia in the brain The mitochondria stress targeting pathway is another promising therapy for PD. The drug inosine is a purine nucleoside (precursor of urate) and showed potential neuroprotective and anti-inflammatory actions. In various clinical trials, inosine and rotenone showed antiparkinsonian effects (Khatri et al., 2021). A few other studies also confirm the neuroprotective potential of antibiotics. Tetracycline and gramicidin S inhibit the Aβ aggregation pathways whereas D-cycloserine, amphotericin, and doxorubicin decrease Aβ-42 fibrillization. Rifampicin also inhibits aggregation of Aβ1-40 and decreases the accumulation of Aβ, tau and acts against AD. Antibiotics like tetracyclines, rifampicin, geldanamycin, ceftriaxone, and rapamycin inhibit α-synuclein fibrils and amyloid aggregation and are effective against PD (Socias et al., 2018). Doxycycline targets both α-synuclein and tau whereas clioquinol targets Zn/Cu and avoids coprecipitation with Aβ deposition (Liu et al., 2022).

3.13. Miscellaneous

A thorough literature search of the repurposed drugs showed that some of the existing drugs such as zidovudine, thalidomide, raloxifene, bupropion, bromocriptine, and allopurinol have the potential to treat AD (Das et al., 2021). The cough linctus drug ambroxol was found to increase glucocerebrosidase activity (mutation of the gene encoding glucocerebrosidase is a major risk factor for PD). The Phase II clinical trials further support ambroxol to be safe and tolerable. Furthermore, it increased α-synuclein in CSF with brain clearance led to its progress to Phase III clinical trials (Fletcher et al., 2021). A β_2 adrenoceptor agonist salbutamol showed reduced endogenous α-synuclein gene expression in human neuroblastoma cells. It was also evident from the Norwegian Prescription data bases that there is 3-fold reduction in five-years period in the use of the salbutamol and PD incidence. The data also showed the protective effect of salbutamol in the 6-

OHDA-lesion rat model of PD. Still, the results were rebutted by many other groups (Fletcher et al., 2021; Mittal et al., 2017). Johnston et al., identified 3 drugs acamprosate, ganaxolone, and lorcaserin based on *in silico* screening to treat L-DOPA-induced dyskinesia in PD. Among the three drugs, lorcaserin showed 58% reduction in the rotational asymmetry against the L-DOPA induced hyperactivity in the rodent model. It also showed 82% reduction in L-DOPA-induced dyskinesia in MPTP-lesioned dyskinetic macaques (Johnston et al., 2024). Ursodeoxycholic acid (UDCA), is indicated for the treatment of cholestatic liver disease and primary biliary cirrhosis. UDCA was demonstrated to increase phosphorylation of Akt, antioxidant, and anti-inflammatory actions in hepatocellular models and hence play an important role in rescuing mitochondrial function in both parkin and LRRK2 mutant fibroblasts. It has also shown promising effects in PD experimental models. It improved behavioral function and attenuated dopaminergic cell loss *in vivo* induced by rotenone (Abdelkader et al., 2016). Zileuton, an anti-asthmatic drug, has been found to improve memory by reducing the deposition of Aβ aggregates and tau pathology in aged mice. It was proposed that the beneficial effects of zileuton are due to the blockage of 5-lipooxygenase (5-LO) which is higher in AD patients (Di Meco et al., 2014). The emerging technology such as AI includes ChatGPT and are promising in identifying and repurposing the drugs against AD. The ChatGPT proposed the 20 most promising repurposed drugs from the available literature and then 10 drugs were tested against two large clinical sets comprising of the exposed and unexposed individuals over 65 years of age at risk of incident AD. The drugs metformin, simvastatin, and losartan are found to be associated with lower AD risk in the meta-analysis (Yan et al., 2024). The retrospective cohort studies of the four most prescribed drugs (antihypertensive, statins, SSRI, and proton-pump inhibitors) were conducted involving 2 million patients between 2008 and 2020 in large multi-speciality urban medical centers in New York, USA. The electronic health records (HER) data including the prescription order of mild cognitive impairment (MCI) were measured for the average treatment effect of various drugs and incidence of dementia. Among the 14269 patients of MCI, a total of 2501 progressed to dementia. The drugs were identified that slow down the progression of MCI to dementia include rosuvastatin, citalopram, escitalopram, and omeprazole. The outcome of this study supports for identifying the repurposed drugs from the prescribed ones but needs further investigation (Xu et al., 2023). It is evident that citalopram reduces the CSF Aβ levels and increases the activity of α-secretase (Séguy et al., 2022). There is a linkage between the branched-chain amino acid transferase 1 (BCAT1) dysfunction and the development of PD. Screening of 50 FDA-approved drugs revealed enasidenib, ethosuximide, metformin, and nitisinone to have beneficial role in intervention for late-in-life PD (Sohrabi et al., 2021). A network-based drug repurposing method was used for AD to rank the drug candidates. The scoring for the compounds is based on three different network- based approaches for normalised CT score, CoDReS score, and scoring based on BBB permeability. The approach led to the selection of 26 compounds. Although the drugs were from different categories, they all share common drug targets i.e., ACE, ESR1 and CYP2D6. The angiotensin I converting enzyme (ACE) inhibitors enhanced memory and decreased the memory deficit in mouse model. The ESR1 encodes for estrogen receptor-α and related to AD. The CYP2D6 is involved in the metabolism of drugs and polymorphism of this gene induces alteration and changes the drug efficiency such as donepezil. Perindopril is an ACE inhibitor, triphenyl ethylene stilbene is an estrogen agonist, Iloperidone approved for schizophrenia can be a potential candidate for AD (Savva et al., 2022). The use of the immunotherapy with monoclonal antibodies vaccination was tried to reduce the proteins i.e., α-synuclein and Aβ involved in the pathogenesis of AD and PD. The development and the use of therapeutic vaccines produce a humoral response in their initial stages, but their safety and tolerability are being currently investigated under clinical trials (Arbo et al., 2022). The mammalian suppressor of tauopathy 2 (MSUT2) controls the accumulation of tau and the knocking out of the MSUT2 protects neurons from neurodegenerative diseases. A library of small molecules was screened by the HTS method for their potential in inhibiting poly(A): MSUT2 RNA-protein and to mitigate the tau accumulation. This led to identification of duloxetine, saquinavir, and clofazimine as potential repurposed drugs against tau accumulation, a hallmark for the development of AD and dementia (Baker et al., 2020). Leucine-rich repeat kinase 2 (LRRK2) is one of the potential targets against PD. The drug repurposing against LRRK2 kinase by Tan et al., involves structure-based drug repurposing that includes molecular docking, MD simulation, and binding energy calculation by MM-GBSA; molecular mechanics-generalized Born surface area. Among 13 compounds screened, six compounds viz., abivertinib, aumolertinib, encorafenib, bosutinib, rilzabrutinib, and mobocertinib showed potent activity with an IC_{50} less than 5 μM. Among them the most potent compound Abivertinib showed potent inhibitions against G2019S mutation and wild-type LRRK2 with an IC_{50} of 410.3 nM and 177.0 nM, respectively (Tan et al., 2023). Over the past two decades there has been a surge in the clinical trials of the FDA-approved drugs against AD and PD. The drugs selected for these trials against AD include donepezil, galantamine, mematine, rivastigmine, and suvorexant whereas for PD the list of drugs includes amantadine, apomorphine, carbidopa, entacapone, istradefylline, levodopa, opicapone, pramipexole, rasagiline, ropinirole, rotigotine, safinamide, and selegiline (Mortberg et al., 2022).

Conclusion

Numerous potential repurposed drugs have been projected for the treatment of AD and PD. However, to date, none has emerged as a leading drug candidate. Most repurposed drugs have failed at either phase II or III clinical trials. Interestingly, while some drugs demonstrate preclinical efficacy, they often fail to replicate these results in placebo-controlled clinical trials.

Despite considerable efforts in recent years to advance repurposed drugs, expected health benefits for AD and PD patients have not been achieved. Moreover, in certain cases, the use of repurposed drugs has been associated with an increased risk of developing AD or PD. This highlights the need for further refinement of the structural properties of repurposed drugs to improve their compatibility with target receptors and enhance therapeutic outcomes.

Emerging tools and technologies, such as artificial intelligence (AI), bioinformatics, and computational approaches like network-based and similarity analyses, high-throughput screening, and genome-wide association studies, hold promise for identifying more effective drug candidates. Retrospective cohort studies on commonly prescribed drugs have also shown potential in discovering treatments that may slow disease progression.

AI platforms, like ChatGPT, can rapidly identify drug candidates using real-world clinical data, offering a significant advantage over traditional approaches. Nevertheless, the repurposing of drugs for AD and PD treatment requires further rigorous validation and investigation to identify new, effective drug candidates.

References

Abdelkader, N. F., Safar, M. M., & Salem, H. A. (2016). Ursodeoxycholic acid ameliorates apoptotic cascade in the rotenone model of Parkinson's disease: modulation of mitochondrial perturbations. *Molecular Neurobiology*, *53*, 810-817.

Advani, D., & Kumar, P. (2021). Therapeutic targeting of repurposed anticancer drugs in Alzheimer's disease: Using the multiomics approach. *ACS Omega*, *6*(21), 13870-13887.

Advani, D., & Kumar, P. (2022). Deciphering the molecular mechanism and crosstalk between Parkinson's disease and breast cancer through multi-omics and drug repurposing approach. *Neuropeptides*, *96*, 102283.

Advani, D., Gupta, R., Tripathi, R., Sharma, S., Ambasta, R. K., & Kumar, P. (2020). Protective role of anticancer drugs in neurodegenerative disorders: A drug repurposing approach. *Neurochemistry International*, *140*, 104841.

Akkaya, D., Seyhan, G., Sari, S., & Barut, B. (2024). In vitro and in silico investigation of FDA-approved drugs to be repurposed against Alzheimer's disease. *Drug Development Research*, *85*(3), e22184.

Albert-Gascó, H., Ros-Bernal, F., Castillo-Gómez, E., & Olucha-Bordonau, F. E. (2020). MAP/ERK signaling in developing cognitive and emotional function and its effect on pathological and neurodegenerative processes. *International Journal of Molecular Sciences*, *21*(12), 4471.

Almoyad, M. A. A., Wahab, S., Mohanto, S., & Khan, N. J. (2024). Repurposing Drugs to Modulate Sortilin: Structure-Guided Strategies Against Atherogenesis, Coronary Artery Disease, and Neurological Disorders. *ACS Omega*, *9*(16), 18438-18448.

Alvarado, C. X., Makarious, M. B., Vitale, D., Koretsky, M. J., Bandres-Ciga, S., Iwaki, H., . . . Nalls, M. A. (2023). omicSynth: an open multi-omic community resource for identifying druggable targets across neurodegenerative diseases. *Alzheimer's & Dementia*, *19*, e082359.

Arbo, B. D., Schimith, L. E., Santos, M. G., & Hort, M. A. (2022). Repositioning and development of new treatments for neurodegenerative diseases: Focus on neuroinflammation. *European Journal of Pharmacology*, *919*, 174800.

Atiya, A., Das Gupta, D., Alsayari, A., Alrouji, M., Alotaibi, A., Sharaf, S. E., . . . Shamsi, A. (2023). Linagliptin and empagliflozin inhibit microtubule affinity regulatory kinase 4: Repurposing anti-diabetic drugs in neurodegenerative disorders using *in silico* and *in vitro* approaches. *ACS Omega*, *8*(7), 6423-6430.

Baker, J. D., Uhrich, R. L., Strovas, T. J., Saxton, A. D., & Kraemer, B. C. (2020). Targeting pathological tau by small molecule inhibition of the poly (A): MSUT2 RNA–protein interaction. *ACS Chemical Neuroscience*, *11*(15), 2277-2285.

Ballard, C., Aarsland, D., Cummings, J., O'Brien, J., Mills, R., Molinuevo, J. L., . . . Corbett, A. (2020). Drug repositioning and repurposing for Alzheimer disease. *Nature Reviews Neurology*, *16*(12), 661-673.

Baruah, P., Das, A., Paul, D., Chakrabarty, S., Aguan, K., & Mitra, S. (2021). Sulfonylurea class of antidiabetic drugs inhibit acetylcholinesterase activity: unexplored auxiliary pharmacological benefit toward Alzheimer's disease. *ACS Pharmacology & Translational Science*, *4*(1), 193-205.

Ben-Shushan, S., & Miller, Y. (2021). Neuropeptides: Roles and activities as metal chelators in neurodegenerative diseases. *The Journal of Physical Chemistry B*, *125*(11), 2796-2811.

Bhat, A., Dalvi, H., Jain, H., Rangaraj, N., Singh, S. B., & Srivastava, S. (2021). Perspective insights of repurposing the pleiotropic efficacy of statins in neurodegenerative disorders: An expository appraisal. *Current Research in Pharmacology and Drug Discovery, 2*, 100012.

Birajdar, S. V., Mazahir, F., Alam, M. I., Kumar, A., & Yadav, A. K. (2023). Repurposing and clinical attributes of antidiabetic drugs for the treatment of neurodegenerative disorders. *European Journal of Pharmacology*, 176117.

Bourdakou, M. M., Fernández-Ginés, R., Cuadrado, A., & Spyrou, G. M. (2023). Drug repurposing on Alzheimer's disease through modulation of NRF2 neighborhood. *Redox Biology, 67*, 102881.

Cartas-Cejudo, P., Cortés, A., Lachén-Montes, M., Anaya-Cubero, E., Puerta, E., Solas, M., . . . Santamaría, E. (2024). Neuropathological stage-dependent proteome mapping of the olfactory tract in Alzheimer's disease: From early olfactory-related omics signatures to computational repurposing of drug candidates. *Brain Pathology*, e13252.

Chang, J. R., Ghafouri, M., Mukerjee, R., Bagashev, A., Chabrashvili, T., & Sawaya, B. E. (2012). Role of p53 in neurodegenerative diseases. *Neurodegenerative Diseases, 9*(2), 68-80.

Chen, S.D., Chuang, Y.C., Lin, T.K., & Yang, J.L. (2023). Alternative role of glucagon-like Peptide-1 receptor agonists in neurodegenerative diseases. *European Journal of Pharmacology, 938*, 175439.

Chowdhury, S. (2021). In silico identification of phytocompounds as potential inhibitors of Glycogen synthase kinase 3 beta (GSK-3β). *Alzheimer's & Dementia, 17*, e058744.

Das, A. K. K., Sharma, D., & Sharma, L. (2021). Drug repurposing strategy for treating Alzheimer's disease. *Alzheimer's & Dementia, 17*, e058503.

Das, V., Miller, J. H., Alladi, C. G., Annadurai, N., De Sanctis, J. B., Hrubá, L., & Hajdúch, M. (2024). Antineoplastics for treating Alzheimer's disease and dementia: Evidence from preclinical and observational studies. *Medicinal Research Reviews, 44*(5), 2078-2111.

Devos, D., Moreau, C., Delval, A., Dujardin, K., Defebvre, L., & Bordet, R. (2013). Methylphenidate: a treatment for Parkinson's disease? *CNS Drugs, 27*, 1-14.

Di Meco, A., Lauretti, E., Vagnozzi, A. N., & Praticò, D. (2014). Zileuton restores memory impairments and reverses amyloid and tau pathology in aged Alzheimer's disease mice. *Neurobiology of Aging, 35*(11), 2458-2464.

Dighe, S. N., Tippana, M., Van Akker, S., & Collet, T. A. (2020). Structure-based scaffold repurposing toward the discovery of novel cholinesterase inhibitors. *ACS Omega, 5*(48), 30971-30979.

Fletcher, E. J., Kaminski, T., Williams, G., & Duty, S. (2021). Drug repurposing strategies of relevance for Parkinson's disease. *Pharmacology Research & Perspectives, 9*(4), e00841.

Furkan, M., Khan, M. S., Shahwan, M., Hassan, N., Yadav, D. K., Anwar, S., . . . Shamsi, A. (2024). Identifying repurposed drugs as potential inhibitors of Apolipoprotein E: A bioinformatics approach to target complex diseases associated with lipid metabolism and neurodegeneration. *International Journal of Biological Macromolecules, 259*, 129167.

Grabowska, M. E., Huang, A., Wen, Z., Li, B., & Wei, W.-Q. (2023). Drug repurposing for Alzheimer's disease from 2012–2022—A 10-year literature review. *Frontiers in Pharmacology, 14*, 1257700.

Gravandi, M. M., Abdian, S., Tahvilian, M., Iranpanah, A., Moradi, S. Z., Fakhri, S., & Echeverría, J. (2023). Therapeutic targeting of Ras/Raf/MAPK pathway by natural products: a systematic and mechanistic approach for neurodegeneration. *Phytomedicine, 115*, 154821.

Halliday, M., Radford, H., Zents, K. A., Molloy, C., Moreno, J. A., Verity, N. C., . . . Bushell, M. (2017). Repurposed drugs targeting eIF2α-P-mediated translational repression prevent neurodegeneration in mice. *Brain, 140*(6), 1768-1783.

Haneczok, J., Delijewski, M., & Moldzio, R. (2023). AI molecular property prediction for Parkinson's Disease reveals potential repurposing drug candidates based on the increase of the expression of PINK1. *Computer Methods and Programs in Biomedicine, 241*, 107731.

Harding, E. C. (2023). Repurposing anti-diabetic drugs for neuroprotection. *Alzheimer's & Dementia, 19*, e075767.

He, H., Duo, H., Hao, Y., Zhang, X., Zhou, X., Zeng, Y., . . . Li, B. (2023). Computational drug repurposing by exploiting large-scale gene expression data: Strategy, methods and applications. *Computers in Biology and Medicine, 155*, 106671.

Hira, S., Saleem, U., Anwar, F., Raza, Z., Rehman, A. U., & Ahmad, B. (2020). *In silico* study and pharmacological evaluation of Eplerinone as an Anti-Alzheimer's drug in STZ-induced Alzheimer's disease model. *ACS Omega, 5*(23), 13973-13983.

Jana, A., Bellver-Sanchis, A., Griñán-Ferré, C., & Banerjee, D. R. (2023). Repurposing of raltitrexed as an effective G9a/EHMT2 inhibitor and promising anti-Alzheimer's agent. *ACS Medicinal Chemistry Letters, 14*(11), 1531-1536.

Johnston, T. H., Lacoste, A. M., Ravenscroft, P., Su, J., Tamadon, S., Seifi, M., . . . Visanji, N. P. (2024). Using artificial intelligence to identify drugs for repurposing to treat l-DOPA-induced dyskinesia. *Neuropharmacology, 248*, 109880.

Karelina, T. (2023). Application of quantitative systems pharmacology model of Alzheimer's disease for simulation of efficacy of tyrosine kinase inhibitors. *Alzheimer's & Dementia, 19*, e065408.

Khan, M. S., Shahwan, M., Shamsi, A., Alhumaydhi, F. A., Alsagaby, S. A., Al Abdulmonem, W., . . . Yadav, D. K. (2022). Elucidating the interactions of fluoxetine with human transferrin employing spectroscopic, calorimetric, and in silico approaches: implications of a potent Alzheimer's drug. *ACS Omega, 7*(10), 9015-9023.

Khatoon, R., Kaushik, P., & Parvez, S. (2022). Mitochondria-related apoptosis regulation by minocycline: a study on a transgenic drosophila model of Alzheimer's disease. *ACS Omega, 7*(23), 19106-19112.

Khatri, D. K., Kadbhane, A., Patel, M., Nene, S., Atmakuri, S., Srivastava, S., & Singh, S. B. (2021). Gauging the role and impact of drug interactions and repurposing in neurodegenerative disorders. *Current Research in Pharmacology and Drug Discovery, 2*, 100022.

Kopp, K. O., Glotfelty, E. J., Li, Y., & Greig, N. H. (2022). Glucagon-like peptide-1 (GLP-1) receptor agonists and neuroinflammation: Implications for neurodegenerative disease treatment. *Pharmacological Research, 186*, 106550.

Krahn, A. I., Wells, C., Drewry, D. H., Beitel, L. K., Durcan, T. M., & Axtman, A. D. (2020). Defining the neural kinome: strategies and opportunities for small molecule drug discovery to target neurodegenerative diseases. *ACS Chemical Neuroscience, 11*(13), 1871-1886.

Kulkarni, V., Alagarsamy, V., Solomon, V., Jose, P., & Murugesan, S. (2023). Drug repurposing: an effective tool in modern drug discovery. *Russian Journal of Bioorganic Chemistry, 49*(2), 157-166.

Li, L., Wu, Q., Lundin, S. K., Naj, A. C., Gur, R. C., Rascovsky, K., . . . Chen, Y. (2023). Investigating drug repurposing for the prevention of Alzheimer's disease and related dementias using electronic health records in the University of Pennsylvania Health System. *Alzheimer's & Dementia, 19*, e079714.

Liu, W., Wang, G., Wang, Z., Wang, G., Huang, J., & Liu, B. (2022). Repurposing small-molecule drugs for modulating toxic protein aggregates in neurodegenerative diseases. *Drug Discovery Today, 27*(7), 1994-2007.

Long, H.Z., Cheng, Y., Zhou, Z.W., Luo, H.Y., Wen, D.D., & Gao, L.C. (2021). PI3K/AKT signal pathway: a target of natural products in the prevention and treatment of Alzheimer's disease and Parkinson's disease. *Frontiers in Pharmacology, 12*, 648636.

Mateev, E., Kondeva-Burdina, M., Georgieva, M., & Zlatkov, A. (2023). Repurposing of FDA-approved drugs as dual-acting MAO-B and AChE inhibitors against Alzheimer's disease: An *in silico* and *in vitro* study. *Journal of Molecular Graphics and Modelling, 122*, 108471.

Medesis Pharma SA (2024). Clinical safety and efficacy evaluation of nanolithiumRNP03 in patients with mild-to-severe Alzheimer's disease. *ClinicalTrials.gov, NCT05423522*(Phase 2).

Mittal, S., Bjørnevik, K., Im, D. S., Flierl, A., Dong, X., Locascio, J. J., . . . Xu, B. (2017). β2-Adrenoreceptor is a regulator of the α-synuclein gene driving risk of Parkinson's disease. *Science, 357*(6354), 891-898.

Mortberg, M. A., Vallabh, S. M., & Minikel, E. V. (2022). Disease stages and therapeutic hypotheses in two decades of neurodegenerative disease clinical trials. *Scientific Reports, 12*(1), 17708.

Padhi, D., & Govindaraju, T. (2022). Mechanistic insights for drug repurposing and the design of hybrid drugs for Alzheimer's disease. *Journal of Medicinal Chemistry, 65*(10), 7088-7105.

Palmas, M. F., Ena, A., Burgaletto, C., Casu, M. A., Cantarella, G., Carboni, E., . . . Cardia, M. C. (2022). Repurposing pomalidomide as a neuroprotective drug: Efficacy in an alpha-synuclein-based model of Parkinson's disease. *Neurotherapeutics, 19*(1), 305-324.

Pan, X., Yun, J., Akdemir, Z. H. C., Jiang, X., Wu, E., Huang, J. H., . . . Yi, S. S. (2023). AI-DrugNet: A network-based deep learning model for drug repurposing and combination therapy in neurological disorders. *Computational and Structural Biotechnology Journal, 21*, 1533-1542.

Parolo, S., Mariotti, F., Bora, P., Carboni, L., & Domenici, E. (2023). Single-cell-led drug repurposing for Alzheimer's disease. *Scientific Reports, 13*(1), 222.

Parvez, S. (2023). Emerging therapeutics agents and recent advances in drug repurposing for Alzheimer's disease. *Ageing Research Reviews, 85*, 101815.

Plun-Favreau, H., Lewis, P. A., Hardy, J., Martins, L. M., & Wood, N. W. (2010). Cancer and neurodegeneration: between the devil and the deep blue sea. *PLoS Genetics*, *6*(12), e1001257.

Quan, P., Wang, K., Yan, S., Wen, S., Wei, C., Zhang, X., . . . Yao, L. (2021). Integrated network analysis identifying potential novel drug candidates and targets for Parkinson's disease. *Scientific Reports*, *11*(1), 13154.

Rani, N., Kaushik, A., Kardam, S., Kag, S., Raj, V. S., Ambasta, R. K., & Kumar, P. (2024). Reimagining old drugs with new tricks: Mechanisms, strategies and notable success stories in drug repurposing for neurological diseases. *Progress in Molecular Biology and Translational Science*, *205*, 23-70.

Rao, P. P. N., Tin, G., Mohamed, T., Shakeri, A., & Pham, A. T. (2020). Drug repurposing for Alzheimer's disease: Selective serotonin reuptake inhibitors: Nonhuman/Target identification and validation studies: Amyloid. *Alzheimer's & Dementia*, *16*, e039163.

Riederer, P., & Müller, T. (2018). Monoamine oxidase-B inhibitors in the treatment of Parkinson's disease: Clinical–pharmacological aspects. *Journal of Neural Transmission*, *125*(11), 1751-1757.

Sahu, M., Vashishth, S., Kukreti, N., Gulia, A., Russell, A., Ambasta, R. K., & Kumar, P. (2024). Synergizing drug repurposing and target identification for neurodegenerative diseases. *Progress in Molecular Biology and Translational Science*, *205*, 111-169.

Sanz, F. J., Solana-Manrique, C., & Paricio, N. (2023). Disease-modifying effects of vincamine supplementation in drosophila and human cell models of Parkinson's disease based on DJ-1 deficiency. *ACS Chemical Neuroscience*, *14*(12), 2294-2301.

Savva, K., Zachariou, M., Bourdakou, M. M., Dietis, N., & Spyrou, G. M. (2022). Network-based stage-specific drug repurposing for Alzheimer's disease. *Computational and Structural Biotechnology Journal*, *20*, 1427-1438.

Séguy, L., Groo, A.-C., & Malzert-Freon, A. (2022). How nano-engineered delivery systems can help marketed and repurposed drugs in Alzheimer's disease treatment? *Drug Discovery Today*, *27*(6), 1575-1589.

Shalaby, K., Nguyen-Tran, V., de la Fuente, A., Thompson, I., Abdesselem, H., Ouararhni, K., & Karim, N. (2023). Exploring the potential of fenamate NSAID analogs for the treatment of Alzheimer's disease. *Alzheimer's & Dementia*, *19*, e076985.

Sharma, P., Sharma, D., & Sharma, L. (2021). Exploring NSAIDs as anti-amyloidal agents: A drug repurposing approach to treat Alzheimer's disease. *Alzheimer's & Dementia*, *17*, e058504.

Siddiqi, F. H., Menzies, F. M., Lopez, A., Stamatakou, E., Karabiyik, C., Ureshino, R., . . . Lai, L. (2019). Felodipine induces autophagy in mouse brains with pharmacokinetics amenable to repurposing. *Nature Communications*, *10*(1), 1817.

Socias, S. B., Gonzalez-Lizarraga, F., Avila, C. L., Vera, C., Acuña, L., Sepulveda-Diaz, J. E., . . . Chehin, R. N. (2018). Exploiting the therapeutic potential of ready-to-use drugs: Repurposing antibiotics against amyloid aggregation in neurodegenerative diseases. *Progress in Neurobiology*, *162*, 17-36.

Sohrabi, S., Mor, D. E., Kaletsky, R., Keyes, W., & Murphy, C. T. (2021). High-throughput behavioral screen in C. elegans reveals Parkinson's disease drug candidates. *Communications Biology*, *4*(1), 203.

Storm, C. S., Kia, D. A., Almramhi, M. M., Bandres-Ciga, S., Finan, C., Hingorani, A. D., & Wood, N. W. (2021). Finding genetically-supported drug targets for Parkinson's disease using Mendelian randomization of the druggable genome. *Nature Communications*, *12*(1), 7342.

Tan, S., Lu, R., Yao, D., Wang, J., Gao, P., Xie, G., . . . Yao, X. (2023). Identification of LRRK2 inhibitors through computational drug repurposing. *ACS Chemical Neuroscience*, *14*(3), 481-493.

Taubes, A., Nova, P., Zalocusky, K. A., Kosti, I., Bicak, M., Zilberter, M. Y., . . . Pineda, S. (2021). Experimental and real-world evidence supporting the computational repurposing of bumetanide for APOE4-related Alzheimer's disease. *Nature Aging*, *1*(10), 932-947.

Tayebi, J., & BabaAli, B. (2024). EKGDR: An End-to-End knowledge graph-based method for computational drug repurposing. *Journal of Chemical Information and Modeling*, *64*(6), 1868-1881.

Urbina, F., Puhl, A. C., & Ekins, S. (2021). Recent advances in drug repurposing using machine learning. *Current Opinion in Chemical Biology*, *65*, 74-84.

Vignaux, P. A., Minerali, E., Lane, T. R., Foil, D. H., Madrid, P. B., Puhl, A. C., & Ekins, S. (2021). The antiviral drug tilorone is a potent and selective inhibitor of acetylcholinesterase. *Chemical Research in Toxicology*, *34*(5), 1296-1307.

Wadhwa, G., & Taliyan, R. (2023). Mechanistic insights of meglitinides drug repurposing and delivery to brain for metabolic syndrome linked neurodegenerative disease. *Alzheimer's & Dementia*, *19*, e080252.

Xu, J., Wang, F., Zang, C., Zhang, H., Niotis, K., Liberman, A. L., . . . Luo, Y. (2023). Comparing the effects of four common drug classes on the progression of mild cognitive impairment to dementia using electronic health records. *Scientific Reports*, *13*(1), 8102.

Yan, C., Grabowska, M. E., Dickson, A. L., Li, B., Wen, Z., Roden, D. M., . . . Feng, Q. (2024). Leveraging generative AI to prioritize drug repurposing candidates for Alzheimer's disease with real-world clinical validation. *NPJ Digital Medicine*, *7*(1), 46.

Yang, Q. Q., & Zhou, J. w. (2019). Neuroinflammation in the central nervous system: Symphony of glial cells. *Glia*, *67*(6), 1017-1035.

Yeong, K. Y., Berdigaliyev, N., & Chang, Y. (2020). Sirtuins and their implications in neurodegenerative diseases from a drug discovery perspective. *ACS Chemical Neuroscience*, *11*(24), 4073-4091.

Yu, T.W., Lane, H.Y., & Lin, C.H. (2021). Novel therapeutic approaches for Alzheimer's disease: an updated review. *International Journal of Molecular Sciences*, *22*(15), 8208.

Zhao, L., Li, Z., Chen, G., Yin, Y., & Chen, C. Y.-C. (2024). Multi-perspective neural network for dual drug repurposing in Alzheimer's disease. *Knowledge-Based Systems*, *283*, 111195.

Chapter 3

Innovative Therapeutic Approaches and Advancements in Repurposing Antidiabetic Drugs

Bharti Sharma
and Archana Kapoor*
Department of Pharmaceutical Sciences, Guru Jambheshwar University of Science and Technology, Hisar, Haryana, India

Abstract

Drug repurposing, or drug redirecting, entails uncovering innovative applications of either prominent or discontinued medicines in healthcare. It involves investigating or acquiring newer pharmaceutical ingredients for use in medicine. Using medication repositioning in research and development, pharmaceutical companies have discovered novel biological targets and generated new drugs. Efficiency, failure reduction, andcost and time savings characterize this strategy. Optimizing a medication's therapeutic value boosts its efficacy. Consequently, medication repositioning can be advantageous in recognizing drugs for novel purposes. Using traditional or *de novo* NME (new molecular entity) discovery for drug development is costly, sluggish, and time-consuming. To discover novel pharmaceutical applications, drug repositioning systematically combines activity-based, experimental, and *in silico* computational techniques. Redirecting safe medications to treat uncommon, difficult -to-treat ailments and disorders is a new tactic. Therapeutic repositioning has substantial commercial potential, as evidenced by the numerous drugs repositioned or repurposed in the

* Corresponding Author's Email: archnakapoor@gjust.org

In: Drug Repurposing
Editors: Shah Alam Khan and Md. Jawaid Akhtar
ISBN: 979-8-89530-422-8

approval pipeline. Recent decades have seen an increase in drug repositioning due to computer technologies like machine learning. The use of type 2 diabetes mellitus medications for neurological, cancer, and cardiovascular issues has been the subject of numerous studies. In this study, we provide an overview of the potential uses of the current antidiabetic medications and concentrate on machine learning techniques for the development of new treatments for type 2 diabetes.

Keywords: drug repurposing, *in silico*, antidiabetic, drug discovery, healthcare

1. Introduction

The notion of pharmacological repurposing, often referred to as pharmaceutical repositioning, was initially introduced in 2004 (Ashburn et al., 2004). The corresponding usage of the term is ambiguous and opaque because of its imperfect meaning, which has changed over time in response to innovations in drug repositioning (Langedijk et al., 2015, Jourdan et al., 2020). The process of discovering new applications for drugs outside the parameters of their original indications is known as drug repurposing, drug redirection, drug repositioning, or drug reprofiling. Drug repositioning involves active ingredients such as drugs with outdated patents, medications taken off the market due to safety concerns, and chemicals that have not demonstrated the anticipated level of safety and efficacy in clinical studies. However, medication repositioning does not include the structural alteration of active molecules by *de novo* drug design techniques (Jourdan et al., 2020). The academic and pharmaceutical groups continue to explore pharmacological repositioning despite the absence of a clear definition (Oprea et al., 2011). The pharmaceutical industry supports drug repositioning because, as compared to traditional drug research and development, it may save production time and costs by omitting many time-consuming and expensive steps (Ashburn et al., 2004). The soaring costs of drug research have made drug repurposing a viable approach to drug discovery. It is widely employed in looking for treatments for a variety of conditions, including cancer (Bertolini et al., 2015), neuropathy (Fava et al., 2018), psychosis (Parsons et al., 2019), heart diseases (Ferrari et al., 2016), rare diseases (van den Berg et al., 2021), infectious diseases (Farha et al., 2019), and T2DM (type 2 diabetes mellitus) (Turner et al., 2016). Although not contagious, T2DM is becoming a serious global health concern. Achieving a fair balance between the effectiveness, side effects, and patient tolerance of the main types of antidiabetic medicines is a challenging task (Chaudhury et al., 2017).

2. Negative Aspects of Using Traditional Approaches to Drug Discovery

Traditional strategies in drug discovery often involve techniques such as high throughput screening (HTS) and structure-based drug design (SBDD). These methods are used to identify potential drug candidates, which are then further developed and investigated for their effects on biological material to find new drug targets. Despite garnering significant attention in recent decades, this strategy has yet to acquire the expected recognition in light of its exorbitant cost, poor success rate, high level of threat, and time-consuming nature. Notably, the traditional method of finding new drugs is usually predicated on a "unidirectional-linear" working paradigm. The typical target-based drug discovery life cycle is shown in Figure 1. It includes target recognition, target validation, assay development, lead identification, lead efficiency, and clinical studies. The scientifically prepared hypothesis, target identification, target confirmation, and therefore the postulated biochemical interactions have all contributed to the evolution of this technique (Zhang et al., 2015). Nevertheless, none of the investigated medication candidates were able to effectively enter the market due to many drawbacks encountered across the process of developing drugs. The extremely high attrition rate is a crucial factor contributing to the low conversion rate of the hypothesis-driven method. Many chemicals have excellent efficacy *in ex-vivo* screening studies but do not exhibit the same degree of potency and specificity when they are introduced into complicated live systems. Certain compounds identified during screening proved to be poor candidates for drugs because they were either poorly absorbed from the gastrointestinal system or could not penetrate the targeted cells (Zierath et al., 2019, Liu et al., 2019). In addition, several prominent substances were unable to obtain acceptance to go to the advanced phases of clinical studies due to previously unidentified adverse effects that could only be revealed by testing on an entire biological system (Aschner et al., 2020, Forouhi et al., 2010). Several drug candidates undergo improvements for clinical use, but some of them ultimately fail or are discontinued as a result of their severe long-term side effects. These adverse effects are difficult to predict during the drug development process, as new biological targets may be identified after the drug has already entered the market (Zierath et al., 2019, Liu et al., 2019).

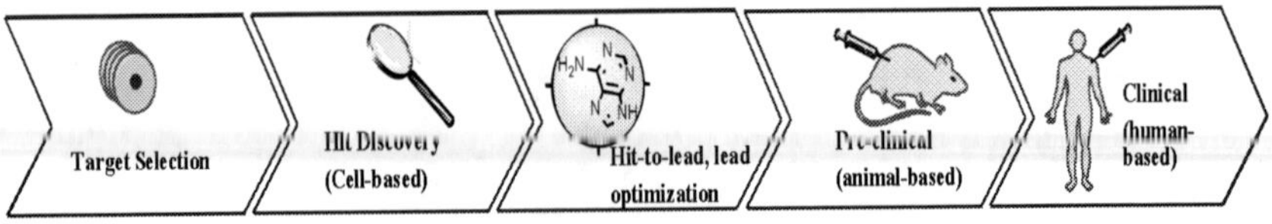

Figure 1. The drug invention process from target identification to the development of novel therapeutic medications.

3. Drug Repurposing: Involves Identifying Alternate Applications for Pre-Existing Medications

Considering the unsatisfactory results of the novel drugs currently on the market, there remains a compelling need to explore new methods to expedite the approval process, reduce costs, and improve outcomes. Drug repurposing, also known as re-profiling, is a systematic approach that aims to find new therapeutic uses for existing medications, different from their original intended use (Zhang et al., 2015). Drug repurposing is a comprehensive approach that speeds up the application of existing knowledge to treat human diseases. It avoids the typical obstacles of drug development, making it more efficient in terms of time and cost. The earlier confirmation of medication safety increases the likelihood of success in repurposing and enables the integration of omics (Pulley et al., 2020). According to Zhang et al., the use of 'omics' data mining to uncover new uses for existing drugs is a useful strategy for discovering potential treatments for diabetes. Moreover, their study findings may connect to other diseases, such as Alzheimer's disease (Zhang et al., 2015). Recently, this strategy has gained popularity due to its numerous uses and the growing discovery of medications that may treat different diseases with comparable clinical symptoms (Mohanty et al., 2020). Drug repurposing offers extra advantages in the field of drug discovery as compared to the conventional strategy (Figure 2). The safety of diabetic treatment is a significant concern in developing new medications. Firstly, a chemical undergoes regulatory scrutiny to become a therapeutic molecule and enters the clinical pipeline. At this stage, the lead drug candidates already have established safety margins and pharmacokinetic characteristics. This contributes to the decrease in expenses, duration, and potential hazards linked to the medical therapy management for emerging medical conditions. Furthermore, the prolonged administration of existing medications in specific patients can result in undesired concurrent side effects that may not be evident during clinical trials. Also, due to their successful completion of all stages of the clinical trial, the current medications possess comprehensive, unambiguous, and easily understandable data on all relevant parameters that may be beneficial for drug repurposing. In addition to this, the current medications have also undergone clinical establishing stages to enhance their bioavailability and *in-vivo* efficacy. This improves the safety margin and reduces the time required to bring the medicine to market (Zhang et al., 2015). The process of selecting the most effective medications for a patient usually involves combining patient data, like genetic or proteomic information, with drug data, including chemical descriptors of the compounds, to assess the therapeutic effectiveness of the pharmaceuticals. Drug interaction prediction frequently depends on similarity metrics, which assume that medications with analogous structures or targets would exhibit comparable behavior or potentially impede one other. The optimization of medication dose schedules is achieved by utilizing mathematical models to analyze pharmacodynamic and pharmacokinetic data (Rommet al., 2020).

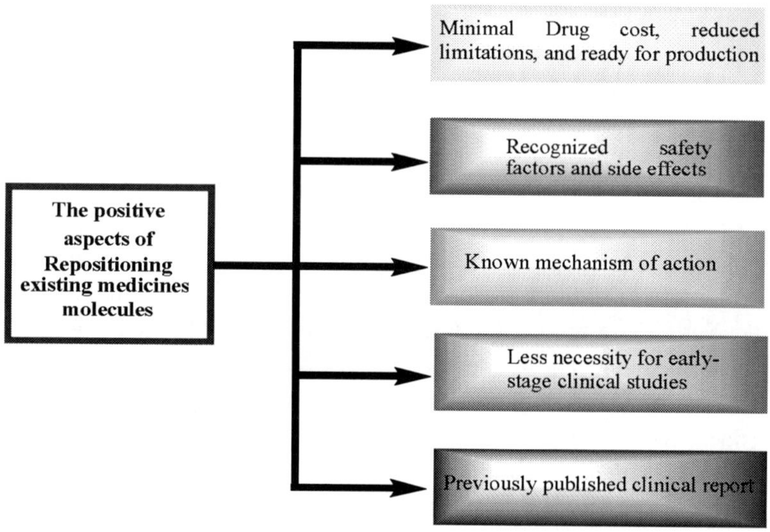

Figure 2. Repurposing existing medications has several advantages.

4. Current Data and Strategies for Managing T2DM

A growing number of people worldwide are suffering from diabetes mellitus (DM), a non-communicable illness. The number of people with diabetes is expected to increase to 643 million by 2030 and 783 million by 2045, and the International Diabetes Federation (IDF) reported that 537 million individuals (20–79 years old) had the disease in 2021 (Sunet al., 2022). Diabetes represents a major threat to the wellness and health of people, their families, and society at large due to its growing number of potential high-risk populations (541 million adults) and elevated mortality (6.7 million deaths in 2021, or 12.2% of adult deaths worldwide from all causes), as well as its disproportionate global health expenditure (USD 966 billion in 2021 for adults). Approximately 90% of all instances of diabetes globally are specifically related to T2DM (Holman et al., 2015, Bruno et al., 2005). T2DM is a long-term condition brought on by either peripheral tissue's impaired insulin sensitivity or malfunctioning pancreatic β-cells' failure to secrete enough insulin to sustain normal body utilization. Several genes (Kwak et al., 2016) and environmental variables play a major role in developing T2DM, as shown in Figure 3. Living a sedentary lifestyle and consuming excessive amounts of calories can lead to obesity and being overweight, which are directly linked to T2DM (Stinkens et al., 2015). Nephropathy, retinopathy, diabetic neuropathy, hypertension, diabetic foot, and atherosclerotic cardiovascular disease (ASCVD) are the conditions most frequently associated with T2DM comorbidities (Faselis et al., 2020, Viigimaa et al., 2020). Medication therapy, regular blood glucose testing, and lifestyle modification are all parts of managing diabetes (Chaudhury et al., 2017). Biguanides, sulfonylureas, agonists of the glucagon-like peptide 1 (GLP-1) receptor, thiazolidinedione (TZD), inhibitors of dipeptidyl peptidase 4 (DPP-4), inhibitors of sodium-glucose cotransporter 2 (SGLT-2), and inhibitors of α-glucosidase are among the anti-diabetic drugs that are frequently used in medical treatment. The age, progression of illness, patient tolerance, physical state, adverse effects, and medication efficacy of antidiabetic medications are considered while determining polypharmacy or monotherapy. Therefore, researchers have been working hard to produce new T2DM medications with improved effectiveness and fewer adverse effects. Bioinformatics and computational methods offer a fresh viewpoint on drug repositioning to identify novel resemblance based on an array of illnesses, genes, targets, medications, and existing information (Pushpakomet al., 2019).

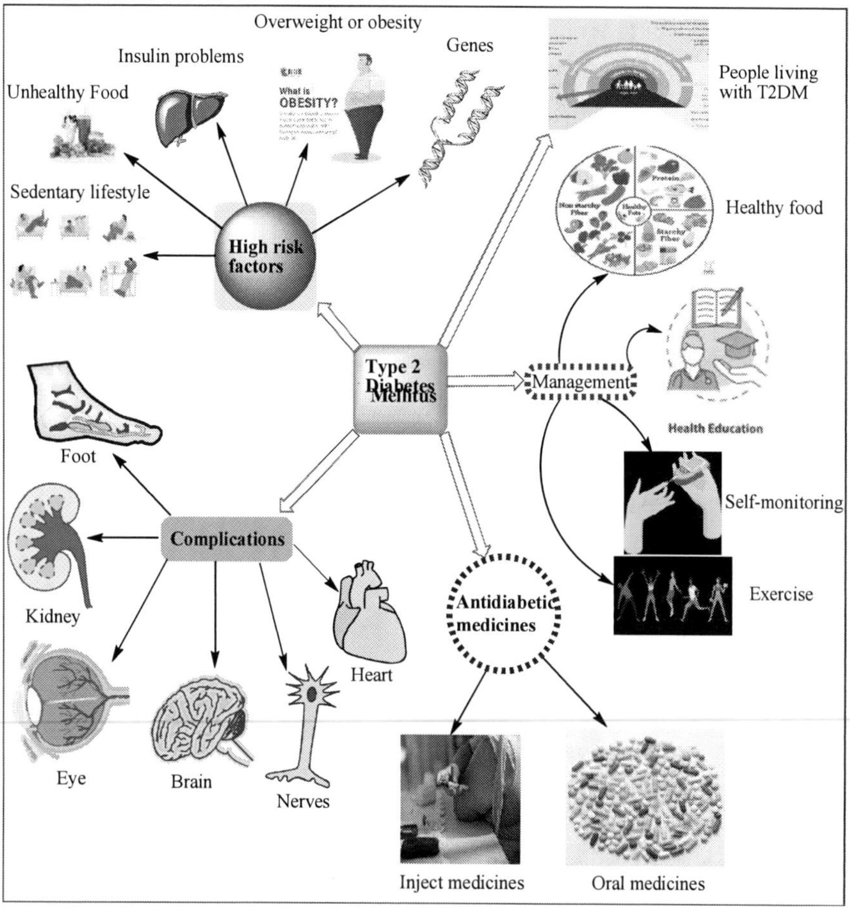

Figure 3. Schematic diagram highlighting the problems, high-risk variables, and T2DM management.

5. Drug Repurposing Using Current Computational Intelligence Models

Modern technology usage greatly enhances medicine development. Repurposing drugs is advantageous as it avoids toxicity studies and first clinical trials (Mohanty et al., 2020). Combining various methodologies and data can significantly progress computational drug repositioning.

5.1. Modern Drug Repurposing Strategies

Modern developments including AI (Artificial Intelligence), visualization methods, Big Data analytics, molecular dynamics modeling, and network-based methodologies (Figure 4) are being adopted by pharmaceutical firms for efficient medication repurposing (Park et al., 2019).

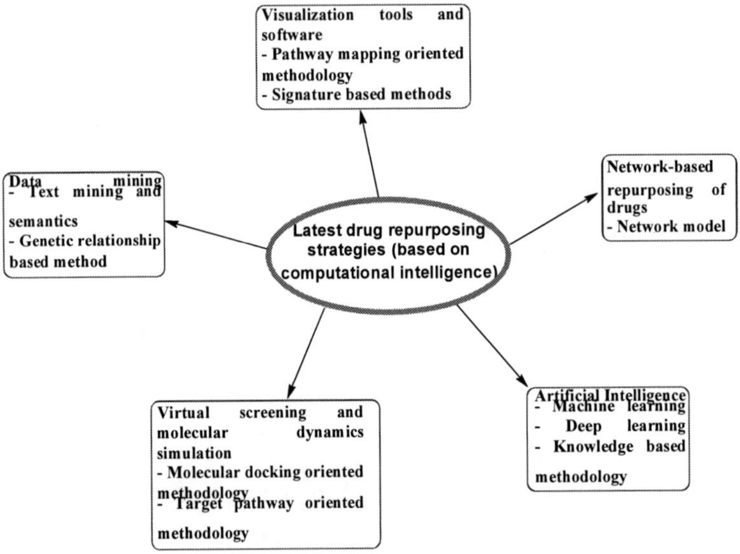

Figure 4. The chart outlines the latest technologies for repurposing drugs.

Considering the availability of large data sets for therapeutic candidates, current drug development has moved towards the big data epoch. Developing AI-based methodologies for ingenious modelling is crucial due to the dynamic, varied, and vast nature of pharmacological information sets. Recent AI methods, like deep learning (DL) and modelling, offer novel techniques for evaluating drug candidates' effectiveness and safety using big data assessment (Zhu et al., 2020). Drug prospects and their potential adverse drug responses or novel uses are sometimes hard to forecast since their basic principle is vague, convoluted, or hidden in information reservoirs. The latest subject in research is multi-domain pharmaco-modules, which are effective for smaller networks of biomolecules and are extracted from numerous datasets in a compilation of databases. Jaswanth and his colleagues have provided a comprehensive overview and analysis of the existing and developing field of computational medicine repurposing (Yella et al., 2018). Liu et al., emphasized the significance of combining AI and large data in recognizing targeted medications (Liu et al., 2019). Younhee Ko et al., offered public medication repositioning databases and outlined methodologies. Their study also emphasized and contrasted the traits and obstacles that are considered for pharmacological repositioning (Ko et al., 2020). Generative adversarial systems create drug structure visual representations from actual data (Park et al., 2019, Levin et al., 2020). A novel structure-based virtual screening method called consensus virtual screening, computer-aided drug design (CADD), can improve accuracy and reduce errors. Virtual databases like the Protein Data Bank hold molecular three-dimensional structures. Homology modelling can predict a protein's three-dimensional structure from its amino acid sequence (Maia et al., 2020).

In computer-aided therapeutic repurposing, data mining is effective for evaluating large amounts of both organized and unorganized biomedical data from many sources, which requires an integrated strategy (Masoudi-Sobhanzadeh et al., 2020, Yang et al., 2020). Visualization technologies aid in building 3D drug models, whereas programming languages enable high-quality drug structure pictures. Virtual evaluation and molecular dynamics models aid structure-based drug development by predicting molecular configuration (Yang et al., 2020). Drug repurposing using network strategies includes Gene Regulatory Systems, Metabolic Systems, Diffusion Sensing Microscopy Systems, and integrated techniques. This approach forecasts drug molecular structure using a graphical depiction of borders and nodes (Zeng et al., 2020).

5.2. Forecasting Novel Interactions for Medication Repurposing

The drug repurposing approach typically involves three steps before bringing a potential drug to market: identifying new drug-target (or drug-drug, drug-disease) groups, assessing the mechanism of the drug's effects in early-stage animal studies, and evaluating the therapeutic impact in clinical studies (Pushpakom et al., 2019). Out of these various phases, it is essential to discover a very reliable new connection between pharmaceutical compounds, medicinal targets, and illnesses. This is vital to the next phase of drug repositioning. Medications frequently combat diseases by targeting certain critical factors that have a distinct function in the condition. Medication-target interactions may be assessed using binding or competing cellular tests. The effectiveness of the medication affecting the target is subsequently determined by analyzing the dose-effect graph and calculating the EC_{50} or IC_{50} values (Keiser et al., 2007). Nevertheless, cellular tests cannot replicate the intricate physiological conditions of an individual's body, potentially leading to the subsequent failure of novel medicinal products (Langhans et al., 2018). Drugs that modify phenotypes in complicated biology can be found by phenotypic analysis in cellular or animal studies. This process involves assessing model actions, identifying the disease, and figuring out the mechanisms of action (Zheng et al., 2013). Phenotypic screening is usually confined to one disease experiment and is time-consuming (Hodos et al., 2016). Large volumes of complex chemical and biomedical data, including drug targets, gene expression facts, clinical medical data, signaling mechanisms, electronic health information, and data on adverse drug effects, can be used by machine learning (ML) techniques to uncover novel probably unanticipated drug-disease (or drug-drug, drug-target) interactions (Bellazzi et al., 2011).

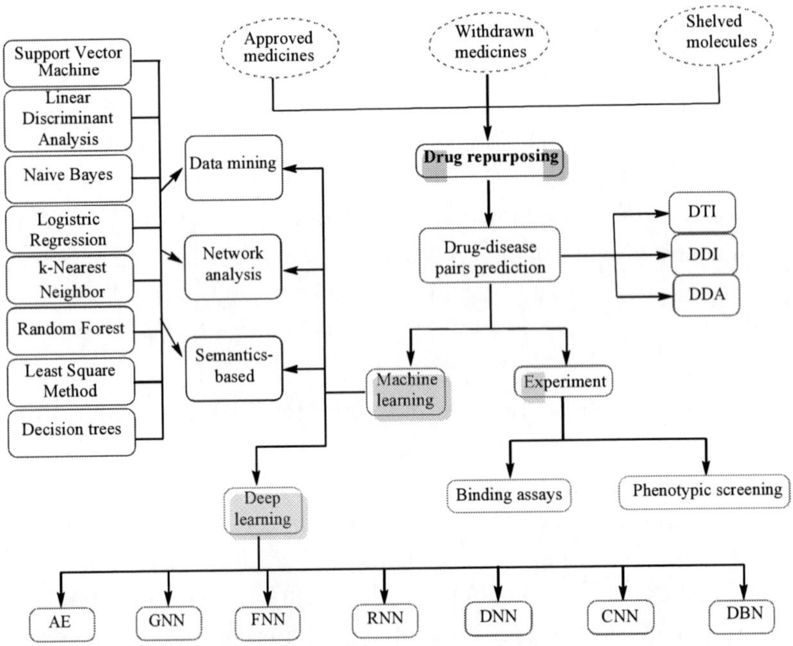

Figure 5. ML techniques for predicting medication-disease pairings in drug repurposing. DTI (drug-target interaction), DDI (drug-drug interaction), DDA (drug-disease association), AE (autoencoder), GNN (graph neural network), FNN (feedforward neural network), RNN (recurrent neural network), DNN (deep neural network), CNN (convolutional neural network), DBN (deep belief network).

ML techniques, including random forest (RF), Naive Bayes, and support vector machine (SVM) (Figure 5), demonstrate exceptional predictive accuracy and have contributed to significant technical advancements. Data mining, network visualization, and semantic investigation are often employed techniques in ML to forecast DTI (drug-target interactions), DDI (drug-drug interactions), and DDA (drug-disease associations). In recent decades, DL (deep learning), a branch of ML, has acquired greater curiosity for its application in medication repositioning (Vamathevan et al., 2019, Ekins et al., 2016, Gawehn et al., 2016, Pinoli et al., 2021, and Karimi et al., 2019). DL utilizes AI neural systems to simulate and create advanced learning structures, aiming to acquire more valuable characteristics and inevitably enhance the precision of categorization or forecasting. GNNs (Graph neural networks), AEs (autoencoders), and CNNs (convolutional neural networks) are examples of DL techniques that are more imposing than conventional approaches to ML at consequently extracting characteristics and effectively representing features to accumulate multiple levels of data (Schmidhuber et al., 2015, Korotcov et al., 2017, and Aliper et al., 2016).

5.3. Drug Repositioning Resources

Many biological information and pharmaceutical genomics resources serve as the primary source of data for medication repositioning. A variety of outstanding reviews have provided detailed descriptions of the information reports, algorithmic tools, and research techniques, all of which are relevant to T2DM drug repurposing. Drug-target connection, illness, organic, and biomolecular databases are the four main categories in which Tanoli et al., (Tanoli et al., 021) provided a summary of 102 sources of databases that are accessible for drug repurposing and demonstrated their uses for drug repositioning. A thorough compilation of databases for drug repositioning was given by Zhang et al., (Zhang et al., 019), who also listed the different kinds of data found within every database, listed techniques for representing data, and listed servers or software that performed data modification. Finally, they concluded with a list of popular ML techniques for drug repositioning. The three best algorithmic structures for pharmaceutical repositioning – similarity-based approaches, networking strategies, and matrix factorization – were listed by Hodos et al., (Hodos et al., 016) along with an explanation of different data kinds that can be used for medication repositioning and the encoding of various data types for AI input. In addition to summarizing the advances in

DL frameworks for medication repurposing in recent times, Pan et al., (Pan et al., 022) also included a description of frequently utilized internet open databases and graph-based and sequence-based data representation techniques. The six most commonlyutilized DL structures – DNNs, DBNs, CNNs, AEs, RNNs, and GNNs – were thoroughly outlined by Yu et al., (Yu et al., 022). Besides this, the authors covered some indicative models for DL built on input information with a wide range of complexity. Further reviews concentrated on ML techniques for handling various data kinds in medication repositioning (Qiu et al., 021, Ding et al., 014, Kim et al., 021).

5.4. ML-Based Techniques for Estimating DTI, DDI, and DDA

The process of forecasting an innovative DTI, DDA, or DDI *via* ML algorithms consists of three steps. To collect training data, the provided information as medication adverse effects, molecule structures, and illness genes – are first preliminary processed utilizing extracted features and feature preference; second, pertinent ML techniques are then used for training; and lastly, predictive algorithms are applied to produce drug repurposing outcomes on an experiment collection. Before entering a model that uses ML for learning representations, data are converted to a standard normalized manner, for instance, a computer-readable direction vector or matrix. A set of methods for converting unprocessed data into a form that ML can use efficiently is representation learning)Bengio et al., 013) (also known as feature learning). The two primary sequence-based (D'Souza et al., 020) information representations comprise the SMILES (Simplified Molecular Input Line Entry System) and the InChI (International Chemical Identifier), which describe three-dimensional structures of chemicals *via* text strings using a specific syntax (Weininger et al., 988 and 1989). Proteins and macromolecules' spatial and structural connections, in addition to the interconnectivity with other biological data, may be represented using graph-based visualizations (Chuang et al., 020, Fasoulis et al., 2021, and Zamora-Resendiz et al., 019). The main division of representation learning is between supervised and unsupervised techniques for obtaining characteristics from the input information for use in subsequent training assignments.

5.5. Drug Repositioning ML Algorithms

5.5.1. DTIs Prediction

Novel targets may provide novel indicators. Earlier literature reviews offer different viewpoints on ML techniques for predicting DTIs. Chen et al., (Chen et al., 018) classified machine learning techniques into supervised and semi-supervised categories and provided a detailed explanation of each subclass. In the framework of ligand-based simulation screening, Lavecchia explained (Lavecchia et al., 015) the theoretical underpinnings and practical uses of ML methods for DTI prediction. A thorough summary of resources for databases and genomics tools for ML methods for drug repositioning was given by Serçinoğlu et al., 019. Bagherian et al., (Bagherian et al., 021) elaborated on the four domains of similarity/distance-based techniques, matrix factorization strategies, algorithms for DL, and feature-based techniques, emphasizing ML approaches for DTI prediction. To repurpose drug molecules, Maryam et al., (Bagherian et al., 021) recommended an algorithm called the CTMC (Coupled Tensor-Matrix Completion), which merged a matrix factorization-based technique in drug-drug and target-target tensors. A program called MolAICal was designed by Bai et al., (Bai et al., 020 and 2021) to investigate the relationship between ligands and targets using conventional and DL techniques (Bai et al., 022). Ceddia et al., (Ceddia et al., 020) suggested a method to improve the shortest-path algorithm. The NMTF (Non-negative Matrix Tri-Factorization) strategy utilizes protein-protein connection systems to evaluate the shortest path between drug-protein pairs. This approach enhances the number and precision of predictions and demonstrates the benefits of integrating diverse network data for predicting DTIs.

5.5.2. DDIs Prediction

To treat the same disorders, drugs with relatively similar characteristics will probably target the same physiological effects. The researchers described information sources for DDIs forecasting in a review article. They included a thorough overview of computational techniques for DDIs finding, incorporating literature-based, pharmacovigilance-based data mining, and ML-based strategies (Qiu et al., 021). The most common use of ML is often in DDI prediction. To forecast potential DDIs, Wang et al., (Wang et al., 022) presented a GCNMK (graph convolution network with multi-kernel), which employed fully linked layers and two existing DDI graph kernels applied to the graph convolutional layers. Mei et al., (Mei et al., 021) trained a basic DDI prediction model using medication target genes and signaling networks instead of data integration in a biological context. Drug GIP (Gaussian interaction profile) kernel resemblance and RLS (regularized least squares) predictor were utilized by Yan et al., (Yan et al., 019) to create a similarity-based system called DDIGIP, which predicted DDIs for both novel and existing therapies. In performance evaluations based on a DNN, Yan et al., (Yan et al., 021) introduced a system called NMDADNN that effectively predicted DDIs by integrating many heterogeneous data sources and collecting uniform drug mapping characteristics.

5.5.3. DDAs Prediction

Researchers can now more easily uncover previously unknown DDAs within complicated networks, potentially leading to innovative approaches for medication repositioning. This progress is driven by the collection of vast amounts of biomedical data and advancements in computational techniques. Low-rank tensors and matrix complement technologies were utilized by Hu et al., (Hu et al., 021) to develop the LMFDA prediction model, which integrates multiple similar networks to identify potential DDAs. Yang and colleagues (Yang et al., 021) introduced the MSBMF (multi-similarities bilinear matrix factorization) model, which effectively assesses the association between medications and illnesses by combining matrix factorization with the alternating direction approach of the multiplier's algorithm. This model evalutes similarities between medications and illnesses at multiple levels.

6. Repurposing Abilities of Antidiabetic Medications

Due to the similar pathophysiological pathways of T2DM with several additional illnesses, the repurposing of antidiabetic medications has been thoroughly investigated for novel purposes. Furthermore, the substantial prevalence of patients with T2DM and the extensive history of pharmaceutical usage provide antidiabetic medicines with reasonably comprehensive clinical safety data, establishing them as the preferred candidates for repositioning. The repurposing effectiveness of antidiabetic medications mostly pertains to cancer, cardiovascular illnesses, and neurological disorders.

6.1. Repurposing Antidiabetic Medications for the Treatment of Cancer

T2DM and carcinoma share risk variables such as gender, age, weight gain, regular exercise, food, alcohol, inhaling smoke, etc. (Giovannucci et al., 010). Pathological conditions such as hyperglycemia, inflammation, hyperinsulinemia, and oxidative stress also link T2DM to several malignancies (Ferguson et al., 012, Guo et al., 012, Reuter et al., 010, Abudawood et al., 019). Despite most clinical investigations being done in cancer survivors with diabetes, antidiabetic medicines have been shown to reduce cancer prevalence and death. Research studies conducted on antidiabetic medications for cancer therapy are summarized in Table 1.

Metformin is the foremost extensively researched anti-diabetic medication renovated for carcinoma use. Lately, based on data meta-analyses of case-control and cohort investigations indicated that metformin has been linked to a reduced risk of across cancer rate and death (Zhu et al., 022). The result was similarly observed in site-specific carcinomas, including colorectal carcinoma, breast carcinoma, pancreatic carcinoma, liver carcinoma, lung carcinoma, and endometrial carcinoma. Retrospective investigations have found conflicting or negligible associations between sulfonylurea medications and cancer risk and death. Glyburide, a sulfonylurea, is widely used to treat T2DM. Glyburide may help fight cancer by modulating ATP-binding circuit receptor subfamily and ATP-dependent potassium channels (Zhu et al., 022). The latest mouse model research revealed that glyburide may suppress the NLRP3 (NOD-like receptor family pyrin domain containing 3) inflammasome to diminish inflammation-linked lung carcinogenesis. An additional research investigation found gliclazide reduced liver cancer risk (Lee et al., 2019 and Li et al., 021). Numerous laboratory investigations involving animals, cancer cells, and clinical information provide evidence that GLP-1 receptor agonists, a class of antidiabetic medications, may offer significant clinical advantages for colorectal carcinoma, breast carcinoma, endometrial carcinoma, prostate carcinoma, ovarian carcinoma, polycystic ovary syndrome, and pancreatic carcinoma (Eftekhari et al., 020, Kosowska et al., 017, Kanda et al., 018, Cao et al., 020, Wenjing et al., 017, Koehler et al., 011, and Han et al., 019). Active glucose absorption in cancer cells involves SGLT-2. SGLT-2 antagonist drugs reduce glucose absorption in SGLT-2-expressing tumors, preventing malignancy. Canagliflozin mitigates intracellular glucose absorption and suppresses SGLT-2-expressing liver tumors in animals. SGLT-2 blockers diminished breast cancer cell survival by causing G1 phase arrest and apoptosis via the AMPK/mTOR (adenosine-monophosphate activated protein kinase/mammalian target of rapamycin) signaling cascade (Zhou et al., 020).

Table 1. Repurposing of antidiabetic medications for oncological therapy

Antidiabetic drug Types	Medicines	Research approaches	Cancer Type
Biguanides	Liraglutide	Meta-analysis	Breast carcinoma (Zhu et al., 022) Liver carcinoma (Zhu et al., 022) Lung carcinoma (Zhu et al., 022) Colorectal carcinoma (Zhu et al., 022) Endometrial carcinoma (Zhu et al., 022) Pancreatic carcinoma (Zhu et al., 022)
GLP-I receptor agonists	Liraglutide Exendin-4	Meta-analysis Cell study Cell study Cell/Animal study Cell study Meta-analysis	Breast carcinoma (Cao et al., 020) Prostate carcinoma (Eftekhari et al., 020) Endometrial carcinoma (Kanda et al., 018) Colorectal carcinoma (Wenjing et al., 017 and Koehler et al., 011) Ovarian carcinoma (Kosowska et al., 017) Polycystic ovary syndrome (Han et al., 019)
Sulfonylurea	Glyburide Gliclazide	Animal study Case study	Lung carcinoma (Li et al., 021) Liver carcinoma (Lee et al., 019)
DPP4 inhibitors	Linagliptin Vildagliptin	Computational/Meta-analysis/Animal study Cell study Animal study	Colorectal carcinoma (Zhu et al., 022) Gastric carcinoma (Zhu et al., 022) Colorectal lung carcinoma (Zhu et al., 022)
SGLT-2 inhibitors	Canagliflozin Dapagliflozin	Cell study Cell study Cell study Retrospective cohort study	Breast carcinoma (Zhou et al., 020) Lung carcinoma (Yamamoto et al., 021) Breast carcinoma (Zhou et al., 020) Sitagliptin Prostate carcinoma (Xu et al., 020)

Further research suggested that SGLT-2 might reduce lung and pancreatic carcinoma cell growth (Yamamoto et al., 021 and Xu et al., 020). By blocking the action of the DPP-4 enzyme, DPP-4 inhibitors can improve glycaemic management by extending and increasing the function of endogenous secreted hormones, such as GLP-1 and GIP (glucose-dependent insulinotropic polypeptide). Research has demonstrated that DPP-4 inhibitors are linked to a lower incidence of breast carcinoma, oral carcinoma, and prostate carcinoma.

According to a repurposing investigation that utilized gene expression monitoring and molecular docking techniques, linagliptin exerted a therapeutic impact on colorectal carcinoma. Another cell investigation found that sitagliptin activated AMPK and inhibited YAP (Yes-associated protein) and MAGE-A3 (Melanoma-associated antigen-A3) to limit gastric carcinoma cell development and clonality. Additionally, sitagliptin lowered colorectal cancer risk. Vildagliptin inhibited colorectal lung metastases in animals (Zhu et al., 022). Although antidiabetic medications have not been proven to lessen cancer risk, their anti-cancer effects are being widely explored. If the approach of repurposing antidiabetic medications for combating cancer is viable, it will prove economically advantageous for both consumers as well as pharmaceutical firms.

6.2. Repurposing Antidiabetic Medications for the Treatment of Cardiovascular Illness

CVD (cardiovascular disease) is more common in diabetics than in non-diabetics (https://www.idf.org/our-activities/care-prevention/cardiovascular-disease.html). Insulin resistance is a key characteristic of T2DM and has been identified as a separate risk factor for cardiovascular illnesses. Furthermore, a pro-inflammatory and hyperlipidemic atmosphere is induced by lipid metabolism blockage linked to T2DM. This is a significant contributing factor to cardiovascular events. Furthermore, chronic hyperglycemia arising from T2DM has been attributed to the sedimentation of AGEs (advanced glycation end-products), which further degrade macrovascular and microvascular structures, eventually triggering CVD-like atherosclerosis. Due to the similar biological processes between T2DM and CVD, research has increasingly focused on the association between antidiabetic medication usage and CVD risk (Zhu et al., 022). SGLT-2 blockers may reduce the incidence of cardiac death and MACE (major adverse cardiovascular events). SGLT-2 blockers and agonists of GLP-1 declined myocardial infarction and stroke mortality in T2DM patients with ASCVD (atherosclerotic cardiovascular disease), according to a meta-analysis. Receptor agonists of GLP-1 lowers the probability of stroke, whereas SGLT-2 antagonists significantly reduce the need for hospitalization due to heart failure and the progression of renal disease. The agonists of the GLP-1 receptor and SGLT-2 inhibitors were also shown to lower cardiac events and improve renal outcomes in other lines of studies (Zhu et al., 022). DPP-4 blockers failed to substantially decrease the risk of MACE in individuals with T2DM in clinical studies intended to evaluate the cardiovascular benefits and safety of the drug. However, the results of another clinical research showed that patients treated with DPP-4 blocking drugs for T2DM had a higher risk of heart failure hospitalization. International recommendations prescribe SGLT-2 blockers or agonists of GLP-1 receptors for ASCVD-diagnosed T2DM patients' glycemic control. Inhibitors of SGLT-2 have been proposed for T2DM patients having CKD (chronic kidney disease) and ASCVD adolescents with cardiac failure (Zhu et al., 022). Antidiabetic medication repositioning studies improve T2DM patients' quality life span and management by diminishing cardiovascular-related impairment and early death.

6.3. Repurposing Antidiabetic Medications for the Treatment of Neurodegenerative Disorders

Insulin resistance causes issues with neuroinflammation, oxidative damage, poor glucose metabolism, and dysfunctional insulin metabolism in those with T2DM. Neuritic plaque accumulation, intracellular neurofibrillary tangles, vascular disease, inflammation-related impairment, and alterations in the structure and function of mitochondria are the principal clinical symptoms of neurodegenerative disorders. The two main neurodegenerative illnesses are AD (Alzheimer's disease) and PD (Parkinson's disease). Several investigations reveal that antidiabetic medication repositioning may have the potential for neurological conditions. Some studies offer a thorough analysis of the epidemiological data about the connections between T2DM and neurological conditions (Zhu et al., 022), and the possible therapeutic approach of antidiabetic medications for the management of neurodegenerative illnesses. The agonists of the GLP-1 receptor have a major neurotrophic function and neuroprotective effects by interfering with the signaling mechanisms of PI3K/AKT (phosphatidylinositol 3-kinase/protein kinase B) and cAMP/PKA (cyclic adenosine monophosphate/protein kinase A), along with controlling intracellular calcium equilibrium and inflammatory events. Exenatide has shown promise as a treatment for PD patients' motor symptoms, as evidenced by a meta-analysis and other clinical studies. To find out how exenatide affects PD patients' decreased motor degeneration over time, a bigger phase 3, placebo-controlled, 96-week clinical research is now underway. A clinical study in phase I identified as NCT03456687 (https://clinicaltrials.gov) aims to investigate how exenatide affects PD patients' brain and motor function. Phase

Table 2. The repurposing of antidiabetic drugs for oncology, cardiovascular disorders, and neurological illnesses in clinical studies

Study Title	Conditions/Interventions	Phase	NCT identifier	Sponsor
Cancer in Patients on Dapagliflozin and Other Antidiabetic Treatment	Breast cancer and Bladder cancer (Dapagliflozin)	Recruiting	NCT02695121	AstraZeneca
Altered Tumor Oxygenation by Metformin, a Potential Step in Overcoming Radiotherapy Resistance in LACC (METOXY-LACC)	Advanced Cervical Cancer (Metformin, Cisplatin)	Phase 2 (Recruiting)	NCT04275713	Oslo University Hospital
A Study to Evaluate the Effect of Multiple Doses of JNJ-56021927 on the Pharmacokinetics of Multiple Cytochrome P450 and Transporter Substrates in Participants with Castration-Resistant Prostate Cancer	Castration-Resistant Prostate Cancer (JNJ-56021927, Drug Cocktail, Pioglitazone, Rosuvastatin)	Phase 1 (Recruiting)	NCT02592317	Aragon Pharmaceuticals, Inc.
A Phase 1 Study to Evaluate the Potential Drug Interactions Between Repotrectinib and Metformin, Digoxin, and Rosuvastatin in Patients with Advanced Solid Tumors	Advanced Solid Tumors Harboring ROS1 or NTRK1-3 (TPX-0005, Metformin, Digoxin, Rosuvastatin)	Phase 1 (not yet Recruiting)	NCT05828303	Turning Point Therapeutics, Inc.
The Efficacy of SGLT-2 Inhibitor in Patients with CAD and DM Undergoing PCI	Coronary Heart Disease and Diabetes Mellitus (SGLT2 inhibitor)	(Recruiting)	NCT05333159	Beijing Anzhen Hospital
Effect of Empagliflozin on Left Atrial Function in Adults at Risk for Heart Failure	Cardiovascular Diseases (Empagliflozin, Placebo tablet)	Early Phase 1 (not yet Recruiting)	NCT06507657	University of Minnesota
Second-line Therapies for Patients with T2DM and Moderate Cardiovascular Disease Risk	T2DMand Moderate Cardiovascular Disease Risk (Glucagon-like peptide 1 receptor agonist, Sodium-glucose cotransporter 2 inhibitor, Dipeptidyl Peptidase 4 Inhibitor, Sulfonylurea)	Active, not recruiting	NCT05214573	Mayo Clinic

Table 2. Continued.

Study Title	Conditions/Interventions	Phase	NCT identifier	Sponsor
Semaglutide and Vascular Regeneration (SEMA-VR)	T2DM, Cardiovascular Diseases (Semaglutide Pen Injector)	Phase 4 (Recruiting)	NCT05870462	Canadian Medical and Surgical Knowledge Translation Research Group
Anti-Diabetic Medications to Fight PD and LBD	Parkinson's Disease and Lewy Body Dementia (Sitagliptin, Dapagliflozin, Placebo)	Phase 4 (Recruiting)	NCT06263673	Mayo Clinic
Clinical Study to Evaluate the Possible Efficacy of Metformin in Patients with Parkinson's Disease	Parkinson's Disease (Levodopa-carbidopa, Metformin)	Phase 2 (Recruiting)	NCT05781711	Tanta University
A Research Study Investigating Semaglutide in People with Early Alzheimer's Disease (EVOKE Plus)	Early Alzheimer´s Disease (Semaglutide, Placebo)	Phase 3 (Active, not recruiting)	NCT04777409	Novo Nordisk A/S
Intranasal Insulin in Parkinson's Disease (INI-PD)	Parkinson's Disease (Regular Novolin R, Placebo)	Phase 2 (Recruiting)	NCT04251585	HealthPartners Institute

IIa, placebo-controlled, parallel comparative NCT04269642 is a clinical trial designed to assess the safety and effectiveness of SR (sustained-release) exenatide (PT320) in managing early PD patients. The randomly assigned, parallel-controlled, phase 2 research investigation NCT04305002 examines how exenatide affects the progression of illness in early-stage PD sufferers with T2DM. In PD mice, liraglutide stimulates p-AMPK expression and suppresses NFjB protein to reduce neuroinflammation. According to a phase II clinical investigation that was both randomized and placebo-controlled (NCT02953665), liraglutide drastically enhanced the motor and cognitive signs experienced by individuals with PD. In conclusion, a great deal of research has been conducted on PD and AD as potential areas of medication repurposing of antidiabetic medicines, highlighting the tight connections between the pathogenic processes of diabetes and neurodegenerative disorders. Exenatide and liraglutide, two agonists of the GLP-1 receptor, are now undergoing late-stage clinical studies. It is very desirable to discover novel therapeutic approaches for neurodegenerative illnesses with significant unmet medical requirements. However, for instance, for the repositioned medications to be utilized efficiently in managing novel indications, extended follow-up times, more strategically designed studies, and a considerably bigger recruited population are still required for effectiveness and safety verification (Zhu et al., 022).

Clinical trials are currently underway to investigate the repurposing of antidiabetic medications for cancer, cardiovascular diseases, and neurodegenerative conditions, as detailed in Table 2 (available at clinicaltrials.gov).

Conclusion

Considering that the traditional methods of developing medicines are a venture that is not only time-consuming but also expensive and fraught with danger, the introduction of contemporary computational technologies that are based on drug repurposing is a viable alternative in the global pharmaceutical sector. A computational intelligence system that has a wider range of applications can be adapted to the process of repurposing anti-diabetic medications for use in treating a variety of conditions such as cancer, cardiovascular diseases, and neurodegenerative conditions. This encompasses Artificial Intelligence, Deep Learning, Visualization Tools, Machine Learning, Data Mining, Virtual Screening and Molecular Dynamics Simulation, and Network-Based Drug

Repurposing, all of which have a significant influence on the research that is being done to repurpose anti-diabetic medications. The present work focuses on exploring knowledge-based computational methodologies for repositioning the existing medications for T2DM to improve the success rate in treating various diseasess. According to the findings of this study, the drug-repositioning strategy is more cost effective, practical, and safe than the traditional de novo drug design technique.

References

Abudawood, M., (2019). Diabetes and cancer: a comprehensive review. *J. Res. Med. Sci. 24*, 94.

Aliper, A., Plis, S., Artemov, A., Ulloa, A., Mamoshina, P., & Zhavoronkov, A., (2016). Deep learning applications for predicting pharmacological properties of drugs and drug repurposing using transcriptomic data. *Mol. Pharm. 13*, 2524-2530.

Aschner, P., (2020). Insulin therapy in type 2 diabetes. *Am. J. Therapeut.27*, e79-e90.

Ashburn, T. T., & Thor, K. B., (2004). Drug repositioning: identifying and developing new uses for existing drugs. *Nat. Rev. Drug Discov. 3*, 673-683.

Bagherian, M., Kim, R. B., Jiang, C., Sartor, M. A., Derksen, H., & Najarian, K., (2021). Coupled matrix–matrix and coupled tensor–matrix completion methods for predicting drug–target interactions. *Brief. Bioinform. 22*, 2161-2171.

Bagherian, M., Sabeti, E., Wang, K., Sartor, M. A., Nikolovska-Coleska, Z., & Najarian, K., (2021). Machine learning approaches and databases for prediction of drug–target interaction: a survey paper. *Brief. Bioinform.22*, 247-269.

Bai, Q., (2020). Research and development of MolAICal for drug design via deep learning and classical programming. *arXiv preprint arXiv:2006.09747*.

Bai, Q., Liu, S., Tian, Y., Xu, T., Banegas-Luna, A. J., Pérez-Sánchez, H., ... & Yao, X., (2022). Application advances of deep learning methods for de novo drug design and molecular dynamics simulation. *WIREs Comput. Mol. Sci.12*, e1581.

Bai, Q., Tan, S., Xu, T., Liu, H., Huang, J., & Yao, X., (2021). MolAICal: a soft tool for 3D drug design of protein targets by artificial intelligence and classical algorithm. *Brief. Bioinform. 22*, bbaa161.

Bellazzi, R., Diomidous, M., Sarkar, I. N., Takabayashi, K., Ziegler, A., & McCray, A. T., (2011). Data analysis and data mining: current issues in biomedical informatics. *Methods Inf. Med. 50*, 536-544.

Bengio, Y., Courville, A., & Vincent, P., (2013). Representation learning: A review and new perspectives. *IEEE Trans. Pattern Anal. Mach. Intell.35*, 1798-1828.

Bertolini, F., Sukhatme, V. P., & Bouche, G., (2015). Drug repurposing in oncology—patient and health systems opportunities. *Nat. Rev. Clin. Oncol.12*, 732-742.

Bruno, G., Runzo, C., Cavallo-Perin, P., Merletti, F., Rivetti, M., Pinach, S., ... & Piedmont Study Group for Diabetes Epidemiology., (2005). Incidence of type 1 and type 2 diabetes in adults aged 30–49 years: the population-based registry in the province of Turin, Italy. *Diabetes Care. 28*, 2613-2619.

Cao, C., Yang, S., & Zhou, Z., (2020). GLP-1 receptor agonists and pancreatic safety concerns in type 2 diabetic patients: data from cardiovascular outcome trials. *Endocrine*, *68*, 518-525.

Ceddia, G., Pinoli, P., Ceri, S., & Masseroli, M., (2020). Matrix factorization-based technique for drug repurposing predictions. *IEEE J. Biomed. Health Inform. 24*, 3162-3172.

Chaudhury, A., Duvoor, C., Reddy Dendi, V. S., Kraleti, S., Chada, A., Ravilla, R., ... & Mirza, W., (2017). Clinical review of antidiabetic drugs: implications for type 2 diabetes mellitus management. *Front. Endocrinol.8*, 6.

Chen, R., Liu, X., Jin, S., Lin, J., & Liu, J., (2018). Machine learning for drug-target interaction prediction. *Molecules. 23*, 2208.

Chuang, K. V., Gunsalus, L. M., & Keiser, M. J., (2020). Learning molecular representations for medicinal chemistry: miniperspective. *J. Med. Chem.63*, 8705-8722.

D'Souza, S., Prema, K. V., & Balaji, S., (2020). Machine learning models for drug–target interactions: current knowledge and future directions. *Drug Discov. Today. 25*, 748-756.

Ding, H., Takigawa, I., Mamitsuka, H., & Zhu, S., (2014). Similarity-based machine learning methods for predicting drug–target interactions: a brief review. *Brief. Bioinform.15*, 734-747.

Eftekhari, S., Montazeri, H., & Tarighi, P., (2020). Synergistic anti-tumor effects of Liraglutide, a glucagon-like peptide-1 receptor agonist, along with Docetaxel on LNCaP prostate cancer cell line. *Eur. J. Pharmacol. 878*, 173102.

Ekins, S., (2016). The next era: deep learning in pharmaceutical research. *Pharm. Res.33*, 2594-2603.

Farha, M. A., & Brown, E. D., (2019). Drug repurposing for antimicrobial discovery. *Nat. Microbiol.4*, 565-577.

Faselis, C., Katsimardou, A., Imprialos, K., Deligkaris, P., Kallistratos, M., & Dimitriadis, K., (2020). Microvascular complications of type 2 diabetes mellitus. *Curr. Vasc. Pharmacol. 18*, 117-124.

Fasoulis, R., Paliouras, G., &Kavraki, L. E., (2021). Graph representation learning for structural proteomics. *Emerg. Top. Life Sci.5*, 789-802.

Fava, M., (2018). The promise and challenges of drug repurposing in psychiatry. *World Psychiatry. 17*, 28.

Ferguson, R. D., Novosyadlyy, R., Fierz, Y., Alikhani, N., Sun, H., Yakar, S., & LeRoith, D., (2012). Hyperinsulinemia enhances c-Myc-mediated mammary tumor development and advances metastatic progression to the lung in a mouse model of type 2 diabetes. *Breast Cancer Res. 14*, 1-12.

Ferrari, R., & Lüscher, T. F., (2016). Reincarnated medicines: using out-dated drugs for novel indications. *Eur. Heart* J. *37*, 2571-2576.

Forouhi, N. G., & Wareham, N. J., (2010). Epidemiology of diabetes. *Medicine. 38*, 602-606.

Gawehn, E., Hiss, J. A., & Schneider, G., (2016). Deep learning in drug discovery. *Mol. Inform.35*, 3-14.

Giovannucci, E., Harlan, D. M., Archer, M. C., Bergenstal, R. M., Gapstur, S. M., Habel, L. A., ... & Yee, D., (2010). Diabetes and cancer: a consensus report. *CA Cancer J. Clin. 60*, 207-221.

Guo, Y., Xu, F., Lu, T., Duan, Z., & Zhang, Z., (2012). Interleukin-6 signaling pathway in targeted therapy for cancer. *Cancer Treat. Rev. 38*, 904-910.

Han, Y., Li, Y., & He, B., (2019). GLP-1 receptor agonists versus metformin in PCOS: a systematic review and meta-analysis. *Reprod. Biomed. Online. 39*, 332-342.

Hodos, R. A., Kidd, B. A., Khader, S., Readhead, B. P., & Dudley, J. T., (2016). Computational approaches to drug repurposing and pharmacology. *Wiley Interdiscip. Rev. Syst. Biol. Med. 8*, 186.

Holman, N., Young, B., & Gadsby, R., (2015). Invited Editorial, Current prevalence of Type 1 and Type 2 diabetes in adults and children in the UK. *Diabet. Med. 32*, 1119-1120.

Hu, P., Huang, Y. A., Mei, J., Leung, H., Chen, Z. H., Kuang, Z. M., ... & Hu, L., (2021). Learning from low-rank multimodal representations for predicting disease-drug associations. *BMC Med. Inform. Decis. Mak. 21*, 1-13.

Jourdan, J. P., Bureau, R., Rochais, C., & Dallemagne, P., (2020). Drug repositioning: a brief overview. *J. Pharm. Pharmacol.72*, 1145-1151.

Kanda, R., Hiraike, H., Wada-Hiraike, O., Ichinose, T., Nagasaka, K., Sasajima, Y., ... & Ayabe, T., (2018). Expression of the glucagon-like peptide-1 receptor and its role in regulating autophagy in endometrial cancer. *BMC Cancer, 18*, 1-11.

Karimi, M., Wu, D., Wang, Z., & Shen, Y., (2019). DeepAffinity: interpretable deep learning of compound–protein affinity through unified recurrent and convolutional neural networks. *Bioinformatics. 35*, 3329-3338.

Keiser, M. J., Roth, B. L., Armbruster, B. N., Ernsberger, P., Irwin, J. J., & Shoichet, B. K., (2007). Relating protein pharmacology by ligand chemistry. *Nat. Biotechnol.25*, 197-206.

Kim, J., Park, S., Min, D., & Kim, W., (2021). Comprehensive survey of recent drug discovery using deep learning. *Int. J. Mol. Sci.22*, 9983.

Ko, Y., (2020). Computational drug repositioning: current progress and challenges. *Appl. Sci. 10*, 5076.

Koehler, J. A., Kain, T., & Drucker, D. J., (2011). Glucagon-like peptide-1 receptor activation inhibits growth and augments apoptosis in murine CT26 colon cancer cells. *Endocrinology. 152*, 3362-3372.

Korotcov, A., Tkachenko, V., Russo, D. P., & Ekins, S., (2017). Comparison of deep learning with multiple machine learning methods and metrics using diverse drug discovery data sets. *Mol. Pharm. 14*, 4462-4475.

Kosowska, A., Gallego-Colon, E., Garczorz, W., Kłych-Ratuszny, A., Aghdam, M. R. F., Woz, M., ... & Francuz, T., (2017). Exenatide modulates tumor–endothelial cell interactions in human ovarian cancer cells. *Endocr. Connect. 6*(8), 856-865.

Kwak, S. H., & Park, K. S., (2016). Recent progress in genetic and epigenetic research on type 2 diabetes. *Exp. Mol. Med. 48*, e220-e220.

Langedijk, J., Mantel-Teeuwisse, A. K., Slijkerman, D. S., & Schutjens, M. H. D., (2015). Drug repositioning and repurposing: terminology and definitions in literature. *Drug Discov. Today. 20*, 1027-1034.

Langhans, S. A., (2018). Three-dimensional *in vitro* cell culture models in drug discovery and drug repositioning. *Front. Pharmacol. 9*, 6.

Lavecchia, A., (2015). Machine-learning approaches in drug discovery: methods and applications. *Drug Discov. Today. 20*, 318-331.

Lee, J. Y., Jang, S. Y., Nam, C. M., & Kang, E. S., (2019). Incident hepatocellular carcinoma risk in patients treated with a sulfonylurea: a nationwide, nested, case-control study. *Sci. Rep. 9*, 8532.

Levin, J. M., Oprea, T. I., Davidovich, S., Clozel, T., Overington, J. P., Vanhaelen, Q., ... & Zhavoronkov, A., (2020). Artificial intelligence, drug repurposing and peer review. *Nat. Biotechnol. 38*, 1127-1131.

Li, M., Liu, H., Shao, H., Zhang, P., Gao, M., Huang, L., ... & Feng, F., (2021). Glyburide attenuates B (a) p and LPS-induced inflammation-related lung tumorigenesis in mice. *Environ. Toxicol. 36*, 1713-1722.

Liu, B., He, H., Luo, H., Zhang, T., & Jiang, J., (2019). Artificial intelligence and big data facilitated targeted drug discovery. *Stroke Vasc. Neurol. 4*, 206e13.

Liu, Z., & Yang, B., (2019). Drug development strategy for type 2 diabetes: Targeting positive energy balances. *Curr. Drug Targets. 20*, 879-890.

Maia, E. H. B., Assis, L. C., De Oliveira, T. A., Da Silva, A. M., & Taranto, A. G., (2020). Structure-based virtual screening: from classical to artificial intelligence. *Front. Chem. 8*, 343.

Masoudi-Sobhanzadeh, Y., (2020). Computational-based drug repurposing methods in COVID-19. *BioImpacts: BI. 10*, 205.

Mei, S., & Zhang, K., (2021). A machine learning framework for predicting drug–drug interactions. *Sci. Rep. 11*, 17619.

Mohanty, S., Rashid, M. H. A., Mridul, M., Mohanty, C., & Swayamsiddha, S., (2020). Application of Artificial Intelligence in COVID-19 drug repurposing. *Diabetes Metab. Syndr. 14*, 1027-1031.

Oprea, T. I., Bauman, J. E., Bologa, C. G., Buranda, T., Chigaev, A., Edwards, B. S., ... & Sklar, L. A., (2011). Drug repurposing from an academic perspective. *Drug Discov. Today Ther. Strateg. 8*, 61-69.

Pan, X., Lin, X., Cao, D., Zeng, X., Yu, P. S., He, L., ... & Cheng, F., (2022). Deep learning for drug repurposing: Methods, databases, and applications. *WIREs Comput. Mol. Sci. 12*, e1597.

Park, K., (2019). A review of computational drug repurposing. *Transl. Clin. Pharmacol. 27*, 59-63.

Parsons, C. G., (2019). CNS repurposing-Potential new uses for old drugs: examples of screens for Alzheimer's disease, Parkinson's disease and spasticity. *Neuropharmacology. 147*, 4-10.

Pinoli, P., Ceddia, G., Ceri, S., & Masseroli, M., (2021). Predicting drug synergism by means of non-negative matrix tri-factorization. *IEEE/ACM Trans. Comput. Biol. Bioinform. 19*, 1956-1967.

Pulley, J. M., Rhoads, J. P., Jerome, R. N., Challa, A. P., Erreger, K. B., Joly, M. M., ... & Aronoff, D. M., (2020). Using what we already have: uncovering new drug repurposing strategies in existing omics data. *Annu. Rev. Pharmacol. Toxicol. 60*, 333-352.

Pushpakom, S., Iorio, F., Eyers, P. A., Escott, K. J., Hopper, S., Wells, A., ... & Pirmohamed, M., (2019). Drug repurposing: progress, challenges and recommendations. *Nat. Rev. Drug Discov. 18*, 41-58.

Qiu, Y., Zhang, Y., Deng, Y., Liu, S., & Zhang, W., (2021). A comprehensive review of computational methods for drug-drug interaction detection. *IEEE/ACM Trans. Comput. Biol. Bioinform. 19*, 1968-1985.

Reuter, S., Gupta, S. C., Chaturvedi, M. M., & Aggarwal, B. B., (2010). Oxidative stress, inflammation, and cancer: how are they linked?. *Free Radic. Biol. Med. 49*, 1603-1616.

Romm, E. L., &Tsigelny, I. F., (2020). Artificial intelligence in drug treatment. *Annu. Rev. Pharmacol. Toxicol.60*, 353-369.

Schmidhuber, J., (2015). Deep learning in neural networks: An overview. *Neural Netw. 61*, 85-117.

Serçinoğlu, O., & Sarica, P. O., (2019). *In silico* databases and tools for drug repurposing. In *In Silico Drug Design.* 703-742. Academic Press.

Stinkens, R., Goossens, G. H., Jocken, J. W., & Blaak, E. E., (2015). Targeting fatty acid metabolism to improve glucose metabolism. *Obes. Rev. 16*, 715-757.

Sun, H., Saeedi, P., Karuranga, S., Pinkepank, M., Ogurtsova, K., Duncan, B. B., ... & Magliano, D. J., (2022). IDF Diabetes Atlas: Global, regional and country-level diabetes prevalence estimates for 2021 and projections for 2045. *Diabetes Res. Clin. Pract.183*, 109119.

Tanoli, Z., Seemab, U., Scherer, A., Wennerberg, K., Tang, J., & Vähä-Koskela, M., (2021). Exploration of databases and methods supporting drug repurposing: a comprehensive survey. *Brief. Bioinform. 22*, 1656-1678.

Turner, N., Zeng, X. Y., Osborne, B., Rogers, S., & Ye, J. M., (2016). Repurposing drugs to target the diabetes epidemic. *Trends Pharmacol. Sci. 3*, 379-389.

Vamathevan, J., Clark, D., Czodrowski, P., Dunham, I., Ferran, E., Lee, G., ... & Zhao, S., (2019). Applications of machine learning in drug discovery and development. *Nat. Rev. Drug Discov. 18*, 463-477.

van den Berg, S., de Visser, S., Leufkens, H. G., & Hollak, C. E., (2021). Drug repurposing for rare diseases: a role for academia. *Front. Pharmacol. 12*, 746987.

Viigimaa, M., Sachinidis, A., Toumpourleka, M., Koutsampasopoulos, K., Alliksoo, S., & Titma, T., (2020). Macrovascular complications of type 2 diabetes mellitus. *Curr. Vasc. Pharmacol. 18*, 110-116.

Wang, F., Lei, X., Liao, B., & Wu, F. X., (2022). Predicting drug–drug interactions by graph convolutional network with multi-kernel. *Brief. Bioinform. 23*, bbab511.

Weininger, D., (1988). SMILES, a chemical language and information system. 1. Introduction to methodology and encoding rules. *J. Chem. Inf. Comput. Sci. 28*, 31-36.

Weininger, D., Weininger, A., & Weininger, J. L., (1989). SMILES. 2. Algorithm for generation of unique SMILES notation. *J. Chem. Inf. Comput. Sci. 29*, 97-101.

Wenjing, H., Shuang, Y., Weisong, L., & Haipeng, X., (2017). Exendin-4 does not modify growth or apoptosis of human colon cancer cells. *Endocr. Res. 42*, 209-218.

Xu, D., Zhou, Y., Xie, X., He, L., Ding, J., Pang, S., ... & Zhou, C., (2020). Inhibitory effects of canagliflozin on pancreatic cancer are mediated via the downregulation of glucose transporter-1 and lactate dehydrogenase A. *Int. J. Oncol. 57*, 1223-1233.

Yamamoto, L., Yamashita, S., Nomiyama, T., Kawanami, T., Hamaguchi, Y., Shigeoka, T., ... & Iwasaki, A., (2021). Sodium-glucose cotransporter 2 inhibitor canagliflozin attenuates lung cancer cell proliferation in vitro. *Diabetol. Int. 12,* 389–398.

Yan, C., Duan, G., Pan, Y., Wu, F. X., & Wang, J., (2019). DDIGIP: predicting drug-drug interactions based on Gaussian interaction profile kernels. *BMC Bioinf. 20*, 1-10.

Yan, X. Y., Yin, P. W., Wu, X. M., & Han, J. X., (2021). Prediction of the drug–drug interaction types with the unified embedding features from drug similarity networks. *Front. Pharmacol. 12*, 794205.

Yang, J., Li, Y., Liu, Q., Li, L., Feng, A., Wang, T., ... & Lyu, J., (2020). Brief introduction of medical database and data mining technology in big data era. *J. Evid. Base Med. 13*, 57-69.

Yang, M., Wu, G., Zhao, Q., Li, Y., & Wang, J., (2021). Computational drug repositioning based on multi-similarities bilinear matrix factorization. *Brief. Bioinform. 22*, bbaa267.

Yella, J. K., Yaddanapudi, S., Wang, Y., & Jegga, A. G., (2018). Changing trends in computational drug repositioning. *Pharmaceuticals. 11*, 57.

Yu, J. L., Dai, Q. Q., & Li, G. B., (2022). Deep learning in target prediction and drug repositioning: Recent advances and challenges. *Drug Discov. Today. 27*, 1796-1814.

Zamora-Resendiz, R., & Crivelli, S., (2019). Structural learning of proteins using graph convolutional neural networks. *BioRxiv*. 610444.

Zeng, X., Zhu, S., Hou, Y., Zhang, P., Li, L., Li, J., ... & Cheng, F., (2020). Network-based prediction of drug–target interactions using an arbitrary-order proximity embedded deep forest. *Bioinformatics*, *36*, 2805-2812.

Zhang, M., Luo, H., Xi, Z., &Rogaeva, E., (2015). Drug repositioning for diabetes-based on'omics' data mining. *PloS One*. *10*, e0126082.

Zhang, T., Leng, J., & Liu, Y., (2020). Deep learning for drug–drug interaction extraction from the literature: a review. *Brief. Bioinform. 21*, 1609-1627.

Zhang, W., Lin, W., Zhang, D., Wang, S., Shi, J., & Niu, Y., (2019). Recent advances in the machine learning-based drug-target interaction prediction. *Curr. Drug Metab.20*, 194-202.

Zheng, W., Thorne, N., & McKew, J. C., (2013). Phenotypic screens as a renewed approach for drug discovery. *Drug Discov. Today. 18*, 1067-1073.

Zhou, B., Wang, R., Wu, P., & Kong, D. X., (2015). Drug repurposing based on drug–drug interaction. *Chem. Biol. Drug Des. 85*, 137-144.

Zhou, J., Zhu, J., Yu, S. J., Ma, H. L., Chen, J., Ding, X. F., ... & Zhang, Q., (2020). Sodium-glucose co-transporter-2 (SGLT-2) inhibition reduces glucose uptake to induce breast cancer-cell growth arrest through AMPK/mTOR pathway. *Biomed. Pharmacother. 132*, 110821.

Zhu, H., (2020). Big data and artificial intelligence modeling for drug discovery. *Annu. Rev. Pharmacol. Toxicol. 60*, 573-589.

Zhu, S., Bai, Q., Li, L., & Xu, T., (2022). Drug repositioning in drug discovery of T2DM and repositioning potential of antidiabetic agents. *Comput. Struct. Biotechnol. J. 20*, 2839-2847.

Zierath, J. R., (2019). Major advances and discoveries in diabetes-2019 in review. *Curr. Diabetes Rep.19*, 1-9.

Chapter 4

Repositioning of Drugs in Non-Small Cell Lung Cancer: Old Weapons for a New War

Rohit Pal[1,*]
Gurubasavaraja Swamy Purawarga Matada[1,†]
Ghanshyam Teli[2]
Ketan R. Hosamani[1]
Kumaraswamy B.[1]
Aayishamma I.[1]
MD Ashadul Sk.[1]
and Viji M. P.[1]
[1]Integrated Drug Discovery Centre, Department of Pharmaceutical Chemistry, Acharya & BM Reddy College of Pharmacy, Bengaluru, Karnataka, India
[2]School of Pharmacy, Sangam University, Atoon, Bhilwara, Rajasthan, India

Abstract

Lung cancer is the second-most prevalent cancer and the leading cause of cancer-related mortality in 2022. Lung cancers are classed as either small-cell (SCLC) or non-small-cell (NSCLC). Non-small cell lung cancer (NSCLC) accounts for the vast majority (84%) of lung cancer cases whereas SCLC accounts for roughly 16%. Despite increasing screening efforts, metastatic NSCLC has a miserable median five-year survival rate of about 3%. Surgery, heterogeneous chemotherapy, simultaneous radiotherapy, and chemotherapy are all typical treatment

* Corresponding Author's Email: rohitpal.rp096@gmail.com
† Corresponding Author's Email: gurubasavarajaswamy@gmail.com

In: Drug Repurposing
Editors: Shah Alam Khan and Md. Jawaid Akhtar
ISBN: 979-8-89530-422-8

options for NSCLC. Over the last several years, there have been numerous breakthroughs in the management of NSCLC regarding screening, diagnosis, and therapy. Unfortunately, most NSCLCs resist existing therapies and grow to later stages. This results in an elevated incidence of treatment resistance, attributed to its heterogeneity and the adaptability of cancer stem cells. Drug repurposing provides a more rapid and less expensive method of developing novel antineoplastic uses for current drugs, to eliminate pharmaceutical resistance. The availability of the pharmacokinetic and pharmacodynamic profiles of FDA-approved pharmaceuticals that can be repurposed reduces the time and money required. This chapter outlines the drug-repurposing techniques and modes of action of possible candidate drugs used to treat NSCLC, such as anti-hypertensives, anti-hyperlipidemic, anti-inflammatory, anti-diabetics, and antimicrobials.

Keywords: lung cancer, NSCLC, resistance, drug repurposing, FDA-approved

1. Introduction

Lung cancer is the leading cause of cancer-related death worldwide, accounting for the greatest mortality rates among men and women (Murray & Levy, 2019; Siegel, Miller, Fuchs, & Jemal, 2022). According to the International Agency for Research on Cancer's (IARC) GLOBOCAN 2020 cancer incidence and mortality statistics, lung cancer is the leading cause of cancer death, accounting for an estimated 1.8 million deaths (18%) in 2020 (Kumbhar et al., 2022). It has the second-highest incidence rate among all malignancies, with 236,740 expected new cases in 2022, second only to breast cancer in women and prostate cancer in men. Tobacco usage (including cigarettes, cigars, and pipes) is the primary cause of lung cancer, but it can also occur in non-smokers. Other risk factors include second-hand smoke, occupational hazards (including asbestos, radon, and certain chemicals), air pollution, hereditary cancer syndromes, and a history of chronic lung disease (Travis et al., 2011). Therefore, lung cancer is exceedingly diverse and can grow in various sites of the bronchial tree, resulting in widely varying symptoms (Deligiorgi & Trafalis, 2020; Paliogiannis et al., 2021).

Lung cancer is divided into two types: Small cell lung carcinoma (SCLC) and Non-small cell lung carcinoma (NSCLC), with NSCLC accounting for nearly 85% of cases. Adenocarcinoma is the most prevalent subtype of NSCLC, accounting for ~60% of cases (Figure 1) (Byers & Rudin, 2015;

Santarpia et al., 2016; Travis et al., 2011). Other NSCLCs include squamous, large squamous, pleomorphic, spindle, and giant cell carcinomas. SCLS is a deadly and aggressive form of lung cancer that often metastasizes to other types of cancer (Islami et al., 2018; Rosell & González-Cao, 2022). It accounts for 10-15% of all lung cancer occurrences. Patients with SCLS are usually identified at an advanced stage and have a short median survival time (N1 year with chemotherapeutic intervention) (Pal, Teli, Sengupta, Maji, & Purawarga Matada, 2023; Pal, Teli, Swamy Purawarga Matada, & Sanjay Dhiwar, 2023).

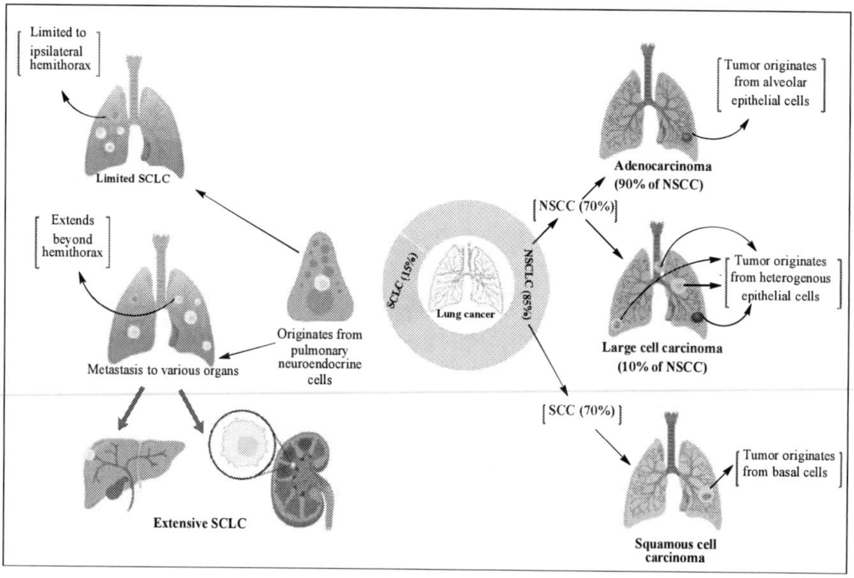

Figure 1. Lung cancer is of two types: small cell lung cancer (SCLC) (15%) and non-small cell lung cancer (NSCLC), which accounts for the other 85%. SCLC is caused by pulmonary neuroendocrine cells and can be restricted or extensive. Restricted SCLC involves just the ipsilateral haemithorax, whereas extensive SCLC involves the contralateral haemithorax or beyond. NSCLC is classified into two types: squamous cell carcinoma (SCC), which originates from lung basal cells and accounts for 30% of all cases, and non-squamous cell carcinoma (NSCC), which accounts for 70%. Most NSCC patients (90%) are adenocarcinoma from alveolar epithelial cells, whereas only 10% are big cell carcinoma with a highly heterogeneous origin. Non-squamous cell carcinoma (NSCC) accounts for 70% of all NSCLC occurrences. Ninety percent of NSCC cases are adenocarcinoma originating from alveolar epithelial cells, whereas only 10% are big cell carcinoma, which has a highly heterogeneous origin (Choudhary et al., 2023; Sánchez-Ortega, Carrera, & Garrido, 2021; Wistuba & Gazdar, 2006).

NSCLC pathogenesis involves many molecular pathways influencing oncogenes and tumor suppressor genes. The epidermal growth factor receptor (EGFR) gene is a critical cancer-related gene that affects NSCLCs (B. Li et al., 2020). Mutations in EGFR genes are commonly detected in adenocarcinoma patients; they are prevalent among people with never-smoked status, female gender, and East Asian ethnicity. Another gene typically seen in adenocarcinoma is the anaplastic lymphoma kinase (ALK) gene, especially the EML4-ALK fusion in young, non-smoking individuals. The Kirsten rat sarcoma viral oncogene homolog gene (KRAS) is also often identified in smokers' adenocarcinoma (Kaur & Suresh, 2023; Paliogiannis et al., 2021; Parvathaneni, Kulkarni, et al., 2020). The KRAS and EGFR pathways in adenocarcinoma are mutually exclusive, implying that separate molecular pathways are involved in adenocarcinoma formation between smokers and nonsmokers. The MET (Mesenchymal Epithelial Transition) gene, found in 7% of NSCLC cases, encodes a receptor tyrosine kinase that activates many signaling pathways implicated in lung cancer development and metastasis. The FGFR1 gene codes for a tyrosine kinase receptor (TKR) found on the cell surface, belonging to the FGFR (Fibroblast growth factor receptor) family. FGFR1 is shown to be amplified in 20% of SCC and 1-3% of adenocarcinomas. Mutations in the discoidin domain receptor 2 (DDR2) gene, a tyrosine kinase receptor, were identified in 4% of squamous cell carcinomas (Kaur & Suresh, 2023). Figure 2 illustrates the various pathways involved in NSCLC.

The National Cancer Institute (2024) devised different processes for detecting and diagnosing lung cancer stages to find the best therapies for each stage (Bernhardt & Jalal, 2016; Kadara, Scheet, Wistuba, & Spira, 2016). Treatment options differ depending on the stage of lung cancer, the patient's susceptibility and health circumstances, and the disease's molecular profile. So far, the treatment options include surgery, chemotherapy, external radiation therapy, radiofrequency ablation, stereotactic radiosurgery, targeted therapy, immunotherapy, and palliative care. Although surgery is the chosen treatment choice for early-stage lung cancer, some individuals' limited tolerance for surgery prevents this intervention. Furthermore, most lung cancer patients are diagnosed at an advanced stage, therefore chemotherapy alone, or in conjunction with other therapies, is the primary line of treatment (Cristea & Sage, 2016; Johnson, Pennell, & Borghaei, 2018). Despite breakthroughs in diagnostic procedures and the discovery of new chemotherapeutics, platinum-based chemotherapeutics like cisplatin remain the most effective medicines in treating lung cancer, and they are frequently used in conjunction with another chemotherapeutic. Cisplatin and other platinum-core compounds (such as carboplatin) are used as first-line treatments for NSCLC. Paclitaxel (PTX), docetaxel (DTX), pemetrexed, and gemcitabine (GCB) are some more chemotherapeutic drugs used for the same purpose. In lung cancer caused by gene alterations, tyrosine kinase inhibitors (TKIs) are suggested as first-line therapy. When first-line medications fail, patients are typically given second-line medicines such as GCB, DTX, and pemetrexed (Rajasegaran, How, Saud, Ali, & Lim, 2023).

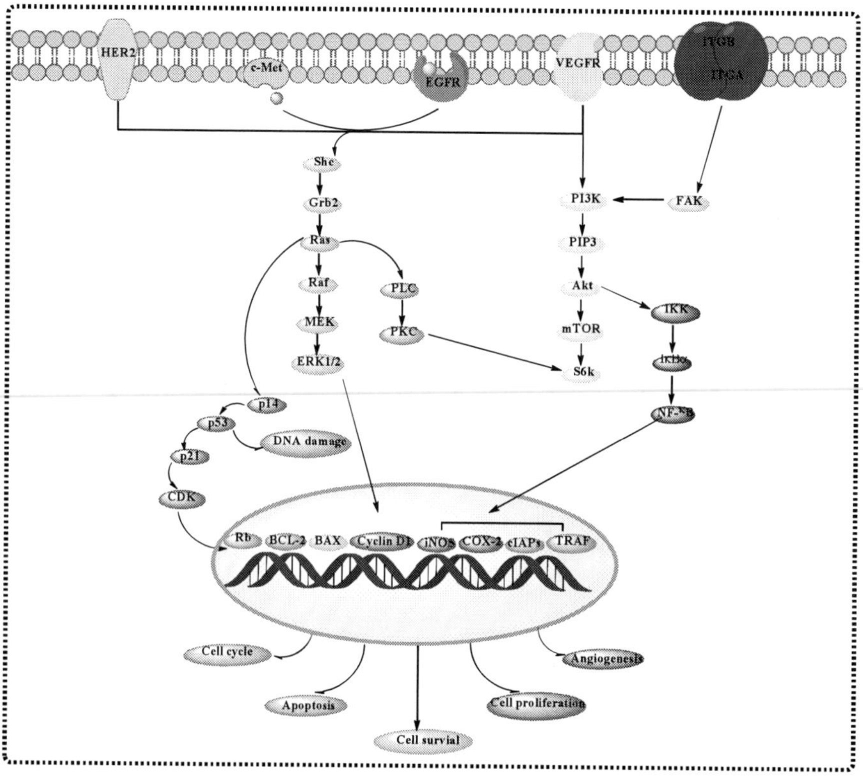

Figure 2. Various molecular pathways are involved in lung cancer.

Despite advances in cancer biology and therapy, overall survival rates have remained modest. Lung cancer therapies have a lengthy development path from discovery to clinical usage, high costs, regulatory barriers, and high failure rates. A newly approved medicine typically takes 13 years and costs US$1.8 billion to develop and enter clinical practice (Rajasegaran et al., 2023). To overcome the hurdles of therapeutic development for lung cancer, repurposing existing medicines for additional indications is a viable option. Drug repurposing allows for faster entry into clinical trials compared to newly produced medications, as existing pharmaceuticals have already been thoroughly tested for safety, dose, and toxicity (W.-H. Lee et al., 2018; Thunnissen et al., 2012). Drug repurposing refers to using existing drugs to treat a new medical health condition that was not its original indication. The new indication is based on existing drug safety, pharmacokinetics, and manufacturing data, including permitted, discontinued, shelved, and experimental therapies. There are two types of repurposed drugs: those that are no longer utilized due to toxicity or adverse effects but can be used in reduced dosages or for a new demographic (Rajasegaran et al., 2023). For example, thalidomide, previously used to treat morning sickness in pregnant women, was subsequently proven to be safe and effective in treating refractory multiple myeloma. Some repurposed medications, such as sildenafil, were once designed as an anti-anginal treatment but are now used to treat erectile dysfunction (ED). Additional classes of drugs are effective in addressing numerous etiologies but are only utilized for one. The synergistic effect refers to drugs that enhance the effects of traditional treatments (Kim, 2012; Kim & Scialli, 2011; Thunnissen et al., 2012).

Albeit improvement in screening efforts has led to decreased incidence rates, NSCLC continues to present significant treatment hurdles (Frisone, Friedlaender, Addeo, & Tsantoulis, 2022; Meador & Hata, 2020; Rotow & Bivona, 2017). High resistance to standard chemotherapy makes single-drug therapy ineffective, requiring double therapy or a combination of chemotherapy and radiation. NSCLC can benefit from medication repurposing due to the need for more treatment alternatives. Therefore, drug repositioning is an effective way to identify new indications of the drugs (Horvath, Thienpont, Zhao, Wolf, & Pircher, 2020; Maione et al., 2015). Using several repurposing methodologies, scientists identified prospective targets based on their mode of action. However, only a few medicines have reached the clinical trial stage. The data presented in Table 1 illustrates the various drugs under clinical trial that are repositioned for NSCLC.

Table 1. Current clinical trials investigating repurposed drugs for the treatment of NSCLC

S. No.	Study Title	ClinicalTrials.gov Identifier	Intervention	Status References
1.	Atovaquone With Radical ChemorADIotherapy in Locally Advanced NSCLC	NCT04648033	Drug: Atovaquone Oral Suspension Drug: Standard of care chemotherapy	Phase 1 (ClinicalTrials.gov)
2.	A Study of Canakinumab with Chemotherapy, Radiation Therapy, and Durvalumab in People with Lung Cancer (CHORUS)	NCT04905316	Drug: Canakinumab Drug: Durvalumab Drug: Chemotherapy	Phase 2 (ClinicalTrials.gov)
3	Pyrotinib Plus Thalidomide in Advanced NSCLC Patients Harboring HER2 Exon 20 Insertions	NCT04382300	Drug: pyrotinib combined with thalidomide	Phase 2 (ClinicalTrials.gov, 2024h)
4.	A Study of Abemaciclib (LY2835219) in Participants with NSCLC or Breast Cancer	NCT02779751	Drug: Abemaciclib Drug: Pembrolizumab Drug: Anastrozole	Phase 1 (ClinicalTrials.gov, 2024i)
5.	Combination of osimertinib and Aspirin to Treat EGFR Mutation NSCLC Patients	NCT04184921	Drug: Aspirin 100mg Drug: Osimertinib 80 MG	Observational (ClinicalTrials.gov, 2024a)
6.	Imatinib Mesylate and Paclitaxel in Treating Older Patients with Stage IIIB or Stage IV Non-Small Cell Lung Cancer	NCT00408460	Drug: imatinib mesylate Drug: paclitaxel	Phase 2 (ClinicalTrials.gov, 2024c)
7.	A Phase Ib Study to Evaluate the Safety and Preliminary Efficacy of IL6-receptor Antibody Sarilumab in Combination With antiPD1 Antibody Cemiplimab for Patients with NSCLC.	NCT05704634	Drug: Cemiplimab Drug: Kevzara (Sarilumab)	Phase 1 (ClinicalTrials.gov, 2024g)
8.	Tocilizumab and Atezolizumab in Adults with Locally Advanced or Metastatic Non-Small Cell Lung Cancer Refractory to 1st Line Immune Checkpoint Inhibitor-Based Therapy	NCT04691817	Drug: Atezolizumab Drug: Tocilizumab	Phase 1 and 2 (ClinicalTrials.gov, 2024k)

Table 1. (Continued)

S. No.	Study Title	ClinicalTrials.gov Identifier	Intervention	Status References
9.	A Study of Multiple Immunotherapy-Based Treatment Combinations in Participants with Metastatic NSCLC (Morpheus- Non-Small Cell Lung Cancer) (Morpheus Lung)	NCT03337698	Drug: Atezolizumab Drug: Cobimetinib Drug: RO6958688 Drug: Docetaxel Drug: CPI-444 Drug: Pemetrexed Drug: Carboplatin Drug: Gemcitabine Drug: Linagliptin Drug: Tocilizumab Drug: Ipatasertib Drug: Bevacizumab Drug: Sacituzumab Govitecan Other: Radiation Drug: Evolocumab Drug: Tiragolumab Drug: XL092 Drug: Camonsertib	Phase 1 and 2 (ClinicalTrials.gov, 2024j)
10.	Pembro and Vorinostat for Patients with Stage IV NSCLC	NCT02638090	Drug: Vorinostat Drug: Pembrolizumab	Phase 1 and 2 (ClinicalTrials.gov, 2024e)
11.	A Phase 2 Study of CRLX101(NLG207) in Patients with Advanced NSCLC	NCT01380769	Drug: CRLX101	Phase 2 (ClinicalTrials.gov, 2024f)
12.	Combined Statin and PD-1/PD-L1 Inhibitors in Treating NSCLC	NCT05636592	Drug: immunotherapy	Observational (ClinicalTrials.gov, 2024b)
13.	Metformin Plus Tyrosine Kinase Inhibitors for Treatment of Patients with NSCLC with EGFR Mutations (METLUNG)	NCT05445791	Drug: Metformin Hydrochloride Other: Placebo	Phase 3 (ClinicalTrials.gov, 2024d)

2. Advancement of Approved Drug Repositioning Against NSCLC

2.1. Repurposing of Anti-Malarial Drugs Against NSCLC

Drugs such as atovaquone, chloroquine, amodiaquine, and dihydroartemisinin (Figure 3) are well-known antimalarial drugs that act by contrasting mechanisms as anticancer agents. In 2021, Skwaski and coworkers evaluated the atovaquone (Figure 3) for NSCLC. In the cohort study atovaquone was given orally to 2 cohort groups. They tested for tumor hypoxic volume (HV) which was examined by the hypoxia PET-CT analysis. The study showed that atovaquone significantly reduced tumor hypoxia in treated patients, with a median HV reduction of 28% compared to a 15.5% increase in untreated patients and considerably downregulated hypoxia-regulated genes without adverse events. Linear regression estimated a 55% lower mean HV in the atovaquone group ($P = 0.004$) (Skwarski et al., 2021). In 2020, amodiaquine was also repositioned against NSCLC cell lines. Repurposing atovaquone as a radiosensitizer might enhance radiation outcomes in NSCLC patients by addressing tumor hypoxia, a frequent barrier to radiation, chemotherapy, and immunotherapy. Parvathaneni and co-workers formulated amodiaquine-loaded nanoparticles and evaluated them for cytotoxic activity. Anticancer activity of the formulation and plain drug was estimated against a series of cancer cell lines and results revealed that amodiaquine and amodiaquine-loaded nanoparticles displayed significant anticancer activity against the lung cancer cell lines with IC_{50} values of 21.8, and 9.4 μM (A549), 14.4, and 1.9 μM (H2122), 22.5 and 16.3 μM (H157), 28.8 and 10.3 μM (H460), 28.8 and 13.5 μM (H4006) (Parvathaneni, Kulkarni, et al., 2020). Similarly, Quinacrine-loaded (Figure 2) nanoparticles (QA-NP) were formulated and evaluated by Vaidya and co-authors for anticancer activity against NSCLC cell lines. Cytotoxicity activity of QA solution, QA-NP (PEI stabilized), and QA-NP (BSA coated) were tested against a group of cell lines. The study demonstrated that QA solution possesses moderate activity with the IC_{50} values of 3.0, 3.7, 3.4, and 0.8 μM, QA-NP (PEI stabilized) exhibited good activity with IC_{50} values of 2.1, 2.6, 2.7 and 0.9 μM and BSA coated QA-NP showed potent activity with IC_{50} values of 1.3,1.6,2.0 and 0.9 μM against A549, H4006, H157, and H460 cell lines, respectively (Vaidya et al., 2020). Chloroquine (Figure 3) for paclitaxel-acquired resistance against NSCLC was evaluated in a cell viability study which revealed that pre-treated chloroquine

+ paclitaxel could significantly decrease cell death with 31% in 20 days and 45% in 40 days compared with paclitaxel alone treated cell which showed only 22% cell death in 20 days and no significant death till 40th day. Furthermore, in the MTT cell proliferation assay on A549 cell lines, pretreated chloroquine with pretreatment with paclitaxel showed significant anticancer activity with an IC_{50} value of 29 μM (Datta et al., 2019). Jiang and co-workers elaborated that dihydroartemisinin (DHA) (Figure 3) acts as an anti-cancer agent by inhibiting the NF-kB/GLUT1 axis. The antiproliferative activity of DHA was evaluated against assay on 95C, 95D, H460, and H1650 cell lines with different concentrations of DHA for 72 hrs. The study revealed that DHA decreases cell viability in a dose-dependent manner. There is no cytotoxic activity at < 15μM, whereas a 15-20% decrease in cell viability was noted at the concentration of 30 μM (J. Jiang et al., 2016).

2.2. Repurposing of Anti-Microbial Drugs Against NSCLC

Bedaquiline (Figure 3), initially approved by the FDA as an anti-tuberculosis (TB) medication, has shown promising anti-cancer effects in several experimental studies. Najib Ullah and co-workers formulated the bedaquiline (BDQ) loaded solid lipid nanoparticles (SLN) and evaluated them for cell viability study on the A549 cell line using MTT assay at 24, 48, and 72 hrs. The study revealed that BDQ-loaded SLN showed potent inhibitory activity against the A549 cell line with a 6.121% cell survival and blank SLN demonstrated inhibitory activity with IC_{50} of 120 μg/ml and a percentage of cell viability of 98.01% (Najib Ullah et al., 2023). Parvathaneni and co-researchers also investigated the anti-lung cancer properties of BDQ (Figure 3) and formulated inhalable bedaquiline cyclodextrin-based (BDQ-CD). The molecular inclusion complexe formulation was prepared by a freeze-drying method. BDQ and BDQ-CD complex exhibited significant anticancer activity against the H1299 cell line with IC_{50} values of 37.2 and 43.8 μM after the 72-hour incubation (Parvathaneni et al., 2021). Further, in 2016 Song and coworkers evaluated levofloxacin (Figure 3) for the treatment of lung cancer. Levofloxacin was tested against a panel of NSCLC lines (A549, H3255, NCL-69, and H460), and it was found that levofloxacin inhibits cancer growth in a dose-dependent manner in all the cancer cell lines. Furthermore, levofloxacin showed an antioxidant effect that confirms that levofloxacin acts on lung cancer by inducing oxidative stress (Song et al., 2016). In the same year, tigecycline (Figure 3) was also tested for the treatment of NSCLC. Tigecycline showed cytotoxicity against A549, PC9, H157, and EBC-1 with IC_{50} 16.4 μM, 21.7 μM, 9.6 μM and 2.3 μM, respectively. Tigecycline was also evaluated for the *in-vivo* study with the A549 xenograft model and the result revealed that after 2 weeks of daily treatment with tigecycline, the tumor size and weight were reduced by 60% compared to the control (Jia, Gu, Chen, & Jiao, 2016). Aftab et al., reported a repurposing of itraconazole (Figure 3) for the NSCLC. The anti-cancer activity was evaluated in NSCLC primary xenograft models LX-14 and LX-7. The findings indicated that itraconazole (75 mg/kg) inhibited tumor growth by 72% and 79% in the LX-14 and LX-7 models, respectively which is comparable with cisplatin which resulted in 75% and 48% inhibition of LX-14 and LX-7 models, respectively. Furthermore, hypoxia markers and vascularity were studied for the LX-14 and LX-7 xenograft tissue which revealed an increase in tumor HIF1α and a decrease in the cancer vascular area in NSCLC xenografts (Aftab, Dobromilskaya, Liu, & Rudin, 2011).

2.3. Repurposing of Anti-Viral Drugs Against NSCLC

Nelfinavir and saquinavir are popular protease inhibitors and are employed in the treatment of Human immunodeficiency viruses (HIV). Pereira and co-researchers investigated saquinavir as an anticancer agent. Saquinavir (Figure 3) (SAQ) alone and in combination with 5-fluorouracil (5-FU) were tested for anticancer activity on the A549 cell line by MTT assay. Results revealed that saquinavir (IC_{50} = 41.04 μM) showed less activity than 5-FU (IC_{50} = 6.070 μM) after 72 hr. The synergistic study was conducted using CompuSyn. The study demonstrated that the combination of 25 μM of 5-FU with various concentrations of saquinavir exhibited more synergism effect which decreased with an increase in the concentration of SAQ whereas the combination of 25 μM of SAQ with various concentrations of 5-FU decreased the synergism with an increase in the concentration of 5-FU (Pereira & Vale, 2022).

Parvathaneni et al. reported nelfinavir (NFV) (Figure 3) loaded PLGA nanoparticles. The formulated nanoparticles and plain nelfinavir were assessed against a panel of NSCLC cell lines. The study revealed that NFV and NFV nanoparticles to possess potent activity against A549 (18.1 and 8.3 μM), H4006 (24.3 and 5.0 μM), H460 (20.3 and 4.0 μM) (Parvathaneni, Goyal, Kulkarni, Shukla, & Gupta, 2020). In 2020, Marima et al., reported the repurposing of the anti-retroviral drugs efavienz (Figure 3) and a combination of lopinavir-ritonavir. It was found that Efavienz was not active in the 4 μM concentration but at 13 μM concentration, it showed a slight increase in cell viability. While the lopinavir-ritonavir combination showed a significant increase in the cytotoxic effect at 10 μM the effect decreased with higher concentrations like 32 μM (Marima, Hull, Dlamini, & Penny, 2020).

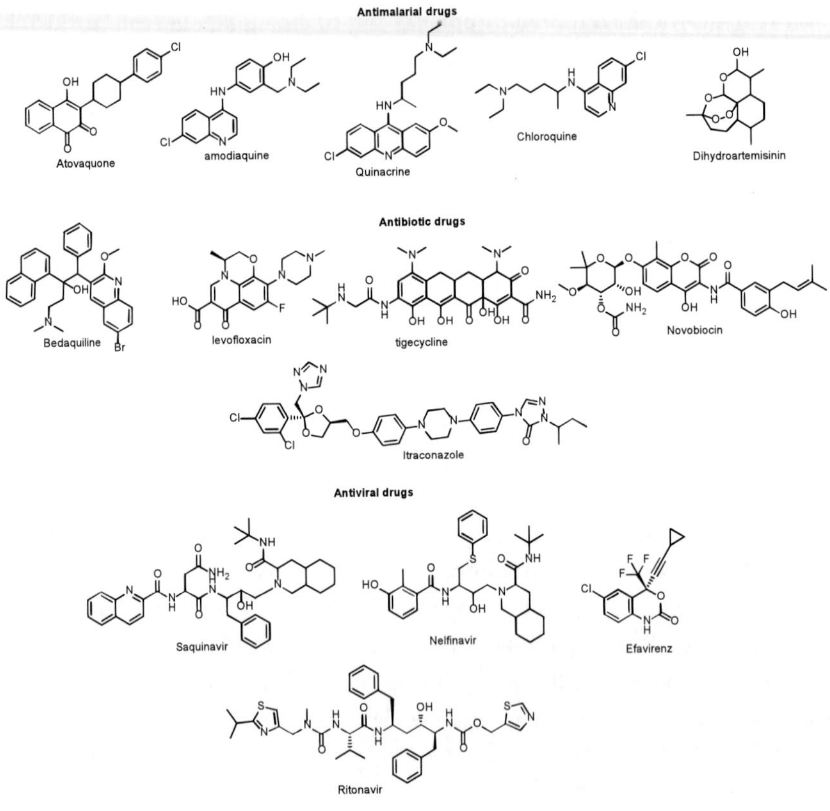

Figure 3. Chemical structures of anti-malarial, anti-microbial, and anti-viral drugs used for the repositioning against NSCLC.

2.4. Repurposing of Cardiovascular Drugs Against NSCLC

In cardiovascular diseases, hyperlipidaemia is a major disease conation and anti-hyperlipidemic drugs such as statins are the most often used medications to treat high blood cholesterol levels. Many preclinical and clinical studies have demonstrated statins' anti-neoplastic properties. Thus, drug repurposing research has concentrated on statins as an off-label cancer therapy. Statins block the enzyme hydroxymethylglutaryl-CoA (HMG-CoA) reductase, which synthesizes cholesterol. This reduces the generation of mevalonated derivatives, which are crucial for growth regulation processes including proliferation, apoptosis, and differentiation (Regulska, Regulski, Karolak, Murias, & Stanisz, 2019). Therefore, Hosseinimehr and co-researchers evaluated atorvastatin as an anti-cancer drug. Atorvastatin (Figure 4) was subjected to a radio-sensitizing effect on the lung cancer cell line (A549). Radiotherapy usually fails in cancer treatment because of their resistance towards cancer cells. Therefore, to overcome this problem, the radio-sensitizing technique is used in cancer therapy. However, atorvastatin 10 μM on X-ray radiation-induced apoptosis at 56.88% in the A549 cell line (Hosseinimehr et al., 2020). Lee and co-authors also evaluated atorvastatin for apoptosis and autophagic mechanism in NSCLC. The apoptosis activity of atorvastatin was evaluated at three different concentrations, 5, 10, and 20 μM. At 5, 10, and 20 μM concentrations, atorvastatin showed 43.41% in the G0/G1 phase, 35.92% in the G0/G1 phase, and 40.18% in the Sub G1 phase apoptosis activity. The authors concluded that atorvastatin induces apoptosis and autophagic cell death in H1299 cells *via* the CNOT2 signaling pathway (J. Lee et al., 2019). Similarly, Otahal et al., also evaluated the anti-cancer activity of pitavastatin and fluvastatin (Figure 4). Pitavastatin, and fluvastatin induced cell death in EGFR-TKI expressing (A549) cell line. The pitavastatin showed EC_{50} values of 6.344, 13.13, and 41.77 μM in A549, Calu6, and H1993 cell lines respectively. When pitavastatin was combined with erlotinib, it induced a synergistic effect in cancer activity (Otahal, Aydemir, Tomasich, & Minichsdorfer, 2020). Chou and co-researchers (2019) investigated the anti-cancer activity of statins *via* p53 gene expression. P53 is a crucial suppressor gene for cancer. The study revealed that simvastatin-induced apoptosis in the caspase-dependent apoptotic pathway (Chou et al., 2019). Walther et al. studied the anticancer effect of lovastatin (Figure 4) against lung cancer cell lines A549 and H358. Cell viability results revealed that lovastatin has IC_{50} values of 76.7 μM and 45.2 μM in A549 and H358 cell lines and its cytotoxic activity was found to be concentration-dependent (Walther, Emmrich, Ramer, Mittag, & Hinz, 2016).

Candesartan (Figure 4) is an antihypertensive medicine that acts as an angiotensin II receptor blocker (ARB). It works by preventing the function of a chemical in the body that causes blood vessels to tighten, allowing the blood vessels to relax. Ji et al. evaluated the anticancer activity of candesartan through an enzyme inhibition assay. It showed good inhibition towards TMEM16A with the IC_{50} value of 24.40 μM. Cell apoptosis studies *via* FITC annexin-V apoptosis assay in the LA795 cancer cell line were also conducted. The results showed that apoptosis was induced via CDST in the LA795 cell line and was found at 10.1% at 30 μM, 16.8% at 100 μM, and 23.8% at 300 μM respectively. While the induced apoptosis through TMEM16A was 23.6% at 30 μM concentration in the LA795 cell line (Ji et al., 2023). Carvedilol (Figure 4) is also an anti-hypertensive drug, which act as a β-Adrenergic agonist. Ikhmais et al. investigated carvedilol for repurposing it in the treatment of NSCLC. An experiment quantified basal aldehyde dehydrogenase (ALDH) activity in A549 and H1299 cell lines using NADH fluorescence spectrophotometric assay. The two tested cell lines showed a significant difference in ALDH activity. A549 cells displayed high activity while the H1299 had very low activity. In the MTT colorimetric assay, carvedilol showed an IC_{50} value of 18 and 13.7μM against A549 and H1299 cell lines respectively (Ikhmais et al., 2023).

2.5. Repurposing of Anti-Inflammatory Drugs Against NSCLC

Nonsteroidal anti-inflammatory drugs (NSAIDs) block cyclooxygenase (COX) enzymes, providing anti-inflammatory, analgesic, and antipyretic effects. Recent research indicates that COX-2 is over-expressed in lung adenocarcinomas. COX-2 promotes cell proliferation and reduces apoptosis, which is essential for aggressive tumor development and metastasis. Thus, COX-2 inhibition serves as a critical checkpoint in anticancer therapy. Overexpression of COX-2 proteins in human lung cancer has been associated with a poor prognosis. Celecoxib (Figure 4), a COX-2 inhibitor, was initially used as an analgesic for osteoarthritis and rheumatoid arthritis pain. However, later studies suggested that celecoxib could inhibit cancer growth by both internal and extrinsic mechanisms, including apoptosis and downregulation of NF-kB, caspase-9, BAX, and BCL-xL. Hida et al., screened the anticancer properties of nimesulide (Figure 4) by the MTT assay using NSCLC cell lines (ACC-LC-33, ACC-LC-91, ACC-LC-319, and ACC-LC-323) and two SCLC cell lines (ACC-LC- ACC-LC-48 and ACC-LC--172). Results revealed that ACC-LC-91 cells were most responsive to the nimesulide followed by ACC-LC-323, -319, and -33 cells whereas ACC-LC-172 and ACC-LC-48 were resistant to the nimesulide. Furthermore, nimesulide was screened for anticancer activity in combination with other anti-cancer agents or radiation therapy. The effect of nimesulide at 0, 10, or 30 mM was evaluated in combination with SM-5887, SN-38, and Taxotere. Results revealed that the

ACC-LC-91 cells treated with nimesulide 30 μM with SN-38, SM-5887, and Taxotere showed 77%, 60%, and 54% reduction, respectively (Hida et al., 2000). Sarvepalli and coworkers formulated a COX-2 inhibitor indomethacin (Figure 4) liposomes (IND-Lip) and evaluated against NSCLC. Indomethacin liposomes (IND-Lip) decreased cell growth with an IC_{50} value of 38.4, 82.2, and 101.4 μM against A549, H460, and H1299 cell lines respectively (Sarvepalli, Parvathaneni, Chauhan, Shukla, & Gupta, 2022).

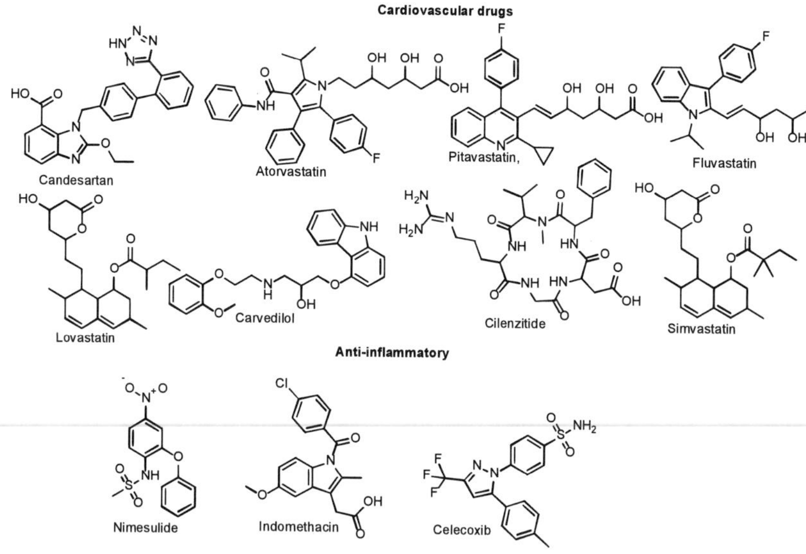

Figure 4. Chemical structures of cardiovascular, and anti-inflammatory drugs used for the repositioning against NSCLC.

2.6. Repurposing of Drugs Acting on the Central Nervous (CNS) Against NSCLC

Sertraline is an antidepressant drug that primarily inhibits presynaptic serotonin reuptake, which leads to a buildup of serotonin. The anticancer properties of sertraline were evaluated by Jiang et al., in comparison with erlotinib by inducing autophagy. Sertraline showed better activity compared to erlotinib with IC_{50} values of 11.08 and 25.23 (A549), 9.40 and 49.07 (H1975) and 10.50 and 30.76 μM (H522). While, erlotinib (IC_{50}: 6.80, and 0.019) exhibited higher activity compared to sertraline (IC_{50}: 9.60, and 4.40) (Figure 5) against PC9/Rand PC9 cell lines (X. Jiang et al., 2018).

Xia and co-workers elaborated repurposing of the antipsychotic trifluoperazine as an anticancer agent. Antipsychotic agents (Haloperidol, Chlorprothixene, Chlorpromazine, Perphenazine, Fluphenazine, Droperidol, Sulpiride, thioridazine, Loxapine, Zuclopenthixol, Trifluoperazine, Pimozide, and Promethazine) (Figure 5) were tested against B16 aggressive cell lines. Amongst all tested antidepressant drugs, trifluoperazine showed the most significant activity with an IC_{50} value of 4.2 μM (Xia et al., 2021).

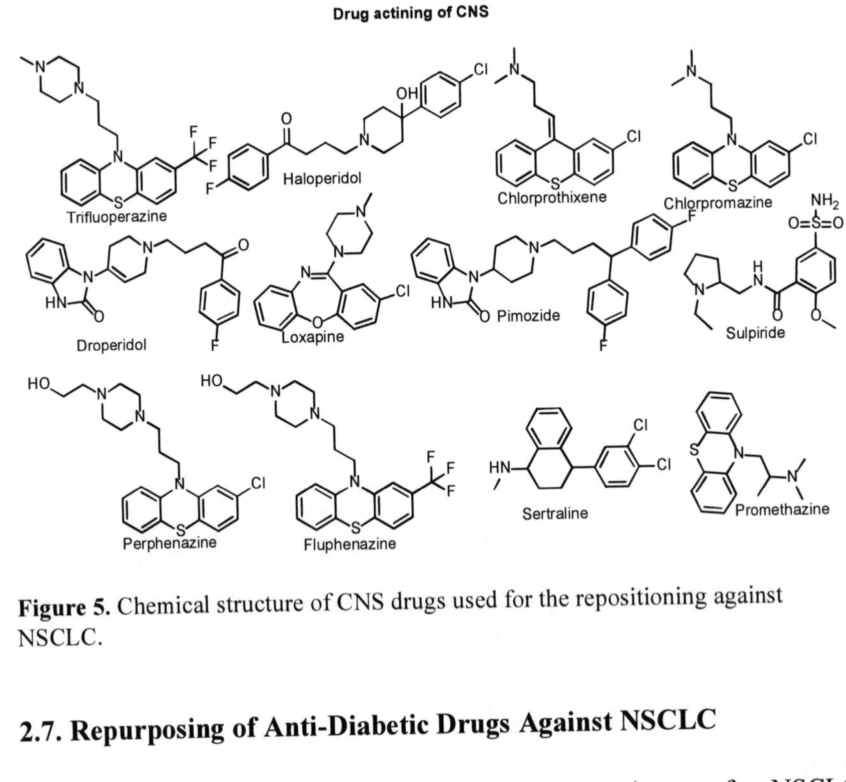

Figure 5. Chemical structure of CNS drugs used for the repositioning against NSCLC.

2.7. Repurposing of Anti-Diabetic Drugs Against NSCLC

Stachnik et al. investigated zoledronic acid as a therapy for NSCLC. Zoledronic acid (Figure 6) was subjected to colony formation assay on HCC827 at different concentrations of 0, 1, 10, and 40 μM while erlotinib was used as a standard drug at a concentration of 50 nM. The study demonstrated that at 1 and 10 μM colony formation occurred with IC_{50} 15.6 and 15.7 μM. At 0 μM concentration colony formation showed an IC_{50} value of 23.7 μM while at 40 μM, no colony formation was observed whereas erlotinib at the concentration 50nM displayed colony formation with IC_{50} 1.17 μM (Stachnik et al., 2014). In 2017, Li et al. tested three drug candidates (indacaterol, canagliflozin, and cis-flupenthixol) (Figure 6) on resistant NSCLC cell lines. The medicines' selectivity for the mutant EGFR was established using a cell-free kinase inhibition experiment. Indacaterol and canagliflozin showed stronger inhibition of L858R/T790M EGFR, consistent with their higher anticancer activity in H1975 cells (with EGFR L858R/T790M mutation) compared to HCC827 cells (with EGFR E746_A750 deletion). The IC_{50} values for these drugs were at least twice as high for wild-type EGFR, suggesting reduced adverse effects on wild-type EGFR. The autophosphorylation of EGFR and its downstream pathway study found that indacaterol and canagliflozin inhibited EGFR autophosphorylation and were more active against H1975 cells than in HCC827 cells (H. Li et al., 2017).

2.8. Repurposing of Anti-Helminthic Drugs Against NSCLC

Pantziarka et al., (2014) have discussed the broad-spectrum anti-cancer activity of mebendazole (MBZ) (Figure 6), which is an anti-helminthic drug. Subsequently, the same authors showed that MBZ causes dose- and time-dependent apoptosis in lung cancer cell lines, with an IC_{50} of ~0. 16 μM arresting the cell cycle at the G2-M phase and inducing apoptosis while sparing normal cells. MBZ was found to suppress the growth of lung, breast, ovary, colon carcinomas, and osteosarcoma cell lines. *Nu\nu* mice were used to infest H460 non-small cell lung cancer cells, and MBZ was discovered to possess dose-related inhibition of tumour improvement and 80% cessation of lung metastasis. MBZ did not kill normal cells, but it was very effective against chemo-resistant cancers such as melanoma, adrenocortical cancer, glioblastoma, colon, breast cancer, and canine osteosarcoma. Despite the lack of clinical trials using MBZ as an anti-cancer drug, two case reports can be noted. In 2011, a patient with metastatic adrenocortical cancer treated only with MBZ (100 mg twice a day) reported stable disease for 19 months with some tumor regression. At six weeks, a 74-year-old with metastatic colon carcinoma and lung and lymph node and liver metastases achieved near complete regression of lung and nodal and partial regression hepatic metastases on MBZ 100 mg. These cases prove the efficacy of MBZ in cancer treatment and warrant further research (Pantziarka, Bouche, Meheus, Sukhatme, & Sukhatme, 2014).

2.9. Other Drugs Used for Repositioning Against NSCLC

Baby et al., in 2023 performed computational drug repurposing of Akt-1 allosteric inhibitors for NSCLC. For Akt-1 allosteric inhibitors molecular docking was performed using the PDB ID: 4EJN. Most hits showed significant interaction with residues such as Trp-80, or Tyr-272. The docking score was in the range of 12.121 to 6.911 kcal/mol. Validation of molecular docking was done and found that the interactions with Trp-80, Val-270, Val-271, Tyr-272, and Asp-292 appeared common among the docked compounds (Table 2) (Baby et al., 2023).

Table 2. Docking score (kcal/mol), H-bonds, and RMSD of Valganciclovir, Dasatinib, and Novobiocin

Top hit compounds	Docking score (kcal\mol)	H-bonds	RMSD value (Å)
Valganciclovir (antiviral)	-8.225	Thr-82, Ser-205, Thr-211, Tyr-272	2.007
Dasatinib (Philadephia chromosome (Ph) positive leukemia)	-7.409	Asn-54, Tyr-272	1.573
Indacaterol (COPD)	-10.413	Thr-211	0.941
Novobiocin (antibiotic)	-9.401	Asp-292	1.907

Albert et al., (2006) investigated the impact of radiation on αvβ3 integrin at the messenger RNA (mRNA) and protein levels in HUVEC (human umbilical vein endothelial cells) and two NSCLC cell lines, H157 and H460. They noticed that before radiation, αvβ3 integrin was present in 99.8% of HUVEC cells, 77.2% of H157 cells, and 27.6% of H460 cells. Radiation enhanced αvβ3 integrin expression in all these cell lines. Radiation also significantly increased cell proliferation: 7-fold in HUVEC and H460 cells, and 1.8-fold in H157 cells after 5 hours. Cilengitide (Figure 6) is an αvβ3 integrin antagonist that was found to suppress cell viability and survival when used in combination with radiation, primarily affecting HUVEC cells. The combined treatment further enhanced the apoptotic effect, more significantly in HUVEC cells. In HUVEC cells, apoptosis increased 2.2-fold, and in H157 and H460 cells with a 1.19-fold increase in apoptosis. Moreover, H460 cells overexpression amphiregulin, which might promote endothelial cell apoptosis and has a moderate effect on NSCLC cells (Albert et al., 2006).

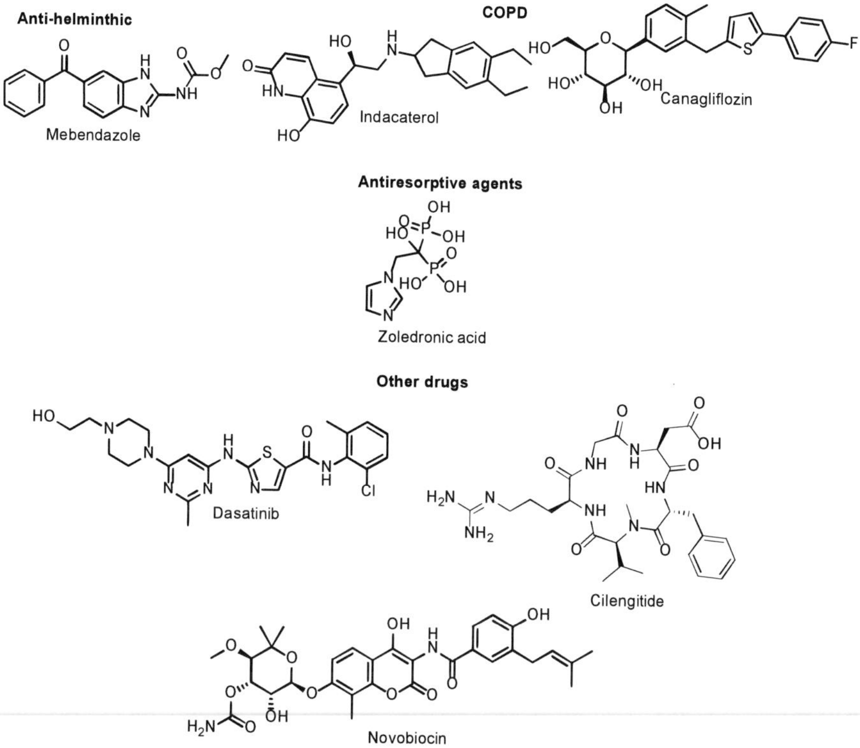

Figure 6. Chemical structure of anthelmintic, COPD, and other miscellaneous drugs used for the repositioning against NSCLC.

Conclusion

Lung cancer is a life-threatening disease impacted by several factors, including genetics and the environment. The intricacy and quick development of therapeutic resistance in lung cancer highlight the need for the identification of novel medications that can effectively kill, eradicate, and reduce tumor progression without developing resistance. There has been an increase in new ways for early diagnosis, treatment, and prognosis of lung cancer. Early detection biomarkers, innovative therapeutics targeting EGFR and ALK, immunological activation using antibodies and interleukin (IL), and novel chemotherapy analogues/derivatives are among the focus areas. However, the effectiveness of innovative medicines in pre-clinical studies does not always translate into clinical outcomes. Drug repurposing is a promising technique for treating lung cancer. Understanding systemic safety, pharmacokinetics, and metabolism of licensed medications can help hasten the development of novel cancer therapeutics through repurposing. Furthermore, medication repurposing is a speedy and cost-effective strategy to offer additional therapy choices to lung cancer patients, potentially enhancing survival rates. Drug repurposing bypasses Phase I and II of clinical trials, making the approval process shorter than usual. With ongoing breakthroughs in precision medicine and artificial intelligence, a broader range of medications may be used in a more focused and strategic manner. This is ultimately about accounting for patient-specific tumor features and increasing survival rates, quality of life, and total expenses through truly personalized therapy. More research, including in vivo tests and clinical trials, is required to investigate possible pharmaceutical repurposing for lung cancer treatment.

Declaration of Competing Interest

The authors have declared that they have no competing financial interests or personal relationships that could have appeared to influence the work reported in this paper.

Acknowledgments

This work was supported by ICMR (No. 5/13/12/2020/NCD-III and No. 5/13/79/2020/NCD-III) projects and Acharya & B M Reddy College of Pharmacy, Bengaluru.

References

Aftab, B. T., Dobromilskaya, I., Liu, J. O., & Rudin, C. M. (2011). Itraconazole inhibits angiogenesis and tumor growth in non–small cell lung cancer. *Cancer Research, 71*(21), 6764-6772.

Albert, J. M., Cao, C., Geng, L., Leavitt, L., Hallahan, D. E., & Lu, B. (2006). Integrin αvβ3 antagonist Cilengitide enhances efficacy of radiotherapy in endothelial cell and non–small-cell lung cancer models. *International Journal of Radiation Oncology* Biology* Physics, 65*(5), 1536-1543.

Baby, K., Maity, S., Mehta, C. H., Nayak, U. Y., Shenoy, G. G., Pai, K. S. R., . . . Nayak, Y. (2023). Computational drug repurposing of Akt-1 allosteric inhibitors for non-small cell lung cancer. *Scientific Reports, 13*(1), 7947.

Bernhardt, E. B., & Jalal, S. I. (2016). Small cell lung cancer. *Lung cancer: Treatment and Research*, 301-322.

Byers, L. A., & Rudin, C. M. (2015). Small cell lung cancer: where do we go from here? *Cancer, 121*(5), 664-672.

Chou, C.W., Lin, C.H., Hsiao, T.H., Lo, C.C., Hsieh, C.Y., Huang, C.C., & Sher, Y.P. (2019). Therapeutic effects of statins against lung adenocarcinoma via p53 mutant-mediated apoptosis. *Scientific Reports, 9*(1), 20403.

Choudhary, N., Bawari, S., Burcher, J. T., Sinha, D., Tewari, D., & Bishayee, A. (2023). Targeting cell signaling pathways in lung cancer by bioactive phytocompounds. *Cancers, 15*(15), 3980.

ClinicalTrials.gov. (2021-01-28). Atovaquone With Radical ChemorADIotherapy in Locally Advanced NSCLC (ARCADIAN). Retrieved from https://clinicaltrials.gov/study/NCT04648033?term=NCT04648033&rank=1.

ClinicalTrials.gov. (2024-06-18). A Study of Canakinumab With Chemotherapy, Radiation Therapy, and Durvalumab in People With Lung Cancer (CHORUS). Retrieved from https://clinicaltrials.gov/study/NCT04905316?term=NCT04905316 &rank=1.

ClinicalTrials.gov. (2024a, 2020-07-23). Combination of Osimertinib and Aspirin to Treat EGFR Mutation NSCLC Patients. Retrieved from https://clinicaltrials.gov/study/NCT04184921?term=NCT04184921&rank=1.

ClinicalTrials.gov. (2024b, 2024-03-12). Combined Statin and PD-1/PD-L1 Inhibitors in Treating Non-small Cell Lung Cancer. Retrieved from https://clinicaltrials.gov/study/NCT05636592?term=NCT05636592&limit=10&rank=1.

ClinicalTrials.gov. (2024c, 2017-04-25). Imatinib Mesylate and Paclitaxel in Treating Older Patients With Stage IIIB or Stage IV Non-Small Cell Lung Cancer. Retrieved from https://clinicaltrials.gov/study/NCT00408460?term=NCT00408460&limit=10 &rank=1.

ClinicalTrials.gov. (2024d, 2023-06-27). Metformin Plus Tyrosine Kinase Inhibitors for Treatment of Patients With Non-small Cell Lung Cancer With EGFR Mutations (METLUNG). Retrieved from https://clinicaltrials.gov/study/NCT05445791?term=NCT05445791&limit=10&rank=1.

ClinicalTrials.gov. (2024e, 2024-02-14). Pembro and Vorinostat for Patients With Stage IV Non-small Cell Lung Cancer (NSCLC). Retrieved from https://clinicaltrials.gov/study/NCT02638090?term=NCT02638090&limit=10&rank=1.

ClinicalTrials.gov. (2024f, 2020-05-28). A Phase 2 Study of CRLX101(NLG207) in Patients With Advanced Non-Small Cell Lung Cancer. Retrieved from https://clinicaltrials.gov/study/NCT01380769?term=NCT01380769&limit=10&rank =1.

ClinicalTrials.gov. (2024g, 2024-07-12). A Phase Ib Study to Evaluate the Safety and Preliminary Efficacy of IL6-receptor Antibody Sarilumab in Combination With antiPD1 Antibody Cemiplimab for Patients With Non-small Cell Lung Cancer. Retrieved from https://clinicaltrials.gov/study/NCT05704634?term=NCT05704634& limit=10&rank=1.

ClinicalTrials.gov. (2024h, 2021-01-28). Pyrotinib Plus Thalidomide in Advanced NSCLC Patients Harboring HER2 Exon 20 Insertions. Retrieved from https://clinicaltrials.gov/study/NCT04382300?term=NCT04382300&rank=1.

ClinicalTrials.gov. (2024i, 2024-03-25). A Study of Abemaciclib (LY2835219) in Participants With Non-Small Cell Lung Cancer or Breast Cancer. Retrieved from https://clinicaltrials.gov/study/NCT02779751?term=NCT02779751&rank=1.

ClinicalTrials.gov. (2024j). A Study Of Multiple Immunotherapy-Based Treatment Combinations In Participants With Metastatic Non-Small Cell Lung Cancer (Morpheus- Non-Small Cell Lung Cancer) (Morpheus Lung). Retrieved from https://clinicaltrials.gov/study/NCT03337698?term=NCT03337698&limit=10&rank=1.

ClinicalTrials.gov. (2024k, 2023-07-28). Tocilizumab and Atezolizumab in Adults With Locally Advanced or Metastatic Non-Small Cell Lung Cancer Refractory to 1st Line Immune Checkpoint Inhibitor-Based Therapy. Retrieved from https://clinicaltrials.gov/study/NCT04691817?term=NCT04691817&limit=10&rank=1.

Cristea, S., & Sage, J. (2016). Is the canonical RAF/MEK/ERK signaling pathway a therapeutic target in SCLC? *Journal of Thoracic Oncology, 11*(8), 1233-1241.

Datta, S., Choudhury, D., Das, A., Mukherjee, D. D., Dasgupta, M., Bandopadhyay, S., & Chakrabarti, G. (2019). Autophagy inhibition with chloroquine reverts paclitaxel resistance and attenuates metastatic potential in human nonsmall lung adenocarcinoma A549 cells via ROS mediated modulation of β-catenin pathway. *Apoptosis, 24*, 414-433.

Deligiorgi, M. V., & Trafalis, D. T. (2020). Repurposing denosumab in lung cancer beyond counteracting the skeletal related events: an intriguing perspective. *Expert Opinion on Biological Therapy, 20*(11), 1331-1346.

Frisone, D., Friedlaender, A., Addeo, A., & Tsantoulis, P. (2022). The landscape of immunotherapy resistance in NSCLC. *Frontiers in Oncology, 12*, 817548.

Hida, T., Kozaki, K.-i., Muramatsu, H., Masuda, A., Shimizu, S., Mitsudomi, T., . . . Takahashi, T. (2000). Cyclooxygenase-2 inhibitor induces apoptosis and enhances cytotoxicity of various anticancer agents in non-small cell lung cancer cell lines. *Clinical Cancer Research, 6*(5), 2006-2011.

Horvath, L., Thienpont, B., Zhao, L., Wolf, D., & Pircher, A. (2020). Overcoming immunotherapy resistance in non-small cell lung cancer (NSCLC)-novel approaches and future outlook. *Molecular Cancer, 19*, 1-15.

Hosseinimehr, S. J., Ghasemi, F., Flahatgar, F., Rahmanian, N., Ghasemi, A., & Asgarian-Omran, H. (2020). Atorvastatin sensitizes breast and lung cancer cells to ionizing radiation. *Iranian Journal of Pharmaceutical Research: IJPR, 19*(2), 80.

Ikhmais, B. A., Hammad, A. M., Abusara, O. H., Hamadneh, L., Abumansour, H., Abdallah, Q. M., . . . Alshehada, R. (2023). Investigating carvedilol's repurposing for the treatment of Non-small Cell Lung Cancer via aldehyde dehydrogenase activity modulation in the presence of β-adrenergic agonists. *Current Issues in Molecular Biology, 45*(10), 7996-8012.

Islami, F., Goding Sauer, A., Miller, K. D., Siegel, R. L., Fedewa, S. A., Jacobs, E. J., . . . Soerjomataram, I. (2018). Proportion and number of cancer cases and deaths attributable to potentially modifiable risk factors in the United States. *CA: A Cancer Journal for Clinicians, 68*(1), 31-54.

Ji, Q., Shi, S., Ma, B., Zhang, W., An, H., & Guo, S. (2023). Drug repurposing and molecular mechanisms of the antihypertensive drug candesartan as a TMEM16A channel inhibitor. *International Journal of Biological Macromolecules, 235*, 123839.

Jia, X., Gu, Z., Chen, W., & Jiao, J. (2016). Tigecycline targets nonsmall cell lung cancer through inhibition of mitochondrial function. *Fundamental & Clinical Pharmacology, 30*(4), 297-306.

Jiang, J., Geng, G., Yu, X., Liu, H., Gao, J., An, H., . . . Wu, X. (2016). Repurposing the anti-malarial drug dihydroartemisinin suppresses metastasis of non-small-cell lung cancer via inhibiting NF-κB/GLUT1 axis. *Oncotarget, 7*(52), 87271.

Jiang, X., Lu, W., Shen, X., Wang, Q., Lv, J., Liu, M., . . . Pang, X. (2018). Repurposing sertraline sensitizes non–small cell lung cancer cells to erlotinib by inducing autophagy. *JCI Insight, 3*(11), e98921.

Johnson, M., Pennell, N. A., & Borghaei, H. (2018). "My patient was diagnosed with nontargetable advanced Non–Small Cell Lung Cancer. What now?" Diagnosis and initial treatment options for newly diagnosed patients with advanced NSCLC. *American Society of Clinical Oncology Educational Book, 38*, 696-707.

Kadara, H., Scheet, P., Wistuba, I. I., & Spira, A. E. (2016). Early events in the molecular pathogenesis of lung cancer. *Cancer Prevention Research, 9*(7), 518-527.

Kaur, R., & Suresh, P. (2023). Chemoresistance mechanisms in Non-Small Cell Lung Cancer—Opportunities for drug repurposing. *Applied Biochemistry and Biotechnology*, 1-57.

Kim, J. H. (2012). Thalidomide: The tragedy of birth defects and the effective treatment of disease. James H. Kim and Anthony R. Scialli. *Toxicological Sciences, 125*(2), 613.

Kim, J. H., & Scialli, A. R. (2011). Thalidomide: the tragedy of birth defects and the effective treatment of disease. *Toxicological Sciences, 122*(1), 1-6.

Kumbhar, P., Manjappa, A., Shah, R., Jha, N. K., Singh, S. K., Dua, K., . . . Patravale, V. (2022). Inhalation delivery of repurposed drugs for lung cancer: approaches, benefits and challenges. *Journal of Controlled Release, 341*, 1-15.

Lee, J., Jung, J. H., Hwang, J., Park, J. E., Kim, J.-H., Park, W. Y., . . . Kim, S.-H. (2019). CNOT2 is critically involved in atorvastatin induced apoptotic and autophagic cell death in non-small cell lung cancers. *Cancers, 11*(10), 1470.

Lee, W.H., Loo, C.Y., Ghadiri, M., Leong, C.R., Young, P. M., & Traini, D. (2018). The potential to treat lung cancer via inhalation of repurposed drugs. *Advanced Drug Delivery Reviews, 133*, 107-130.

Li, B., Dai, C., Wang, L., Deng, H., Li, Y., Guan, Z., & Ni, H. (2020). A novel drug repurposing approach for non-small cell lung cancer using deep learning. *PloS One, 15*(6), e0233112.

Li, H., Tong, C. W.-S., Leung, Y., Wong, M.-H., To, K. K.-W., & Leung, K.-S. (2017). Identification of clinically approved drugs indacaterol and canagliflozin for repurposing to treat epidermal growth factor tyrosine kinase inhibitor-resistant lung cancer. *Frontiers in Oncology, 7*, 288.

Maione, P., Sacco, P. C., Sgambato, A., Casaluce, F., Rossi, A., & Gridelli, C. (2015). Overcoming resistance to targeted therapies in NSCLC: current approaches and clinical application. *Therapeutic Advances in Medical Oncology, 7*(5), 263-273.

Marima, R., Hull, R., Dlamini, Z., & Penny, C. (2020). Efavirenz and lopinavir/ritonavir alter cell cycle regulation in lung cancer. *Frontiers in Oncology, 10*, 1693.

Meador, C. B., & Hata, A. N. (2020). Acquired resistance to targeted therapies in NSCLC: Updates and evolving insights. *Pharmacology & Therapeutics, 210*, 107522.

Murray, J. C., & Levy, B. (2019). Repurposed drugs trials by cancer type: Lung cancer. *The Cancer Journal, 25*(2), 127-133.

Najib Ullah, S. N. M., Afzal, O., Altamimi, A. S. A., Alossaimi, M. A., Almalki, W. H., Alzahrani, A., . . . Shorog, E. M. (2023). Bedaquiline-loaded solid lipid nanoparticles drug delivery in the management of non-small-cell lung cancer (NSCLC). *Pharmaceuticals, 16*(9), 1309.

Otahal, A., Aydemir, D., Tomasich, E., & Minichsdorfer, C. (2020). Delineation of cell death mechanisms induced by synergistic effects of statins and erlotinib in non-small cell lung cancer cell (NSCLC) lines. *Scientific Reports, 10*(1), 959.

Pal, R., Teli, G., Sengupta, S., Maji, L., & Purawarga Matada, G. S. (2023). An outlook of docking analysis and structure-activity relationship of pyrimidine-based analogues as EGFR inhibitors against non-small cell lung cancer (NSCLC). *Journal of Biomolecular Structure and Dynamics, 42*(18), 9795-9811.

Pal, R., Teli, G., Swamy Purawarga Matada, G., & Sanjay Dhiwar, P. (2023). Nitrogen-containing heterocyclic scaffolds as EGFR inhibitors: Design approaches, molecular docking, and structure-activity relationships. *ChemistrySelect, 8*(26), e202301200.

Paliogiannis, P., Fois, S. S., Fois, A. G., Cossu, A., Palmieri, G., & Pintus, G. (2021). Repurposing anticancer drugs for the treatment of idiopathic pulmonary fibrosis and antifibrotic drugs for the treatment of cancer: state of the art. *Current Medicinal Chemistry, 28*(11), 2234-2247.

Pantziarka, P., Bouche, G., Meheus, L., Sukhatme, V., & Sukhatme, V. P. (2014). Repurposing Drugs in Oncology (ReDO)—mebendazole as an anti-cancer agent. *eCancerMedicalScience, 8*.

Parvathaneni, V., Elbatanony, R. S., Goyal, M., Chavan, T., Vega, N., Kolluru, S., . . . Kunda, N. K. (2021). Repurposing bedaquiline for effective non-small cell lung cancer (NSCLC) therapy as inhalable cyclodextrin-based molecular inclusion complexes. *International Journal of Molecular Sciences, 22*(9), 4783.

Parvathaneni, V., Goyal, M., Kulkarni, N. S., Shukla, S. K., & Gupta, V. (2020). Nanotechnology based repositioning of an anti-viral drug for non-small cell lung cancer (NSCLC). *Pharmaceutical Research, 37*, 1-21.

Parvathaneni, V., Kulkarni, N. S., Chauhan, G., Shukla, S. K., Elbatanony, R., Patel, B., . . . Gupta, V. (2020). Development of pharmaceutically scalable inhaled anti-cancer nanotherapy–repurposing amodiaquine for non-small cell lung cancer (NSCLC). *Materials Science and Engineering: C, 115*, 111139.

Pereira, M., & Vale, N. (2022). Repurposing alone and in combination of the antiviral saquinavir with 5-fluorouracil in prostate and lung cancer cells. *International Journal of Molecular Sciences, 23*(20), 12240.

Rajasegaran, T., How, C. W., Saud, A., Ali, A., & Lim, J. C. W. (2023). Targeting inflammation in non-small cell lung cancer through drug repurposing. *Pharmaceuticals, 16*(3), 451.

Regulska, K., Regulski, M., Karolak, B., Murias, M., & Stanisz, B. (2019). Can cardiovascular drugs support cancer treatment? The rationale for drug repurposing. *Drug Discovery Today, 24*(4), 1059-1065.

Rosell, R., & González-Cao, M. (2022). Ablating lung cancer, knowing the tumor better. *The Lancet Regional Health–Europe, 22,* 100494.

Rotow, J., & Bivona, T. G. (2017). Understanding and targeting resistance mechanisms in NSCLC. *Nature Reviews Cancer, 17*(11), 637-658.

Sánchez-Ortega, M., Carrera, A. C., & Garrido, A. (2021). Role of NRF2 in lung cancer. *Cells, 10*(8), 1879.

Santarpia, M., Daffinà, M. G., Karachaliou, N., González-Cao, M., Lazzari, C., Altavilla, G., & Rosell, R. (2016). Targeted drugs in small-cell lung cancer. *Translational Lung Cancer Research, 5*(1), 51.

Sarvepalli, S., Parvathaneni, V., Chauhan, G., Shukla, S. K., & Gupta, V. (2022). Inhaled indomethacin-loaded liposomes as potential therapeutics against non-small cell lung cancer (NSCLC). *Pharmaceutical Research, 39*(11), 2801-2815.

Siegel, R. L., Miller, K. D., Fuchs, H. E., & Jemal, A. (2022). Cancer statistics, 2022. *CA: A Cancer Journal for Clinicians, 72*(1), 7-33.

Skwarski, M., McGowan, D. R., Belcher, E., Di Chiara, F., Stavroulias, D., McCole, M., . . . Chauhan, J. (2021). Mitochondrial inhibitor atovaquone increases tumor oxygenation and inhibits hypoxic gene expression in patients with non–small cell lung cancer. *Clinical Cancer Research, 27*(9), 2459-2469.

Song, M., Wu, H., Wu, S., Ge, T., Wang, G., Zhou, Y., . . . Jiang, J. (2016). Antibiotic drug levofloxacin inhibits proliferation and induces apoptosis of lung cancer cells through inducing mitochondrial dysfunction and oxidative damage. *Biomedicine & Pharmacotherapy, 84*, 1137-1143.

Stachnik, A., Yuen, T., Iqbal, J., Sgobba, M., Gupta, Y., Lu, P., . . . Kim, S.M. (2014). Repurposing of bisphosphonates for the prevention and therapy of nonsmall cell lung and breast cancer. *Proceedings of the National Academy of Sciences, 111*(50), 17995-18000.

Thunnissen, E., Kerr, K. M., Herth, F. J., Lantuejoul, S., Papotti, M., Rintoul, R. C., . . . Bubendorf, L. (2012). The challenge of NSCLC diagnosis and predictive analysis on small samples. Practical approach of a working group. *Lung Cancer, 76*(1), 1-18.

Travis, W. D., Brambilla, E., Noguchi, M., Nicholson, A. G., Geisinger, K. R., Yatabe, Y., . . . Van Schil, P. E. (2011). International association for the study of lung cancer/american thoracic society/european respiratory society international multidisciplinary classification of lung adenocarcinoma. *Journal of Thoracic Oncology, 6*(2), 244-285.

Vaidya, B., Kulkarni, N. S., Shukla, S. K., Parvathaneni, V., Chauhan, G., Damon, J. K., . . . Mitragotri, S. (2020). Development of inhalable quinacrine loaded bovine serum albumin modified cationic nanoparticles: Repurposing quinacrine for lung cancer therapeutics. *International Journal of Pharmaceutics, 577*, 118995.

Walther, U., Emmrich, K., Ramer, R., Mittag, N., & Hinz, B. (2016). Lovastatin lactone elicits human lung cancer cell apoptosis via a COX-2/PPARγ-dependent pathway. *Oncotarget, 7*(9), 10345.

Wistuba, I. I., & Gazdar, A. F. (2006). Lung cancer preneoplasia. *Annu. Rev. Pathol. Mech. Dis., 1*(1), 331-348.

Xia, Y., Xu, F., Xiong, M., Yang, H., Lin, W., Xie, Y., . . . Yu, L. (2021). Repurposing of antipsychotic trifluoperazine for treating brain metastasis, lung metastasis and bone metastasis of melanoma by disrupting autophagy flux. *Pharmacological Research, 163*, 105295.

Chapter 5

Repurposing of Clinically Approved Natural Compounds: A New Frontier in Drug Discovery

Maitrayee Ghosh[1,*]
Debajyoti Roy[2]
and Naresh Kumar Rangra[3]
[1]Department of Pharmaceutics, Amity Institute of Pharmacy, Amity University Uttar Pradesh, Noida, Uttar Pradesh, India
[2]Faculty of Pharmacy, C.V. Raman Global University, Mahura, Bhubaneswar, Odisha, India
[3]Chitkara University School of Pharmacy, Chitkara University, Himachal Pradesh, India

Abstract

Natural sources have been viewed as large reservoirs of biologically active molecules that result in large-scale pharmacological effects and therapeutic potentials. Between 1981 and 2019, two-thirds of all approved drugs were small molecules originating in natural products. Various compounds effective against infections have been established in natural products, including artemisinin one of the best documented for its efficacy against malaria. The application of natural products to living has been fast extrapolated to healthy living in our present-day world. Inherent chemical defences that have been certain in natural products carry a few advantages over their synthetic analogues for effectiveness. These facts would lead to the repurposing of clinically approved natural molecules to be an exciting new research area for the discovery of highly effective

---

* Corresponding Author's Email: maitrayee0225@gmail.com

In: Drug Repurposing
Editors: Shah Alam Khan and Md. Jawaid Akhtar
ISBN: 979-8-89530-422-8

treatments for emerging infectious diseases and common conditions with limited therapeutic options. Repurposing with natural compounds can lead to effective therapies with minimal side effects but at a reduced cost and time. The chapter herein focuses on the overview of different natural compounds under clinical evaluation for repurposing toward various acute and chronic diseases. It further discusses a quick recap of the assorted strategies currently being used in drug repurposing and their practical implications.

Keywords: anti-infective, artemisinin, small molecules, modern medicine, drug repurposing, natural source

1. Introduction

Sources of ancient treatments from herbs, plants, and trees and their parts have been identified for various ailments and diseases. In modern times, such sources have gained increasing interest for the discovery of new small therapeutic molecules for specific conditions (Singh et al., 2024; Harvey et al., 2008). For instance, between 1981 and 2019, two-thirds of clinically approved drugs were small molecules mainly derived from natural resources. The rationale for the design and development of drugs has long been provided by nature. Most drugs used in this modern world are originally derived from natural products. Anticancer drug paclitaxel was initially derived from the Pacific yew tree and penicillin was first isolated from *Penicillium* mold. Similarly, marine organisms are also a source of pharmacologically active molecules which may be used for therapeutic purposes (Rudrapal et al., 2020). Its significance has been further emphasized by the award of the Nobel Prize to W.C. Campbell, S. Omura, and Y. Tu for the discovery of two natural compounds, artemisinin and avermectin, associated with the treatment of parasitic diseases (Efferth et al., 2015; Shen. 2015). Natural source-based or derived drugs have become increasingly significant in our lives, offering a natural alternative to conventional pharmaceuticals (Chin et al., 2006). Therapeutic molecules derived from plants, animals, marine sources, and microorganisms are often considered more compatible with the human body and potentially safer than synthetic drugs (Newman, 2008). These natural and marine drugs have been used in traditional medicine for thousands of years and are increasingly recognized for their medicinal benefits. One of the major advantages these natural compounds have over conventional drugs is that they tend to exhibit fewer side effects. In contrast, synthetic drug molecules show most of the serious adverse reactions and toxicities that can be harmful to our health (Krettli et al., 2009). Drugs of natural origin and of marine origin are very much compliant with nature and the body's chemistry. Natural drugs over the centuries have been taken without having many side effects as noted by Stonik, 2009; Schmidt et al., 2007. The popularity of natural and marine drugs has brought concern in modern society because more and more people are relying on those sources to develop their health and to treat diseases. This transition in trend is powered by an increased level of awareness of the benefits associated with natural and marine drugs, possibly offering targeted and efficient therapies with fewer side effects (Calado et al., 2018). Potential research and development into natural remedies might be the future game-changer for healthcare industries and could offer a healthier, quality lifestyle to people from different walks of life.

In the last two decades, natural and marine sources have emerged as tremendous vehicles for drug discovery and development. Next to this, increasing interest among scientists in the repurposing of already-existent natural products against new targets other than their original use was also observed (Rudrapal et al., 2020). The demand for new and more effective treatments against a variety of diseases, including cancer, infectious diseases, and neurological disorders, has encouraged the quest for novel bioactive molecules from natural sources (Lahlou, 2013). Natural products and marine organisms possess sophisticated chemical defence mechanisms and, as such, they represent a rich source of new drug candidates with therapeutic potential. All these sources have been explored for drug development, yielding a wide range of bioactive compounds, some of which have already been approved for clinical use, while others remain in the preclinical or clinical evaluation process (Martins et al., 2020).

Drug repurposing is the process of exploring new therapeutic uses for already-existing drug molecules. Recently, it has gained significant attention because one-third of newly approved drugs have been repurposed for further l therapeutic use other than their original targets (Pantziarka et al., 2018; Sachs et al., 2017). Very recently, several drug discovery groups have been actively involved in drug repurposing to find remedies against COVID-19 using synthetic and natural sources (Khan et al., 2021). A promising area of drug repurposing from natural and marine sources is the development of anticancer treatments (Rastelli et al., 2020). Cancer represents a significant public health issue, with millions of new diagnoses occurring globally each year. Despite the availability of various treatments, such as chemotherapy, radiation, and surgery, there remains a need for more effective therapies. Rather, compounds from natural and marine sources have been unravelled and identified to possess anticancer properties. Paclitaxel, vinblastine, and vincristine have already gone through clinical approval (El-Sherbeni et al., 2016; Maruca et al., 2020). Anti-infective drugs are another major domain of drug repurposing from natural sources. Infectious diseases remain one of the major global health burdens; for instance, malaria, tuberculosis, and HIV/AIDS can easily kill millions. The isolation of various active principles with potential anti-infective activities has been carried out using natural products and marine organisms. For instance, artemisinin is used in the treatment of malaria; tetrodotoxin is an extremely potent compound in treating a wide range of bacterial and viral infections (Tu et al., 2011; Narahashi. 2008). Advancements in technologies and analytical techniques have empowered natural products and the usage of marine organisms in drug repurposing. For instance, HTS enables fast screening of a large number of compounds against their biological effects. Besides, progress in genomics and metabolomics facilitated the identification and characterization of novel compounds produced by natural and marine sources (Cragg et al., 2009).

2. Historical Perspective

Drug repositioning or drug repurposing refers to using approved drugs for new medical uses. The concept evolved almost with the same timeline as modern biology, pharmacology, and medicine. The following section is an overview of how drug positioning took place from a historical point of view.

2.1. Ancient and Traditional Medicine

Drug repurposing was one of the most common practices in ancient civilizations, empirically formulated and adequately supported by traditional knowledge. Natural products from plants, animals, and minerals were more often used to contribute to therapeutic assistance for various diseases, sometimes offering a wide range of purposes.

2.1.1. Ancient Civilizations

2.1.1.1. Egypt

Ancient Egyptians documented that various natural products were known to treat several illnesses. For example, willow bark was used as an analgesic and anti-inflammatory agent centuries before the finding that the active agent, salicin, belongs to aspirin in terms of its chemical structure (Beltag et al., 2022). Honey, on the other hand, is an antimicrobial agent effective in wound management (Al-Waili et al., 2011).

2.1.1.2. Greece

The father of medicine Hippocrates, used a wide range of natural products for different purposes including vinegar for wounds, opium for pain, and willow bark for fever. It is an early example of repurposing established clinical practice evidence for drug benefits (Mavrogenis et al. 2019).

2.1.2. Ayurveda and Traditional Chinese Medicine (TCM)

- *Ayurveda:* The Ayurvedic system of medicine aimed to reestablish harmony and balance in a diseased state came into existence in India over 3,000 years ago. Ashwagandha (*Withania somnifera*) and turmeric (*Curcuma longa*) were received for centuries in application to traditional medicine. Traditionally, it has helped to reduce inflammation and treat open wounds; today, it is under scrutiny to conclude its potential use in treating chronic diseases like cancer and arthritis (Kunnumakkara et al., 2023).
- *Traditional Chinese Medicine (TCM):* With its history of over a thousand years old, this is the reason TCM uses herbs and natural chemicals so diversely. For instance, ginseng (*Panax ginseng*), has been utilized for ages as an energy booster and immunity enhancer. Investigations into the putative benefits of ginseng on metabolic disorders, cardiovascular illness, and even cancer therapy continue (Yang 2021).

2.1.3. Native American and Indigenous Practices

- *North America:* The indigenous tribes in the North America used the native flora medicinally, finding several uses for one species quite often. For example, echinacea was used as a general tonic as well as to treat snake bites. One of the cornerstones of their healing approach was that local herbs were adapted, based on their proven effectiveness (Meuninck, 2024).

- *South America:* The history of using plants by the natives to control fever, for instance, the cinchona bark, gave a clue from which quinine would be developed much later. Quinine can be considered a very important pharmaceutical intervention in relation to malaria. This traditional knowledge gave a base, to discover drugs from natural sources (Christensen, 2021).

2.2. 19th and Early 20th Century

The 19th and early 20th centuries were a milestone era in the history of medicine during which important developments occurred in the area of drug research, and the philosophy of pharmacology was established as a scientific discipline. Most of the pharmaceuticals developed in that era, in order to treat single health ailments, have since been found for their therapeutic potential to treat manifold other diseases, thus founding the first organized sets of drug repurposing.

2.2.1. Emergence of Modern Pharmacology

- *Scientific foundations:* The emergence of pharmacological institutes, parallel to developing experimental methods of evaluating the effects of drugs, took place in the early 19th century. The emergence of the institute marked the very beginning of pharmacology as an independent field. It is only through emphasizing facts in the medical practices that pioneers like François Magendie and Claude Bernard were able to lay a foundation for the modern-day understanding of the relations between drugs and living organisms (Stahnisch, 2008).
- *Chemical Synthesis and Isolation:* Chemistry has developed to an extent to synthesize and isolate active molecules from natural sources. During this period most of the medicines currently available were prepared in most cases for the first time for contemporary uses.

2.2.1.1. Aspirin

- *Origins:* The most important example of drug repurposing during this era must be aspirin. It's chemical name is acetylsalicylic acid. Aspirin was obtained from the active principle salicin, extracted from willow bark and used since ancient times to relieve pain and inflammation.
- *Repurposing for Cardiovascular Diseases:* Aspirin was first discovered in the early 20th century as an antiplatelet agent that could be used to prevent thrombus formation. It was therefore used in prophylaxis for cardiovascular diseases such as myocardial infarction and cerebrovascular accidents. Since then, aspirin has been widely used in clinical practice because it is cardio-protective and holds great therapeutic potential (Abdelsayed et al., 2022).

2.2.1.2. Quinine

- *Antimalarial Use:* Quinine is an alkaloid extracted from the bark of cinchona, embodying the very first successful treatment for malaria. It was said to stretch back to the 17th century, but general acknowledgement can only be located in the 19th century (D'Acquarica et al., 2023).
- *Additional Therapeutic Uses:* Quinine was also used to treat arthritis, and nocturnal cramps in the legs, among other diversions besides its antimalarial properties. To a very great degree, these broad uses of quinine speak for the general rewriting of the natural product in light of new scientific knowledge (Hariyanti et al., 2022).

2.3. Mid to Late 20th Century

Added to this, the enormous growth of the pharmaceutical industry and research in biomedical institutions spurred considerable advances in most of the mid-to-late 20th century pharmaceutical discovery and development. This has included a variety of drugs with clinically important repurposing to novel therapeutic indications that clearly argue for the strength of drug repurposing in answering new medical needs for the betterment of the patient.

2.3.1. Thalidomide

- *Initial use:* Thalidomide first came to be a common drug during the late 1950s as a sedative and later as a therapeutic medication to treat morning sickness in pregnancy. Within several years, it was strongly linked to causing severe birth defects, after which it was pulled from the market (Vargesson et al., 2021).
- *Repurposing for Leprosy and Cancer:* It was later realized that, although teratogenic, thalidomide was immunomodulatory and anti-angiogenic. It has been reintroduced to treat the very painful complication of leprosy. Its most recent use, because of its anti-angiogenic action, has been against multiple myeloma and many other cancers (Schein, 2020).

2.3.2. All-Trans Retinoic Acid (ATRA)

- *Initial Use:* The derivative of vitamin A, called all-trans retinoic acid, was first studied for dermatological use and its effects on skin health for the treatment of acne and other skin disorders.
- *Repurposing for Acute Promyelocytic Leukemia (APL):* In the 1980s, it was discovered that ATRA could differentiate leukemic cells from acute myeloid leukemia. This consequently made the combination of ATRA with chemotherapy very effective, enhancing remission rates in patients suffering from APL (Nguyen et al., 2020).

2.4. Late 20th Century to Early 21st Century

Drug repositioning is a revolutionary concept, particularly into the late 20th and early 21st centuries, in part due to certain key technological breakthroughs but, more importantly, because of an improvement in our appreciation of molecular biology. This advance in systematic high-throughput approaches developed over the period thus enabled both planned as well as serendipitous new indication discovery for approved drugs.

2.4.1. Artemisinin

- *Initial Use:* Artemisinin (*Artemisia annua*) has been used for ages in traditional Chinese medicine for fever. In fact, in the 1970s, a Chinese scientist named Tu Youyou had proven that artemisinin was effective in the treatment of malaria; this had started the widespread use against malaria.

- *Repurposing for Cancer:* Artemisinin and its derivatives were thus considered for application in therapeutic approaches other than their well-elucidated antimalarial use in cancer. Given that it has been reported to induce apoptosis in some cancer cell lines, lately, artemisinin is being considered as a potential candidate clinical drug in a variety of malignancies. Its capability to selectively target dividing cells makes it an ideal candidate in the area of oncology (Augustin et al., 2020).

2.4.2. Curcumin

- *Initial Use:* The main active compound in turmeric (*Curcuma longa*) is called curcumin. It is used extensively in traditional Chinese and Ayurvedic medicinal practices because of its anti-inflammatory and antioxidant activities.
- *Repurposing for Chronic Diseases:* It was therefore explored for its potential therapeutic activity over the last few decades against a wide variety of chronic diseases that include cancer, neurodegenerative disorders like Alzheimer's disease, and cardiovascular diseases. This protection by curcumin has been considered to occur through its antioxidant and anti-inflammatory action. The several bioactive drug delivery systems of curcumin have derivatized, hence improving bioavailability and therapeutic efficacy (Vaiserman et al., 2021).

2.4.3. Resveratrol

- *Initial Use:* Resveratrol is a natural product that initially gained much interest due to its potential health and antioxidant benefits.
- *Repurposing for Metabolic and Neurodegenerative Diseases:* Natural food sources of resveratrol include peanuts, berries, and grapes; it has been studied reasonably extensively regarding the protection against a wide range of age-related diseases. This effect has been suggested to help protect against neurodegenerative diseases such as Alzheimer's and Parkinson's, as well as metabolic diseases like diabetes and obesity. Most of the research focus has been on its ability to activate sirtuins—proteins characterized by their ability to elongate the lifespan and health of cells (Arbo et al., 2020).

3. Modern Era of Natural Drugs

The renaissance of interest in natural medicines has hugely gained in the modern period, mainly because of the development of genetics and especially biotechnology, with a growing appreciation of the extent to which nature encodes therapeutic knowledge. It is an age of systematic and purposeful drug development and drug repositioning, supported by new technologies in the search for new uses of plant-based materials.

3.1. Biotechnological Advances

- *High-Throughput Screening:* Many natural products are rapidly screened against a host of bioactivities with the help of modern HTS techniques. HTS technology has accelerated the evaluation of the pharmacological properties of large numbers of natural substances and greatly facilitated the identification of candidates for repurposing.
- *Genomic and Proteomic Approaches:* Technological advancements in the areas of proteomics and genomics have not only brought new insight into the molecular causations of many diseases but also, contemporaneously, functions of natural compounds. Being able to appreciate how the genetic and protein components of these molecules interact helps the researcher in the identification and repurposing of natural medications for new therapeutic uses. The most important utilities of this probably are found in oncology, where natural compounds find new application as they can modulate critical pathways related to cancer (Maurya et al., 2021).

3.2. Immunomodulatory Natural Drugs

- *Polysaccharides from Medicinal Mushrooms:* Beta-glucans stand out among polysaccharides extracted from medicinal fungi due to their immunomodulatory properties. It has shown wide-ranging health applications, such as boosting immune responses against infectious diseases, and using immunotherapeutic agents to treat cancer. Several clinical trials are currently underway to improve results in conventional therapy by adding beta glucans as an adjuvant.

- *Marine-Derived Compounds:* Most of these marine organisms also possess unusual classes of bioactive compounds with immunomodulatory activity. Thus, the opportunity for natural products to modulate the activities of immune cells and inhibit inflammatory pathways underscores interest in their use within treatments against cancer and autoimmune diseases (Wainwright et al., 2022).

3.3. Anti-Cancer Natural Drugs

- *Paclitaxel (Taxol):* It was originally derived from the Pacific yew tree (*Taxus brevifolia*), for its highly effective anti-cancer activity. In recent times, it has found application in the treatment of many solid tumors originating from the lung, ovary, and breast. Current studies have been focusing on the use of this compound in combination with other pharmaceuticals to enhance effectiveness and resistance (Barbuti et al., 2015).
- *Camptothecin Derivatives:* Topotecan and irinotecan are two antitumor agents developed from one alkaloid derived from the Chinese tree (*Camptotheca acuminata*), using camptothecin as a common lead pharmacophore. The drugs, among other targeted therapies, are currently being explored about their applications against an even wider range of types of tumors to develop improved therapeutic options (Martino et al., 2017).

3.4. Cardiovascular and Metabolic Natural Drugs

- *Berberine:* This alkaloid, obtained from plants such as Coptis and Berberis, has long been used in Chinese medicine for gastrointestinal ailments. Modern research reinvented berberine into an antidiabetic and cardioprotective agent. The research established berberine as a naturally serviceable drug for metabolic syndrome and cardiovascular diseases by its blood glucose-lowering properties, improvement in lipid profile, and anti-inflammatory actions (Jin et al., 2016).
- *Omega-3 Fatty Acids:* Omega-3 fatty acids were initially identified to be from fish oil and algae, and they had a protective action on the

heart. These days, it is used in the treatment of many more ailments. Studies have shown these reduce inflammation, enhance cognitive functioning, and may even treat some mental health disorders like anxiety and depression (Watanabe et al., 2017).

3.5. Neuroprotective Natural Drugs

- *Ginkgo Biloba: Ginkgo biloba* is an ancient tree whose active ingredients are medicinally used to improve circulation and, in turn, enhance brain function. In recent years, ginkgo extracts have been repurposed for neuroprotective uses in treating neurodegenerative disorders like Parkinson's and Alzheimer's. Current clinical studies on *Ginkgo biloba* are based on its efficacy in the reduction of cognitive decline and enhancement of quality of life in conditions presenting such subjects (Nowak et al., 2021).
- *Curcumin:* The neuroprotective properties of curcumin have also been utilized for purposes other than anti-inflammatory and antioxidant. In fact, a few mechanisms through which curcumin might have an impact on the formation of amyloid plaque, aggregation of tau protein, and neuroinflammatory processes-in a way to delay manifestations of neurodegenerative disease have been identified in various studies. Besides, in therapeutic contexts, its efficacy improves cognitive function and offers neuroprotection against degenerative processes (Farooqui et al., 2019).

4. Methods Used in Drug Repurposing

Pharmaceutical repositioning designates a study of new disease indications of an already approved drug. One of the biggest advantages of such an approach may be possibly drastically reducing development time and cost; repurposed drugs have already been tested at least for safety and their pharmacokinetic profile. Drug repositioning contains a wide array of methodologies, from experimental to computational methods.

4.1. Computational Approaches

4.1.1. In Silico Screening

It is a technique of adopting computational strategies to study the chemical structure, biological target, and interactions of already known pharmaceuticals to predict new uses for these drugs. They may include techniques and tools applied for virtual screening, machine learning algorithms, and molecular docking. This may involve prediction by molecular docking of the binding affinity of a drug to an alternative target; it may relate to the identification of potential inhibitors for emerging viral infections.

4.1.2. Data Mining and Bioinformatics

It studies big data sets—millions of pieces of information on the transcriptome, proteome, and genome—to find potential interactions of drugs with targets. Among the pathway analyses are protein-protein interaction networks and gene expression profiling. Examples include mining clinical trial data and electronic health records for correlations in drug use with unexpected positive effects.

4.1.3. Systems Biology Approaches

In fact, the complex interaction of pharmaceuticals with their target has already been considered and modelled from computational simulations down to systems and network pharmacology. These can be exemplified by pre-defining the system's level of pharmacological effects as opposed to an individual drug target which opens new repositioning chances (Vanhaelen, 2019).

5. Repurposing of Drugs from Various Natural Sources

5.1. Drugs Repurposed from Plants and Herbs

Mankind has conventionally used plants and herbs in the treatment of a large number of diseases and disorders over the centuries. A great deal of these products has been widely researched, during which their pharmacological actions have also been discovered. Some have even undergone further processing as medications to target a specific disease. Drug repositioning implies carrying out further development for a rather new illness based on a molecule originally extracted from some plant or herb. Such development may be less costly and therefore quicker to implement than when developing new drugs. Besides, in general, people tend to tolerate natural substances better than conventional drugs, and sometimes their side effects are lower. Drugs like paclitaxel, artemisinin, resveratrol, curcumin, aspirin, quinine, morphine, taxol, artesunate, etoposide, aconitine, reserpine, salinomycin, and psilocybin were botanical drugs subjected to heavy modification. This class of drugs has traditionally gained approvals for indications in cancer, malaria, pain, and cardiovascular conditions.

For this purpose alone, recently, it was reported that the neuroprotective activity of the chemotherapeutic compound paclitaxel is effective against β-amyloid (Shemesh et al., 2011) and tau proteins (Michaelis et al., 1998), which themselves could serve as possible biomarkers for AD. Another intranasal delivery of paclitaxel formulation was also developed to achieve better penetration through the blood-brain barrier (Cross et al., 2021). Artemisinin is a first-line anti-malarial drug undergoing very intensive research in the ability to cure hepatitis and even cancer (Meng et al., 2021). Resveratrol was identified as a pharmacological agent with an anti-inflammatory effect, and currently, it is under research for treating cardiovascular diseases and cancers (Zhang et al., 2021). A summary of those under research for therapeutic repositioning is given in Table 1.

Another important aspect that involves the use of natural sources in drug repurposing, is the necessity to develop safe and effective pharmacological treatments. The flow of new, technologically advanced and improved methodologies in research, together with new classes of natural products being explored, also continues to develop. This kind of approach would help the patient and indirectly affect the pharmaceutical industry since it reduces the time and costs spent on this kind of development.

5.2. Drugs Repurposed from Marine Sources

Because of its high biodiversity, the ocean is an immense, still poorly explored repository for pharmaceuticals. The variety of marine organisms synthesizes several thousands compounds of potential therapeutic interest (Santos et al., 2020).

Table 1. Potentials of repurposing of drugs obtained from plant and herb sources

S. No.	Natural source	Drug obtained	General pharmacological use	Repurposing area	References
1	*Ananas comosus*	Bromelain	Osteoarthritis, anticancer, muscle soreness, and digestive problem	Anti-inflammation, painkiller	Hong et al., 2021
2	*Taxus brevifolia*	Paclitaxel, Taxol	Chemotherapy drug for treating breast, lung, and ovarian cancer	As Anti-Alzheimer and other neurological disorders	Shemesh, et al., 2011
3	*Psilocybe cubensis*	Psilocybin	Anxiety, migraines, depression, PTSD	Anti-inflammatory	Smedfors et al., 2022
4	*Artemisia annua*	Artesunate	Anti-malarial drug	Anticancer	Meng et al., 2021
5	*Streptomyces albus*	Salinomycin	Antibiotic	Antiviral	Pindiprolu et al., 2020
6	*Vitis vinifera*	Resveratrol	Anti-inflammatory and antioxidant effects	Anticancer, cardiovascular disease	Zhang et al., 2021
7	*Rauwolfia serpentina*	Reserpine	Antihypertensive	Refractory hypertension	Weir, 2020
8	*Curcuma longa*	Curcumin	Management of oxidative and inflammatory conditions, metabolic syndrome, arthritis, anxiety, and hyperlipidaemia	Anticancer, Anti Alzheimer's disease	Roy et al., 2021
9	*Aconitum napellus*	Aconitine	Fear, anxiety, and restlessness, acute sudden fever, influenza, pulsating headaches	COVID-19	Lakhera et al., 2022
10	*Salix alba*	Aspirin	Anti-inflammatory, analgesic, antipyretic	Anticancer and cardiovascular disease	Mohapatra et al., 2019
11	*Podophyllum peltatum*	Etoposide	Anticancer	Alzheimer's disease and other neurological disorders	Ganesan et al., 2022
12	*Papaver somniferum*	Morphine	Painkiller	Antidepressant, Antipsychotic	Sehrawat et al., 2021
13	*Artemisia annua*	Artemisinin	Antimalarial drug	Anticancer, hepatitis	Meng et al., 2021

Table 2. Potentials of repurposing of drugs obtained from marine sources

S. No.	Natural source	Drug obtained	General pharmacological use	Repurposing area	References
1	*Crustacea*	Chitosan	Reduce fat and cholesterol absorption, Wound healing	Treatment of obesity	Lütjohann et al., 2018; Huang et al., 2020
2	*Tetraodontidae*	Tetrodotoxin	To relieve headaches associated with heroin withdrawal	Analgesic and Anticancer	Permana et al., 2013
3	*Salmo salar, Sardina pilchardus*	Omega-3 fatty acids	Prevent heart disease and stroke	Treatment of cardiovascular disease, depression	Chang et al., 2020; Carney 2019
4	*Acremonium falciforme*	Cephalosporins	Treat a variety of bacterial infections, including pneumonia and meningitis	Anti-Alzheimer	Kumari et al., 2021; Tikka et al., 2001
5	*Streptomyces peucetius*	Doxorubicin	Antibiotic	Anticancer	Mandell et al., 2019; Yousefnezhad et al., 2023
6	*Aspergillus terreus*	Lovastatin	Lower cholesterol levels and reduce the risk of heart disease	Anti-Alzheimer	Peyclit et al., 2021
7	*Sarghassum wightii*	Fucoidan	Immune-modulatory and anti-inflammatory	Anticancer	Mustafa et al., 2021
8	Various marine organism	Marine peptides	Antimicrobial, antiviral, antitumor, antioxidative, antihypertensive, anti-atherosclerotic, anticoagulant, immunomodulatory, analgesic, anxiolytic, antidiabetic	Anticancer, anti-inflammation	Kumar et al., 2019

This interest has therefore given birth to further research into utilizing such marine-derived molecules as therapeutic modalities (Jaspars et al., 2016). Such pharmaceutical repurposing from marine organisms has many benefits compared to conventional paths of drug discovery and development, which are not only time-consuming but also expensive. Some of these therapeutic agents - isolated from marine organisms and then modified for use - include trebectin, ziconotide, bryostatin, and discodermolide. These have secured clinical approvals for diseases related to Alzheimer's disease, cancer to neuropathic pain. Over time, there has been increasing interest in the prospect of therapeutic application by compounds of marine origin against the backdrop of an expanding, innovative, and efficacious pharmaceutical market (Püsküllüoğlu et al., 2022). Advancements in technology and research methodologies have made it possible to study and characterize new substances obtained from the aquatic organisms (Fayed et al., 2021). Many of these pharmacological agents, natural or marine-derived, have undergone modifications for new therapeutic applications, and some examples are listed in Table 2.

The marine environment provides a sound resource of chemical entities that can be applied in pharmacology. Some of the better-known pharmaceutical uses outside of recognized indications for clinical use include cephalosporins and tetrodotoxin. The following clinical agents from the marine environment represent agents whose anticancer effects merit further investigation.

5.3. Drugs Repurposed from Microbial Sources

Historically, advancements in pharmaceutics have relied a lot on chemicals produced by microorganisms. They represent a rich source of bioactive compounds with diversified therapeutic applications. Compounds from fungi and bacteria have formed the origin of most medicines in general and antibiotics and other antimicrobial therapies in particular (Abdel-Razek et al., 2020). In practice, however, their application goes beyond this. It has grown rapidly as a result of potential opportunities for repurposing such natural substances in order to discover new therapeutic applications against existing health-related challenges.

One area of immense interest in the development and discovery of drugs currently is in the repositioning of medicines that are already approved since such actions shorten the time it takes for actual effective treatment drugs to reach the market, where they are readily available for consumption. This could be especially important for microbial products, as the mechanisms of action of most of these chemicals are complex and multifaceted.

Any reuse of such compounds allows researchers to take advantage of their preexisting pharmacological properties, hence economizing on time and funds spent in the lengthy process of developing new pharmaceuticals from scratch.

Table 3. Potentials of repurposing of drugs obtained from microbial sources

S. No.	Natural source	Drug obtained	General pharmacological use	Repurposing area	References
1	*Streptomyces hygroscopicus*	Rapamycin (Sirolimus)	antifungal agent	Immunosuppressants and also in certain cancers	Gaumann et al., 2008
2	*Streptomyces griseus*	Streptomycin	antibiotics	Non-tuberculous mycobacterial infections	Gopalaswamy et al., 2020
3	*Streptomyces roseosporus*	Daptomycin	To treat infections caused by Gram-positive bacteria	Complicated infections like endocarditis and osteomyelitis	Dhand et al., 2014
4	*Aspergillus terreus*	Lovastatin	to avoid heart disease and reduce cholesterol	It has been investigated its potential for treating neurological illnesses and a variety of malignancies. It also provides neuroprotection against illnesses such as Alzheimer's.	Jiang et al., 2021; Vitiello et al., 2022
5	*Saccharopolyspora erythraea*	Erythromycin	broad-spectrum antibiotic	Prokinetic agent to treat gastroparesis	Roe et al., 2017
6	*Tolypocladium inflatum*	Cyclosporine	immunosuppressant	It is now used to treat autoimmune conditions like psoriasis and rheumatoid arthritis. It is additionally used to treat persistent dry eye disease in ophthalmology.	Diamanti et al., 2011; Periman et al., 2020
7	*Bacillus brevis*	Gramicidin	antibiotics	Eye infections	Periman et al., 2020

The reuse of pharmaceuticals from microorganisms originated from large medical breakthroughs in neurology, immunosuppression, cardiovascular medicine, oncology, infectious diseases, etc. (Abdella et al., 2021). In light of systematic research related to accidental findings, these studies adapt to complex methodologies such as HTS, high-level bioinformatics, and modernized molecular techniques for characterization. No doubt, these approaches have simplified the ease with which researchers can find new therapeutic uses of already-approved drugs and work out the efficacy and dosing for particular patient populations. Several microbial compounds have also been put to new uses, successfully demonstrating the flexibility and continued application of such microbes in medicine. These agents, including but not limited to the immunosuppressants, antibiotics, anti-cancer therapies and neuroprotective agents considered below, continue to hold great therapeutic promise and further, both improve patient care and push the boundaries of pharmaceutical research. Some representative examples of microbial natural products that have found new clinical uses, including the new indications and clinical benefits that have emerged from this fast-growing area are listed in Table 3.

6. Challenges and Limitations of Natural Drug Repurposing

Repurposing natural drugs entails developing new therapeutic applications for substances derived from natural sources. Even though this sounds very promising, several limitations and challenges are involved at each level, starting from regulations to issues concerning safety and efficacy, as well as identification and isolation processes (Porta et al., 2023).

6.1. Identification and Isolation

6.1.1. Identification Challenges

As such natural sources contain complex chemical mixtures and, therefore, identification of specific active ingredients would pose a challenge.

6.1.2. Isolation Challenges

The extraction of the desired chemicals from natural resources generally poses gigantic technical problems. Presence of hundreds of chemicals in a mixture further complicates the isolation process. Many times natural products are unstable or even deteriorate during extraction.

6.2. Safety and Efficacy

- *Unpredictable Side Effects:* Their ingredients can also interact in unpredictable ways with several pharmaceuticals and can cause unforeseen adverse effects. Comprehensive safety assessment is essential so that all these are ascertained and thus elaborated.
- *Lack of Standardization:* Variations in the preparation used to obtain a natural product can lead to variations in safety. Because repeatable safety testing does not consider the composition consistency of the tested natural product.
- *Bioavailability:* Poor bioavailability is likely to be a problem for the majority of natural products and may limit their use as a therapy. Poor solubility, metabolism, and/or absorption may limit their clinically useful applications.
- *Inconsistent Results:* The origin and preparation of a natural product, along with individual responses, can significantly affect the therapeutic results. This variability can complicate the determination of whether the treatment will consistently lead to a positive outcome.

6.3. Regulatory Hurdles

- *Lack of Precedent:* Most of these new natural compounds would not have clinical precedents for regulatory approval. For new therapeutic uses, abundant data and many clinical trials would be necessary to document safety and efficacy.
- *Complex Documentation:* Regulatory agencies require full documentation and evidence to support new submissions for natural therapeutics. The process—requiring detailed data on aspects such as safety, efficacy, manufacturing processes, and quality assurance—is often long and complex.

6.4. Intellectual Property and Patent Issues

- *Patent Challenges:* Long use by mankind renders the protection of such natural products through intellectual property rights and patenting quite tough. The viability for new patenting might be a matter of if there is already a patented natural product, or it might relate to the particular molecules.
- *Ethical and Legal Concerns:* The utilization of natural resources presents significant ethical and legal challenges, particularly in relation to the indigenous territories or protected regions. Securing agreements and licenses is crucial, yet this process is frequently fraught with difficulties and can be exceedingly time-consuming.

6.5. Market Access and Commercialization

- *Economic Viability:* More crucially, repurposed natural pharmaceuticals are expensive to develop and commercialize, especially when targeting indications with small market sizes. Research and development expenditures, including potential revenue, may easily be outweighed by regulatory and compliance expenditures.
- *Market Acceptance:* Convincing proof of therapeutic advantages and resolving any doubts regarding the efficacy and safety of repurposed natural pharmaceuticals are prerequisites for gaining commercial approval.

These are but a few major steps among many significant things one must do to repurpose natural medicines. The traditional approach to dealing with these concerns typically relies on broad plans through analytical techniques, comprehensive safety and efficacy investigations, and careful management of legal and intellectual property issues. However, this area holds immense promise not only in the perspective of patient benefits but also in the medical treatment modalities. The beauty of it all is finding new uses for already known molecules.

Conclusion

Identification and repurposing of natural chemicals represent a valuable way to further formulate pharmaceuticals that can radically change modern medicine. The discussion in this chapter has considered the historical context, methodology, challenges, and perspectives for future natural drug repurposing—the ins and outs of the discipline. Natural products have been major therapeutic sources because of their complex chemical architecture and manifold bioactivities for several centuries. The historical trajectory of natural products spans from significance assigned to traditional and ancient healing practices to research in the sciences. Different reutilization strategies vary in methodology, starting from high-throughput and phenotypic screening techniques to multiplicity at the level of computational strategies, including data mining and in silico screening. Each technique adds way of seeing, further increasing our perception and devising new applications for existing natural products.

Repurposing natural products is by no means a straightforward challenge. Complex mixtures require sophisticated means of identifying active ingredients and isolating them. As a matter of fact, inherent variability of natural sources along with generally lacking mechanistic understanding obstructs major efforts to employ repurposed natural drugs. Moreover, inconsistent activity, poor absorption, and potential aberrant side effects might reduce the clinical effectiveness of a repurposed natural product; hence, a key issue is their safety and efficacy. Other regulatory challenges of this translation of repurposed natural pharmaceuticals to the clinic are documented in completed documentation, issues concerning intellectual property, and market acceptability. The future is definitely not bleak for the repurposing of natural medicines to base itself on the developments which include analytical technologies, computer methodologies, personalized medicine, synthetic biology, and, finally, collaborative studies. Omics techniques integrated with high-throughput mass spectrometry have led to unprecedented understanding in the identification and characterization of natural products. Revolutionary drug-target interaction and therapeutic potential predictive modelling are fuelled by machine learning, artificial intelligence, and big data. Advanced delivery of medications and patient-derived models would be futuristic personalized medicine approaches ensuring that the right therapy reaches each needy patient at the right time.

Data analysis shows that drug repurposing in the context of natural compounds is a current necessity and exciting field in the discovery of new drugs. Addressing this significant challenge and anticipating potential advancements in science and technology will enable natural medicines to explore new therapeutic avenues, thereby addressing unmet medical needs and ultimately enhancing patient outcomes. This is something that in combination with the advancement of pharmacognosy as a multidisciplinary, collaborative approach to repurposing natural medicines could be literally game changing for world health.

References

Abdella, M., Abdella, B. and Lahiri, C., 2021. Rediscovering and repurposing natural microbial macromolecules through computational approaches. *Microb. Nat. Macromol.,* (pp. 373-400). Academic Press.

Abdel-Razek, A. S., El-Naggar, M. E., Allam, A., Morsy, O. M. and Othman, S. I., 2020. Microbial natural products in drug discovery. *Process.*, *8*(4), pp.470.

Abdelsayed, M., Kort, E. J., Jovinge, S. and Mercola, M., 2022. Repurposing drugs to treat cardiovascular disease in the era of precision medicine. *Nat. Rev. Cardiol.*, 19(11), pp.751-764.

Al-Waili, N., Salom, K. and Al-Ghamdi, A. A., 2011. Honey for wound healing, ulcers, and burns; data supporting its use in clinical practice. *Sci. World J.*, *11*(1), pp.766-787.

Arbo, B. D., André-Miral, C., Nasre-Nasser, R. G., Schimith, L. E., Santos, M. G., Costa-Silva, D., Muccillo-Baisch, A. L. and Hort, M. A., 2020. Resveratrol derivatives as potential treatments for Alzheimer's and Parkinson's disease. *Front. Aging Neurosci.*, *12*, pp.103.

Augustin, Y., Staines, H. M. and Krishna, S., 2020. Artemisinins as a novel anti-cancer therapy: targeting a global cancer pandemic through drug repurposing. *Pharmacol. Therap.*, *216*, pp.107706.

Barbuti, A. M. and Chen, Z. S., 2015. Paclitaxel through the ages of anticancer therapy: exploring its role in chemoresistance and radiation therapy. *Cancers*, *7*(4), pp.2360-2371.

Beltag, D. M., Ali, E. M. M., Sakr, H. M. S. and Radwan, E. H., 2022. Biochemical studies on the effect of prostaglandin inhibitors and willow bark extract on liver cirrhosis induced by acetylsalicylate in rats. *J. Med. Life Sci.*, 4(4), pp.82-104.

Calado, R., Leal, M. C., Gaspar, H., Santos, S., Marques, A., Nunes, M. L. and Vieira, H., 2018. How to succeed in marketing marine natural products for nutraceutical, pharmaceutical and cosmeceutical markets. *Grand Challenges in Marine Biotechnology*, pp.317-403.

Carney, R. M., Freedland, K. E., Rubin, E. H., Rich, M. W., Steinmeyer, B. C. and Harris, W. S., 2019. A randomized placebo-controlled trial of omega-3 and sertraline in depressed patients with or at risk for coronary heart disease. *JCP*, *80*(4), pp.13302.

Chang, J. P. C., Chang, S. S., Yang, H. T., Chen, H. T., Chien, Y. C., Yang, B., Su, H. and Su, K. P., 2020. Omega-3 polyunsaturated fatty acids in cardiovascular diseases comorbid major depressive disorder–Results from a randomized controlled trial. *BBI*, *85*, pp.14-20.

Chin, Y. W., Balunas, M. J., Chai, H. B. and Kinghorn, A. D., 2006. Drug discovery from natural sources. *The AAPS Journal*, *8*, pp.E239-E253.

Christensen, S. B., 2021. Natural products that changed society. *Biomedicines*, 9(5), pp.472.

Cragg, G. M., Grothaus, P. G. and Newman, D. J., 2009. Impact of natural products on developing new anti-cancer agents. *Chem. Rev.*, *109*(7), pp.3012-3043.

Cross, D. J., Huber, B. R., Silverman, M. A., Cline, M. M., Gill, T. B., Cross, C. G., Cook, D. G. and Minoshima, S., 2021. Intranasal paclitaxel alters Alzheimer's disease phenotypic features in 3xTg-AD mice. *J. Alzheimer's Dis.*, *83*(1), pp.379-394.

D'Acquarica, I. and Agranat, I., 2023. Widened scope of drug repurposing/chiral switches, elements of secondary pharmaceuticals: the quinine/quinidine case. *Future Drug Discov.*, 5(3), pp.FDD85.

Dhand, A. and Sakoulas, G., 2014. Daptomycin in combination with other antibiotics for the treatment of complicated methicillin-resistant *Staphylococcus aureus* bacteremia. *Clin. Ther.*, *36*(10), pp.1303-1316.

Diamanti, A. P., Rosado, M., Germano, V., Scarsella, M., Giorda, E., Podestà, E., D'Amelio, R., Carsetti, R. and Laganà, B., 2011. Reversion of resistance to immunosuppressive agents in three patients with psoriatic arthritis by cyclosporine A: Modulation of P-glycoprotein function. *Clin. Immunol.*, *138*(1), pp.9-13.

Efferth, T., Zacchino, S., Georgiev, M. I., Liu, L., Wagner, H. and Panossian, A., 2015. Nobel Prize for artemisinin brings phytotherapy into the spotlight. *Phytomed.*, *22*(13), pp.A1-A3.

El-Sherbeni, A. A. and El-Kadi, A. O., 2016. Repurposing resveratrol and fluconazole to modulate human cytochrome P450-mediated arachidonic acid metabolism. *Mol. Pharmaceutics*, *13*(4), pp.1278-1288.

Farooqui, T. and Farooqui, A. A., 2019. Curcumin: Historical background, chemistry, pharmacological action, and potential therapeutic value. *Curcumin for Neurological and Psychiatric Disorders*, pp.23-44.

Fayed, M. A., El-Behairy, M. F., Abdallah, I. A., Abdel-Bar, H. M., Elimam, H., Mostafa, A., Moatasim, Y., Abouzid, K. A. and Elshaier, Y. A., 2021. Structure-and ligand-based *in silico* studies towards the repurposing of marine bioactive compounds to target SARS-CoV 2. *Arab. J. Chem.*, *14*(4), pp.103092.

Ganesan, V., Meiyazhagan, G., Devaraj, M., Kandasamy, S., Manogaran, P., Suresh Kumar, G., Raji, G., Kattimani, V. S. and Easwaradas Kreedapathy, G., 2022. Repurposing the antibacterial activity of etoposide – a chemotherapeutic drug in combination with eggshell-derived hydroxyapatite. *ACS Biomaterials Science & Engineering*, *8*(2), pp.682-693.

Gaumann, A., Schlitt, H. J. and Geissler, E. K., 2008. Immunosuppression and tumor development in organ transplant recipients: the emerging dualistic role of rapamycin. *Transpl. Int.*, *21*(3), pp.207-217.

Gopalaswamy, R., Shanmugam, S., Mondal, R. and Subbian, S., 2020. Of tuberculosis and non-tuberculous mycobacterial infections–a comparative analysis of epidemiology, diagnosis and treatment. *J. Biomed. Sci.*, *27*, pp.1-17.

Hariyanti, H., Mauludin, R., Sumirtapura, Y. C. and Kurniati, N. F., 2022. A review: Pharmacological activities of quinoline alkaloid of Cinchona sp. *Biointerface Res. Appl. Chem.*, *13*(4), pp.3.

Harvey, A. L., 2008. Natural products in drug discovery. *Drug Discov. Today*, *13*(19-20), pp.894-901.

Hong, J. H., Kim, M. R., Lee, B. N., Oh, W. M., Min, K. S., Im, Y. G. and Hwang, Y. C., 2021. Anti-inflammatory and mineralization effects of bromelain on lipopolysaccharide-induced inflammation of human dental pulp cells. *Medicina*, *57*(6), pp.591.

Huang, H., Liao, D., Zou, Y. and Chi, H., 2020. The effects of chitosan supplementation on body weight and body composition: A systematic review and meta-analysis of randomized controlled trials. *Crit. Rev. Food Sci. Nutr.*, *60*(11), pp.1815-1825.

Jaspars, M., De Pascale, D., Andersen, J. H., Reyes, F., Crawford, A. D. and Ianora, A., 2016. The marine biodiscovery pipeline and ocean medicines of tomorrow. *J. Mar. Biol. Assoc. U. K.*, *96*(1), pp.151-158.

Jiang, W., Hu, J. W., He, X. R., Jin, W. L. and He, X. Y., 2021. Statins: a repurposed drug to fight cancer. *J. Exp. Clin. Cancer Res.*, *40*, pp.1-33.

Jin, Y., Khadka, D. B. and Cho, W. J., 2016. Pharmacological effects of berberine and its derivatives: a patent update. *Expert Opin. Ther. Pat.*, *26*(2), pp.229-243.

Khan, S. A. and Al-Balushi, K., 2021. Combating COVID-19: The role of drug repurposing and medicinal plants. *J. Infect Public Health*, *14*(4), pp.495-503.

Krettli, A. U., Adebayo, J. O. and Krettli, L. G., 2009. Testing of natural products and synthetic molecules aiming at new antimalarials. *Curr. Drug Targets.*, *10*(3), pp.261-270.

Kumar, L. V., Shakila, R. J. and Jeyasekaran, G., 2019. *In vitro* anti-cancer, anti-diabetic, anti-inflammation and wound healing properties of collagen peptides derived from unicorn leatherjacket (*Aluterus monoceros*) at different hydrolysis. *TrJFAS*, *19*(7), pp.551-560.

Kumari, S. and Deshmukh, R., 2021. β-Lactam antibiotics to tame down molecular pathways of Alzheimer's disease. *Eur. J. Pharmacol.*, *895*, pp.173877.

Kunnumakkara, A. B., Hegde, M., Parama, D., Girisa, S., Kumar, A., Daimary, U. D., Garodia, P., Yenisetti, S. C., Oommen, O. V. and Aggarwal, B. B., 2023. Role of turmeric and curcumin in prevention and treatment of chronic diseases: Lessons learned from clinical trials. *ACS Pharmacol. Transl.*, *6*(4), pp.447-518.

Lahlou, M., 2013. The success of natural products in drug discovery. *Pharmacol & Pharmacy*, *4*, 17-31.

Lakhera, S., Devlal, K., Rana, M. and Ghosh, A., 2022. *In-silico* investigation of inhibiting property of phytoconstituents of medicinal herb '*Aconitum heterophyllum*' against Omicron variant of SARS-CoV-2, *Res Sq.*, https://doi.org/10.21203/rs.3.rs-1637533/v1.

Lütjohann, D., Marinova, M., Wolter, K., Willinek, W., Bitterlich, N., Coenen, M., Coch, C. and Stellaard, F., 2018. Influence of chitosan treatment on surrogate serum markers of cholesterol metabolism in obese subjects. *Nutr.*, *10*(1), pp.72.

Mandell, J. B., Lu, F., Fisch, M., Beumer, J. H., Guo, J., Watters, R. J. and Weiss, K. R., 2019. Combination therapy with disulfiram, copper, and doxorubicin for osteosarcoma: in vitro support for a novel drug repurposing strategy. *Sarcoma, 2019*(1), pp.1320201.

Martino, E., Della Volpe, S., Terribile, E., Benetti, E., Sakaj, M., Centamore, A., Sala, A. and Collina, S., 2017. The long story of camptothecin: From traditional medicine to drugs. *BMCL, 27*(4), pp.701-707.

Martins, M., Silva, R., MM Pinto, M. and Sousa, E., 2020. Marine natural products, multitarget therapy and repurposed agents in Alzheimer's disease. *Pharmaceuticals, 13*(9), pp.242.

Maruca, A., Rocca, R., Catalano, R., Mesiti, F., Costa, G., Lanzillotta, D., Salatino, A., Ortuso, F., Trapasso, F., Alcaro, S. and Artese, A., 2020. Natural products extracted from fungal species as new potential anti-cancer drugs: a structure-based drug repurposing approach targeting HDAC7. *Molecules, 25*(23), pp.5524.

Maurya, D. K., Kumar, A., Chaurasiya, U., Hussain, T. and Singh, S. K., 2021. Modern era of microbial biotechnology: opportunities and future prospects. In: *Microbiomes and plant health* (pp. 317-343). Academic Press.

Mavrogenis, A. F., Saranteas, T., Markatos, K., Kotsiou, A. and Tesseromatis, C., 2019. Pharmacies for pain and trauma in ancient Greece. *Int. Orthop., 43*, pp.1529-1536.

Meng, Y., Ma, N., Lyu, H., Wong, Y. K., Zhang, X., Zhu, Y., Gao, P., Sun, P., Song, Y., Lin, L. and Wang, J., 2021. Recent pharmacological advances in the repurposing of artemisinin drugs. *Med. Res. Rev., 41*(6), pp.3156-3181.

Meuninck, J., 2024. Medicinal plants of North America: a field guide. Published by Rowman & Littlefield.

Michaelis, M. L., Ranciat, N., Chen, Y., Bechtel, M., Ragan, R., Hepperle, M., Liu, Y. and Georg, G., 1998. Protection against β-amyloid toxicity in primary neurons by paclitaxel (taxol). *J. Neurochem., 70*(4), pp.1623-1627.

Mohapatra, T. K. and Subudhi, B. B., 2019. Repurposing of aspirin: Opportunities and challenges. *RJPT, 12*(4), pp.2037-2044.

Mustafa, S., Pawar, J. S. and Ghosh, I., 2021. Fucoidan induces ROS-dependent epigenetic modulation in cervical cancer HeLa cell. *Int. J. Biol. Macromol., 181*, pp.180-192.

Narahashi, T., 2008. Tetrodotoxin—A brief history—. *Proceedings of the Japan Academy, Series B, 84*(5), pp.147-154.

Newman, D. J., 2008. Natural products as leads to potential drugs: an old process or the new hope for drug discovery?. *J. Med. Chem., 51*(9), pp.2589-2599.

Nguyen, C. H., Grandits, A. M., Purton, L. E., Sill, H. and Wieser, R., 2020. All-trans retinoic acid in non-promyelocytic acute myeloid leukemia: Driver lesion dependent effects on leukemic stem cells. *Cell Cycle, 19*(20), pp.2573-2588.

Nowak, A., Kojder, K., Zielonka-Brzezicka, J., Wróbel, J., Bosiacki, M., Fabiańska, M., Wróbel, M., Sołek-Pastuszka, J. and Klimowicz, A., 2021. The use of *Ginkgo biloba* L. as a neuroprotective agent in the Alzheimer's disease. *Front. Pharmacol., 12*, pp.775034.

Pantziarka, P., Pirmohamed, M. and Mirza, N., 2018. New uses for old drugs. *BMJ.*,361, pp. k2701.

Periman, L. M., Mah, F. S. and Karpecki, P. M., 2020. A review of the mechanism of action of cyclosporine A: the role of cyclosporine A in dry eye disease and recent formulation developments. *Clin. Ophthalmol.*, pp.4187-4200.

Permana, K. R. and Septian, D. D., 2013, April. Revolutionary therapy for breast cancer (BCa) by using Tetrodotoxin (TTX) extracted from the masked puffer fish arothron diadematus as best investment in community. In: *Proceedings of the 1st Annual International Scholars Conference in Taiwan* (Vol. 1, pp. 601-605).

Peyclit, L., Yousfi, H., Rolain, J. M. and Bittar, F., 2021. Drug repurposing in medical mycology: Identification of compounds as potential antifungals to overcome the emergence of multidrug-resistant fungi. *Pharmaceuticals*, *14*(5), pp.488.

Pindiprolu, S. K. S., Kumar, C. S. P., Golla, V. S. K., Likitha, P., Chandra, S., Busha, S. K. E., and Ramachandra, R. K., 2020. Pulmonary delivery of nanostructured lipid carriers for effective repurposing of salinomycin as an antiviral agent. *Med. Hypotheses*, *143*, pp.109858.

Porta, E. O., Kalesh, K. and Steel, P. G., 2023. Navigating drug repurposing for Chagas disease: advances, challenges, and opportunities. *Front. Pharmacol.*, *14*, pp.1233253.

Püsküllüoğlu, M. and Michalak, I., 2022. An ocean of possibilities: a review of marine organisms as sources of nanoparticles for cancer care. *Nanomed.*, *17*(22), pp.1695-1719.

Rastelli, G., Pellati, F., Pinzi, L. and Gamberini, M. C., 2020. Repositioning natural products in drug discovery. *Molecules*, *25*(5), pp.1154.

Roe, N. A., Sakaan, S., Swanson, H. and Twilla, J. D., 2017. Evaluation of prokinetic agents used in the treatment of gastroparesis. *J. Drug Assess.*, *6*(1), pp.6-9.

Roy, R., Bhowmik, R., Seth, S., Bhattacharyya, S. and Sengupta, S., 2021. Computational repurposing model of curcumin As a drug. *Int. J. Pharm. Sci. Rev. Res.*, *68*(1), pp.165-171.

Rudrapal, M., Khairnar, S. J. and Jadhav, A. G., 2020. Drug repurposing (DR): an emerging approach in drug discovery. In: *Drug repurposing-hypothesis, molecular aspects and therapeutic applications*. Edited by Badria, F.A., IntechOpen. doi:10.5772/intechopen.93193.

Sachs, R. E., Ginsburg, P. B. and Goldman, D. P., 2017. Encouraging new uses for old drugs. *Jama.*, *318*(24), pp.2421-2422.

Santos, J. D., Vitorino, I., Reyes, F., Vicente, F. and Lage, O. M., 2020. From ocean to medicine: Pharmaceutical applications of metabolites from marine bacteria. *Antibiotics*, *9*(8), pp.455.

Schcolnik-Cabrera, A., Juárez-López, D. and Duenas-Gonzalez, A., 2021. Perspectives on drug repurposing. *Curr. Med. Chem.*, *28*(11), pp.2085-2099.

Schein, C. H., 2020. Repurposing approved drugs on the pathway to novel therapies. *Med. Res. Rev.*, 40(2), pp.586-605.

Schmidt, B. M., Ribnicky, D. M., Lipsky, P. E. and Raskin, I., 2007. Revisiting the ancient concept of botanical therapeutics. *Nat. Chem. Biol.*, *3*(7), pp.360-366.

Sehrawat, H., Kumar, N., Sood, D., Kumar, L., Tomar, R. and Chandra, R., 2021. Unraveling the interaction of an opium poppy alkaloid noscapine ionic liquid with human hemoglobin: biophysical and computational studies. *J. Mol. Liq.*, *338*, pp.116710.

Shemesh, O. A. and Spira, M. E., 2011. Rescue of neurons from undergoing hallmark tau-induced Alzheimer's disease cell pathologies by the antimitotic drug paclitaxel. *Neurobiol. Dis.*, *43*(1), pp.163-175.

Shen, B., 2015. A new golden age of natural products drug discovery. *Cell, 163*(6), pp.1297-1300.

Singh, S., Chib, S., Akhtar, M. J., Kumar, B., Chawla, P. A. and Bhatia, R., 2024. Paradigms and success stories of natural products in drug discovery against neurodegenerative disorders (NDDs). *Curr. Neuropharmacol., 22*(6), pp.992-1015.

Smedfors, G., Glotfelty, E., Kalani, N., Hjelle, C. P., Horntvedt, O., Wellfelt, K., Brodin, A., von Kieseritzky, F., Olson, L. and Karlsson, T., 2022. Psilocybin combines rapid synaptogenic and anti-inflammatory effects in vitro. *Research Square. DOI: 10.21203/rs.3.rs-1321542/v1.*

Stahnisch, F. W., 2008. Instrument transfer as knowledge transfer in neurophysiology: François Magendie's (1783–1855) early attempts to measure cerebrospinal fluid pressure. *J. Hist. Neurosci., 17*(1), pp.72-99.

Stonik, V. A., 2009. Marine natural products: A way to new drugs. *Acta Naturae., 1*(2 (2)), pp.15-25.

Tikka, T., Usenius, T., Tenhunen, M., KeinaÈnen, R. and Koistinaho, J., 2001. Tetracycline derivatives and ceftriaxone, a cephalosporin antibiotic, protect neurons against apoptosis induced by ionizing radiation. *J. Neurochem., 78*(6), pp.1409-1414.

Tu, Youyou. 2011. "The discovery of artemisinin (qinghaosu) and gifts from Chinese medicine." *Nat. Med.,* 17(10), pp.1217-1220.

Vaiserman, A., Koliada, A., Lushchak, O. and Castillo, M. J., 2021. Repurposing drugs to fight aging: The difficult path from bench to bedside. *Med. Res. Rev., 41*(3), pp.1676-1700.

Vanhaelen, Q. ed., 2019. Computational methods for drug repurposing (1st Edition, Vol. 1903). Humana Press, New York.

Vargesson, N., Stephens, T., 2021. Thalidomide: history, withdrawal, renaissance, and safety concerns. *Expert Opin. Drug Saf., 20*(12), pp.1455-1457.

Vitiello, A., Troiano, V. and La Porta, R., 2022. Statins in Alzheimer's disease (AD). *Eur. J. Clin. Pharmacol., 78*(7), pp.1201-1202.

Wainwright, C. L., Teixeira, M. M., Adelson, D. L., Braga, F. C., Buenz, E. J., Campana, P. R., David, B., Glaser, K. B., Harata-Lee, Y., Howes, M. J. R. and Izzo, A. A., 2022. Future directions for the discovery of natural product-derived immunomodulating drugs: an IUPHAR positional review. *Pharmacol. Res., 177*, p.106076.

Watanabe, Y. and Tatsuno, I., 2017. Omega-3 polyunsaturated fatty acids for cardiovascular diseases: present, past and future. *Expert Rev. Clin. Pharmacol., 10*(8), pp.865-873.

Weir, M. R., 2020. Reserpine: a new consideration of an old drug for refractory hypertension. *Am. J. Hypertens., 33*(8), pp.708-710.

Yang, M. S., 2021. Chinese ginseng. *Nutraceuticals,* (pp. 853-864). Academic Press.

Yousefnezhad, M., Davaran, S., Babazadeh, M., Akbarzadeh, A. and Pazoki-Toroudi, H., 2023. PCL-based nanoparticles for doxorubicin-ezetimibe co-delivery: A combination therapy for prostate cancer using a drug repurposing strategy. *BI, 13*(3), pp.241.

Zhang, M., Chen, X. and Radacsi, N., 2021. New tricks of old drugs: Repurposing non-chemo drugs and dietary phytochemicals as adjuvants in anti-tumor therapies. *J. Control. Release, 329*, pp.96-120.

Chapter 6

Repurposing Anti-Diabetic Agents: Investigating Their Therapeutic Utility in Managing Co-Occurring Conditions in Diabetes

Aparoop Das*
Urvashee Gogoi
Kalyani Pathak
Riya Saikia
Jon Jyoti Sahariah
Saptasikha Gogoi
Dibyajyoti Das
and Pallab Pramanik
Department of Pharmaceutical Sciences, Dibrugarh University, Dibrugarh, Assam, India

Abstract

Diabetic patients often experience additional health issues such as cardiovascular diseases (CVDs), cancer, obesity, cognitive impairment, gout, and leishmaniasis, among others. Managing these conditions alongside diabetes often requires multiple medications, which can impact quality of life. Insulin and insulin sensitizers are among the most extensively researched candidates for drug repositioning in diabetes. Additionally, several other anti-diabetic medications have also been

* Corresponding Author's Email: aparoopdas@dibru.ac.in

In: Drug Repurposing
Editors: Shah Alam Khan and Md. Jawaid Akhtar
ISBN: 979-8-89530-422-8

repurposed for treating dementia. Some anti-diabetic drugs have shown promise in treating these comorbidities. For example, metformin is linked to improvements in diseases such as cancer, neurodegenerative diseases, and cardiovascular diseases.
Data from preclinical studies, clinical trials, and observational research have shown that drugs from diverse classes beyond those typically used for diabetes treatment exhibit therapeutic effects in managing diabetes associated comorbidities. Some of these drugs demonstrate the ability to mitigate diabetic complications or improve outcomes, while others show potential in preventing the onset of additional complications. The effectiveness of these drugs against comorbidities is likely due to shared underlying mechanisms such as inflammation, oxidative stress, metabolic disruptions, and obesity. This chapter highlights the connections between diabetes and its comorbidities, emphasizing the rationale for using anti-diabetic medications to address these interconnected health challenges.

Keywords: drug repurposing, diabetes, comorbidities, cancer, cardiovascular diseases, Alzheimer's disease, PCOS

1. Introduction

Diabetes mellitus is a rapidly escalating global health issue with significant social, economic, and health impacts. This chronic metabolic disease is marked by high blood glucose levels, which can cause severe damage to the heart, blood vessels, eyes, kidneys, nerves, etc., over time. Type 2 diabetes (T2DM) is the most prevalent form of diabetes, typically affecting adults, and results from the body's resistance to insulin or insufficient insulin production. Over the last 30 years, T2DM rates have surged across countries of all income levels (World Health Organization,2023). According to the International Diabetes Federation (IDF) Diabetes Atlas (2021), 10.5% of adults aged 20-79 have diabetes, with nearly half unaware of their condition. The IDF projects that by 2045, 1 in 8 adults, or about 783 million people, will have diabetes, representing a 46% increase (International Diabetes Federation,2021).

Managing diabetes is made more complex by comorbidities, which refers to the coexistence of two or more chronic diseases or medical conditions in a patient (Valderas et al., 2009). Diabetes not only affects other chronic conditions but is also impacted by them through various interconnected pathways. These comorbidities are closely linked to diabetes, potentially worsening its severity and increasing the risk of complications (Teck, 2022).

This interplay complicates healthcare outcomes, treatment options, care requirements, and associated costs for diabetic patients.

Cardiovascular disease is the leading cause of illness and death in diabetes, and its impact is worsened by hypertension. Diabetes and hypertension are strongly connected due to shared risk factors such as endothelial dysfunction, vascular inflammation, arterial remodeling, atherosclerosis, dyslipidemia, and obesity (Petrie et al., 2018). Obesity is another significant modifiable risk factor for type 2 diabetes (T2D). The link between obesity and T2D involves abdominal fat distribution, dysfunctional adipose tissue, and inflammation, which leads to the secretion of diabetogenic adipokines that impair insulin action in various organs, including skeletal muscle, liver, and brain (Blüher and Stumvoll, 2020). Diabetic patients face a heightened risk of developing various cancers, including those of the urinary tract, liver, biliary tract, pancreas, colon, endometrium, and kidneys (Habib and Rojna, 2013). Additionally, studies suggest a link between polycystic ovary syndrome (PCOS), diabetes, and cardiovascular disease (Zhu et al., 2021). Diabetes mellitus is also frequently associated with advanced liver disease, a connection that significantly increases the risk of morbidity and mortality. A metabolic pathway involving fatty liver and steatohepatitis has been identified, suggesting that cirrhosis might be a late-stage complication of diabetes (Moscatiello et al., 2007). Moreover, the relationship between diabetes and neurological disorders has been extensively studied, revealing a complex connection. Diabetic neuropathy, affecting nearly half of all diabetics, is one of the most common neurological complications (Sachedina, 2013). Diabetes is also linked to a higher risk of psychiatric conditions, including depression and anxiety, as well as neurological disorders such as Alzheimer's disease (AD), Parkinson's disease (PD), and stroke (Thakur et al., 2019).

The rising costs and extended timelines for developing new treatments mean that patients may encounter severe complications or life-threatening conditions if they develop resistance to current medications before new therapies are available. This issue has increased interest in drug repurposing (Pushpakom et al., 2019; Zhang et al., 2020). Drug repurposing, also known as drug repositioning or reprofiling, involves finding new therapeutic uses for existing drugs initially developed for different conditions. This method takes advantage of the known safety profiles and pharmacological data of these drugs, enabling quicker and more cost-effective development compared to traditional drug discovery (Parvathaneni et al., 2019). Investigational molecules that fail to show efficacy for their original use can often be revived through repurposing, making them viable therapies, especially for rare diseases that pose significant challenges in diagnosis, treatment, and resource availability (Delavan et al., 2018; Roessler et al., 2021).

2. Pathophysiology of Diabetes

Depending on the cause of diabetes, the following may cause hyperglycemia:

a) Decreased insulin secretion
b) Reduced body utilization of glucose
c) Increased synthesis of glucose.

2.1. Pathogenesis of Type 1 DM

The primary hallmark of T1DM is the depletion of the β-cell mass, resulting in severe insulin insufficiency. The pathogenesis of type 1A diabetes is attributed to immune-mediated processes, while type 1B diabetes's origins remain idiopathic. Presently, the pathogenesis of type 1A diabetes mellitus is elucidated through a triad of interrelated mechanisms: genetic predisposition, autoimmunity, and specific environmental factors (Baynest, 2015).

2.1.1. Genetic Susceptibility

Multiple genes that contribute to the susceptibility to type 1A DM are inherited.

a) It has been noted that with identical twins, there is a 50% probability that the second twin will also have type 1A DM, but not always. This indicates that under these situations, there are some extra moderating factors involved in the development of DM.
b) The gene linked to susceptibility for type 1A diabetes mellitus is primarily located within the HLA region of chromosome 6 (MHC class II region), particularly at the HLA DR3, DR4, and HLA DQ locus.

2.1.2. Autoimmunity

Numerous immunologic abnormalities have been observed in type 1A DM studies on humans and animal models.

a) Antibodies against GAD (glutamic acid decarboxylase), insulin, and other proteins in the islet cells.
b) β-cell degradation occurs selectively, leaving other islet cell types (such as alpha cells producing glucagon, delta cells producing somatostatin, or PP cells forming polypeptides) unaffected. The underlying mechanisms for this phenomenon involve apoptosis or cytotoxicity mediated by T-cells.
c) Insulitis refers to a state where lymphocytic infiltration occurs within and around the pancreatic islets. Predominantly consisting of CD8+ T cells, insulitis involves varying quantities of CD4+ T lymphocytes and macrophages.
d) Roughly 10–20% of autoimmune conditions like Graves' disease, Addison's disease, Hashimoto's thyroiditis, and pernicious anemia is linked to type 1A diabetes.

2.1.3. Environmental Factors

Epidemiological studies suggest that environmental factors might play a part in the development of type 1A diabetes mellitus, although none have been definitively established. Specific dietary and viral proteins exhibit antigenic characteristics similar to human cell surface proteins, triggering the immune system to mistakenly attack β-cells through a process known as molecular mimicry. These influencing factors encompass the following:

a) Its seasonal and geographical variations in occurrence point to certain common environmental causes.
b) It has been possible to experimentally induce type 1A DM using specific drugs, such as alloxan, streptozotocin, and pentamidine.
c) Several viral infections, such as measles, mumps, coxsackie B virus, cytomegalovirus, and infectious mononucleosis, occur before the beginning of the illness.

2.2. Pathogenesis of T2DM

The core metabolic anomaly in type 2 diabetes is either insulin resistance or compromised insulin secretion, resulting from peripheral tissues' inability to respond effectively to insulin. Despite type 2 diabetes mellitus (DM) being more prevalent than type 1 and having a more intricate origin, its pathogenesis remains less understood (DeFronzo, 2015).

2.2.1. Genetic Factors

Genetic predisposition is more pronounced in type 2 diabetes compared to type 1A diabetes. Despite the absence of specific and consistent genes, the primary factor driving the development of type 2 DM is multifactorial inheritance.

a) If one identical twin develops diabetes, the other twin has a roughly 80% probability of also having the condition.
b) A person who has one parent with type 2 diabetes is more likely to have the disease; however, if both parents have the disease, the risk for the child increases to 40%.

2.2.2. Insulin Resistance

A significant metabolic feature of type 2 diabetes is the reduced responsiveness of peripheral tissues, especially skeletal muscle and the liver, to insulin (DeFronzo, 1997). Insulin resistance and type 2 diabetes are notably connected to obesity. The mechanism behind hyperglycemia in these scenarios is elucidated as follows:

a) Insulin action is inhibited by resistance, which results in impaired glucose utilization and hyperglycemia.
b) There is an elevated production of glucose by the liver.
c) Elevated blood sugar levels in obesity are associated with increased concentrations of free fatty acids and cytokines (such as TNF-α and adiponectin), which influence the responsiveness of peripheral tissues to insulin.

The exact underlying molecular defect responsible for insulin resistance in type 2 diabetes mellitus has not yet been completely pinpointed. Presently, it is suggested that insulin resistance could potentially result from one of the subsequent abnormalities:

a) Polymorphisms in various intracellular signal pathway molecules after the insulin receptor.
b) Increased levels of free fatty acids observed in obesity might contribute, such as impaired glucose utilization in skeletal muscles, heightened liver glucose synthesis, and compromised β-cell function.
c) Insulin resistance syndrome is a constellation of clinical characteristics stemming from insulin resistance and its subsequent metabolic disruptions, including elevated blood sugar and compensatory high insulin levels. These features manifest as accelerated cardiovascular disease and can arise in both obese and non-obese individuals with type 2 diabetes mellitus. The manifestations encompass mild hypertension (linked to endothelial dysfunction) and dyslipidemia (marked by lowered HDL levels, increased triglycerides, and elevated LDL levels).

2.2.3. Dysfunctional Insulin Release

In T2DM, there is a connection between insulin resistance and insulin secretion:

I. In the early stages of the disease, as a response to insulin resistance, there is a compensatory increase in insulin secretion (hyperinsulinemia) aimed at maintaining normal blood glucose levels.
II. Eventually, however, the function of β-cells fails to adequately secrete insulin. While some insulin is still produced, individuals with type 2 DM experience a mild to moderate insulin deficiency (considerably less severe than that seen in type 1 DM), but the complete absence of insulin secretion does not occur.

2.2.4. Elevated Hepatic Glucose Production

One of insulin's typical functions is to enhance the storage of glucose as glycogen in the liver and inhibit gluconeogenesis. However, in type 2 diabetes mellitus, alongside peripheral tissue insulin resistance, the liver also becomes resistant to insulin. Thus, even with elevated insulin levels in the early stages of the disease, the liver's gluconeogenesis is not adequately suppressed. This leads to heightened hepatic glucose synthesis, a contributing factor to the elevated blood sugar levels (hyperglycemia) observed in these instances.

2.2.5. Constitutional Factors

Obesity, hypertension, and degree of physical activity are a few examples of environmental variables that contribute to and influence the phenotyping of the illness.

On the other hand, for both type 1 and type 2 diabetes mellitus, most of the complications like microvascular, macrovascular, and neuropathic are similar. Hyperglycaemia is the key factor for the development of macrovascular and microvascular complications. However macrovascular disease is much less related to glycemia. Lipid abnormalities like elevated levels of small dense low-density lipoprotein cholesterol (LDL-C), low levels of high-density lipoprotein cholesterol (HDL-C), and elevated levels of triglyceride-rich remnant protein are characteristic complications of diabetes mellitus. Besides thrombotic abnormalities like elevated type 1 plasminogen-activated inhibitor (PAI-I), elevated fibrinogen is also a very important factor of diabetes mellitus (Kearney et al., 2017). These all lead to the development of cardiovascular complicacy in the individuals and this risk appears to begin before the development of frank hyperglycemia.

2.3. Oral Anti-Hyperglycaemic Agents or Oral Hypoglycaemic Drugs

Oral hypoglycemic drugs have expanded the therapeutic options for diabetes management, complementing insulin variants. These medications are taken orally and effectively lower blood glucose levels through various mechanisms.

2.3.1. Sulfonylurea

These are a class of antidiabetic drugs used in the management of diabetes. The first clinically acceptable sulfonylureas Tolbutamide was introduced in 1957. Later on, in the 1970s 'second generation' sulfonylureas were developed which are 20-100 times more potent.

E.g.: First generation: Tolbutamide, Chlorpropamide
Second generation: Glyburide, Glipizide etc.

These sulfonylurea drugs typically used for treating T2DM have been explored for repurposing in conditions beyond diabetes due to their pharmacological properties. Chlorpropamide has been studied for its effects on serum lipids and lipoproteins, which are critical factors in cardiovascular health. Research indicates that it can lead to an initial decrease in serum and

VLDL triglycerides, alongside a reduction in LDL cholesterol and an increase in HDL cholesterol over 12 months (Paisey et al., 978). A study using fruit fly models found that glyburide, by stimulating the insulin-signaling pathway, can prevent the formation of harmful protein clumps and improve the structure of genetic material in cells affected by Huntington's disease (Tandon and Sarkar, 2023).

2.3.2. Meglitinides

Sulfonylureas and Meglitinides are also called Insulin Secretagogues. They stimulate insulin release from β cells of the pancreas and thus lower blood sugar levels.

E.g.: Repaglinide, Nateglinide

Recent studies have highlighted the shared pathophysiological mechanisms between diabetes and sarcopenia. Repaglinide, along with other insulin secretagogues, may have roles beyond glycaemic control, potentially influencing muscle metabolism and health (Witham et al., 2023).

2.3.3. Biguanides

Biguanides, primarily represented by metformin, are a class of drugs used primarily for treating type 2 diabetes. Their main mechanism of action involves improving insulin sensitivity and reducing glucose production in the liver.

E.g.: Phenformin, Buformin, Metformin

Metformin shows considerable promise for repurposing in various therapeutic areas, particularly cancer and age-related diseases. Preclinical studies have shown that metformin can inhibit tumor growth through various mechanisms, including the inhibition of mitochondrial oxidative phosphorylation and modulation of cancer immunity (Foretz et al., 2023). Beyond oncology, metformin is being explored for its potential benefits in age-related neurodegenerative diseases and cardiovascular health. Research indicates that metformin may offer cerebroprotective effects in ischemic stroke and improve health span by mitigating age-associated morbidities (Sharma et al., 2021).

2.3.4. Thiazolidinediones

E.g.: Troglitazone, Pioglitazone

These novel classes of drugs are selective agonists for the nuclear peroxisome proliferator-activated receptor γ which enhances the transcription of several insulin-responsive genes. They act by diminishing insulin resistance. They also improve the uptake and metabolism of glucose in the muscles and adipose tissues.

Pioglitazone is being explored for Alzheimer's disease due to its role as a PPARγ agonist, which may help reduce neuroinflammation and improve cognitive function. A systematic review indicated that pioglitazone, administered in doses ranging from 15 to 45 mg over at least six months, showed promise in enhancing memory function in patients with mild to moderate AD. The drug was well tolerated, suggesting a favorable safety profile for long-term use in this population (Basutkar et al., 2023).

2.3.5. α-Glucosidase Inhibitors

They are also called starch blockers. These inhibitors are useful adjuncts to insulin in the treatment of both type 1 and type 2 diabetes mellitus.

E.g.: Acarbose, Miglitol

These inhibitors have attracted interest for potential repurposing in other conditions due to their effects on carbohydrate metabolism and other pathways. Acarbose has shown potential for repurposing in cancer therapy. While its exact mechanism for inhibiting metastasis is unclear, studies have found that acarbose can slow kidney cancer growth and promote protective immune responses (Hijazi et al., 2023).

3. Mechanistic Insights and Pharmacodynamics of Anti-Diabetic Agents

The primary treatments for Type 2 Diabetes (T2DM) involve insulin-based injections and oral hypoglycaemic medications. Despite their essential role in managing T2DM, these therapies are often associated with adverse effects. Insulin has been a cornerstone in controlling severe insulin deficiency since its discovery. Due to the critical shortage of beta cells in T2DM, exogenous insulin administration is indispensable for patient survival. While significant progress has been made in understanding diabetes, developing insulin analogs, and managing the condition, achieving optimal blood sugar control without complications like hypoglycemia and weight gain remains a challenge. Lifestyle interventions including diet and exercise can effectively improve glucose levels, but most individuals with T2DM will eventually need medication to maintain adequate control. Oral hypoglycaemic agents are the choice of drug that adequately improves the condition upon use.

3.1. Insulin

Insulin, a peptide hormone synthesized by pancreatic beta cells, plays a crucial role in regulating blood glucose levels. When the body's insulin production is inadequate or insulin resistance develops, medical intervention with insulin therapy becomes necessary (Mathieu et al., 2017).

Insulin is a primary treatment for diabetes mellitus, a condition characterized by abnormal glucose levels in the blood. In type 1 diabetes, the body's immune system destroys the insulin-producing beta cells in the pancreas, leading to a complete or near-complete absence of insulin (Smith et al., 017; Van Belle et al., 2011; DiMeglio et al., 2018). Consequently, individuals with type 1 diabetes require lifelong insulin therapy.

In Type 2 diabetes, the body's cells become resistant to insulin's effects, causing the pancreas to overwork. Ultimately, the pancreas may fail to produce sufficient insulin, leading to hyperglycemia. While lifestyle modifications and oral medications are often initial treatments, insulin therapy may become necessary for advanced stages of type 2 diabetes (Donner et al., 2012; Howard-Thompson et al., 2018).

Insulin, as the oldest therapeutic agent in the management of diabetes mellitus (DM), has always ranked high among different pharmacologic products in all clinical practice guidelines on standard care of DM. Insulin is the most effective agent for glycaemic control and glycated hemoglobin (HbA1c) reduction that can be used in all stages of type 2 DM as monotherapy or combined with other glucose-lowering agents (Wallia et al., 2015; Abdi et al., 2018).

Insulin exerts its effects by directly interacting with specific receptors embedded in the outer layer of cells. While these receptors are present in all cells, their abundance varies significantly, with liver and fat cells exhibiting the highest concentrations. The insulin receptor is a complex molecule composed of two distinct types of subunits: alpha and beta. The alpha subunits, located on the exterior of the cell, are responsible for binding insulin. The beta subunits, which span the cell membrane, possess an enzyme-like activity known as tyrosine kinase (Kahn, 1985). When insulin attaches to the alpha subunits, it triggers the activation of the tyrosine kinase within the beta subunits. This activation initiates a series of events that result in the relocation of glucose transport proteins from the cell's interior to its surface (Posner, 2017). These transport proteins facilitate the uptake of glucose from the bloodstream into the cell, consequently lowering blood glucose levels (Jaldin-Fincati et al., 2017).

Nowadays, in addition to old human insulin products, several rapid- and long-acting insulin analogs such as pens are available. Arguments, however, continue about the cost-effectiveness of using these new products with pharmacologic profiles closer to normal physiology (Davidson, 2014; Grunberger, 2014). Based on the American Diabetes Association algorithm for the use of glucose-lowering medications in type 2 DM, where cost is a major concern, insulin with the lowest cost would be considered (American Diabetes Association, 2019). According to the World Health Organization guidelines, in low-resource settings, human insulin products, both short-acting regular insulin and intermediate-acting NPH insulin, are strongly recommended as the insulin of choice in adults with type 1 DM and as the third-line treatment for individuals with type 2 DM; long-acting insulin analogs is suggested for patients with frequent severe hypoglycemia with human insulin (Roglic and Norris, 2018).

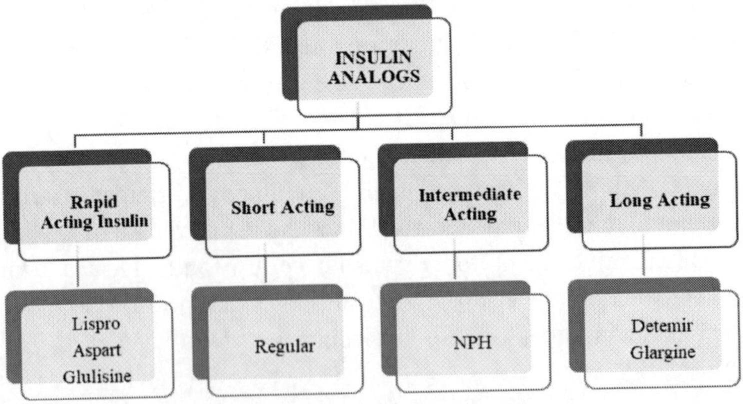

Figure 1. Some types of Insulin preparation with example.

Table 1. Pharmacokinetic properties of various Insulin preparations (Donnor and Sarkar, 2023)

Insulin	Onset of action (hr.)	Peak (hr.)	Duration (hr.)
Fast-acting Insulin Aspart	16 min	~1	~5
Insulin Lispro	23-27 min	~ 1-2	~5
Insulin Lispro-AABC	15-18 minutes	~ 1-2	~4
Insulin Aspart	21 min	1-3	~5
Insulin Glulisine	0.25-0.5	0.5-1	~ 4
Technosphere	within 5 min	15 min	~ 3
Regular	~ 1	2-4	5-8
NPH	1-2	4-10	14+
Insulin Detemir	3-4	6-8 (though relatively flat)	up to 20-24
Insulin Glargine	1.5	Flat	24
Insulin Degludec	1	9	42
Lispro Mix 50/50	0.25-0.5	0.5-3	14-24
Lispro Mix 75/25	0.25-5	0.5-2.5	14-24
Aspart Mix 70/30	0.1-0.2	1-4	18-24

3.2. Oral Antidiabetic Drugs

Effective diabetes management necessitates a collaborative approach incorporating lifestyle changes, such as diet and exercise, and, when necessary, medication to achieve individualized blood sugar targets. As type 2 diabetes progresses, with the pancreas gradually losing its ability to produce insulin, combining lifestyle modifications with oral medications becomes crucial for optimal blood sugar control (Cersosimo et al., 2021).

Table 2. Classification of oral hypoglycaemic drugs

Broad Class	Examples
Biguanides	Metformin
Thiazolidinediones	Pioglitazone, Rosiglitazone
Sulfonylureas	1st generation- Chlorpropamide, Tolbutamide 2nd generation- Glipizide, Glyburide
Meglitinides	Nateglinide, repaglinide
DPP-4 inhibitors	Linagliptin, saxagliptin, sitagliptin
Sodium-glucose cotransporter 2 inhibitors (SGLT-2 inhibitors)	Canagliflozin, Dapagliflozin
Alpha-glucosidase inhibitors	Acarbose, Miglitol
Amylin Analogues	Pramlintide

The complex nature of type 2 diabetes, involving multiple factors contributing to high blood sugar, necessitates a variety of medications with different mechanisms of action to effectively lower glucose levels.

3.2.1. Biguanides

Metformin, a synthetic compound derived from the natural substance guanidine, has been a cornerstone in type 2 diabetes management since its introduction over half a century ago. Its efficacy in controlling blood sugar levels both alone and in combination with other treatments has led to its widespread adoption as the most prescribed oral medication for this condition globally (Tahrani et al., 2016).

This prominence stems from the understanding that metformin's risk of causing lactic acidosis, a serious side effect, is minimal in individuals with normal kidney function. In contrast, phenformin, a chemically similar drug, was withdrawn from the market in many countries due to its higher propensity for this complication (Tahrani et al., 2016).

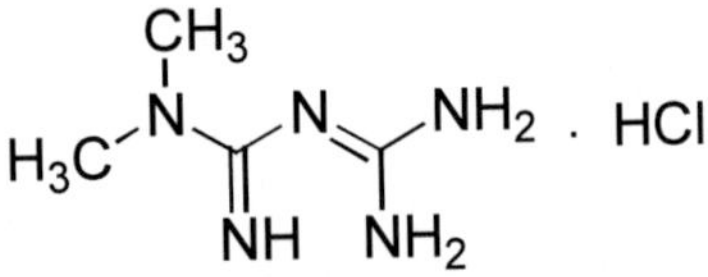

Figure 2. Chemical structure of Metformin hydrochloride.

Metformin primarily reduces glucose production in the liver and enhances the liver's response to insulin. Its impact on improving insulin sensitivity in other tissues is more modest, likely attributed to reductions in blood sugar, triglycerides, and free fatty acids (Rena et al., 2017; Foretz et al., 2014). This drug decreases insulin levels and fasting blood sugar by curbing glucose output from the liver (Tahrani et al., 2016). Additionally, metformin increases glucose utilization in the intestines and stimulates the release of GLP-1 (Rena et al., 2017; Foretz et al., 2014). However, it does not increase insulin production (Tahrani et al., 2016). The precise cellular and molecular mechanisms underlying these effects remain to be fully elucidated.

3.2.2. Thiazolidinediones

Troglitazone, pioglitazone, and rosiglitazone belong to a drug class called thiazolidinediones (TZDs) that enhance insulin sensitivity by activating PPAR gamma. Troglitazone was removed from the market in the United States, Europe, and Japan in 2000 due to rare but severe liver damage leading to liver failure and death in some patients (Tahrani et al., 2016; Yki-Järvinen, 2004). Unlike troglitazone, pioglitazone and rosiglitazone have not been associated with this serious liver problem. TZDs improve the body's response to insulin, both insulin produced by the body and insulin administered as medication, by reducing insulin resistance (Yki-Järvinen, 2004).

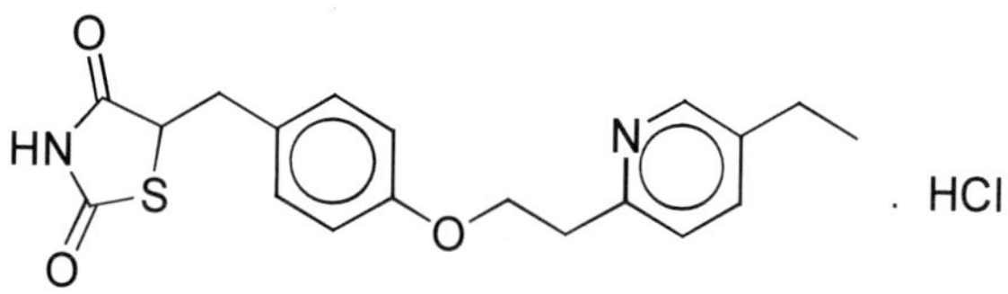

Figure 3. Chemical structure of Pioglitazone hydrochloride.

Pioglitazone and rosiglitazone primarily enhance insulin sensitivity by reducing insulin resistance (Tahrani et al., 2016; Yki-Järvinen, 2004; Willson et al., 2001). These drugs selectively activate the PPAR gamma receptor, a nuclear receptor that regulates gene expression (Yki-Järvinen, 2004; Willson et al., 2001). In its inactive state, PPAR gamma forms a complex with the retinoid X receptor (RXR) and a co-repressor molecule, binding to specific DNA sequences called PPREs. Upon activation by a ligand like pioglitazone or rosiglitazone, the co-repressor is released and replaced by a co-activator complex, leading to increased gene transcription (Yki-Järvinen, 2004; Willson et al., 2001). PPAR gamma can also indirectly inhibit gene expression by interfering with other signaling pathways, such as NF-kB (Yki-Järvinen, 2004). The genes controlled by PPAR gamma are involved in regulating lipid and carbohydrate metabolism, as well as inflammation (Yki-Järvinen, 2004; Willson et al., 2001).

3.2.3. Sulfonylureas

Sulfonylureas, introduced in the 1950s, has been extensively employed in the treatment of type 2 diabetes (Thule and Umpierrez, 2014; Khunti et al., 2018). Early sulfonylureas, such as acetohexamide, chlorpropamide, tolazamide, and tolbutamide, exhibited a lower binding affinity for their target, the ATP-sensitive potassium channel, necessitating higher dosages for therapeutic effects (Thule and Umpierrez, 2014; Khunti et al., 2018). These first-generation agents are now rarely prescribed. Subsequent development in the 1980s led to the introduction of second-generation sulfonylureas including glyburide, glipizide, gliclazide, and glimepiride, which are significantly more potent (approximately 100-fold) (Thule and Umpierrez, 2014). These agents are currently the most commonly used sulfonylureas. While sulfonylureas can be used independently, they are generally not combined with meglitinides due to overlapping mechanisms of action that lower blood glucose levels (Tahrani et al., 2016; Thule and Umpierrez, 2014).

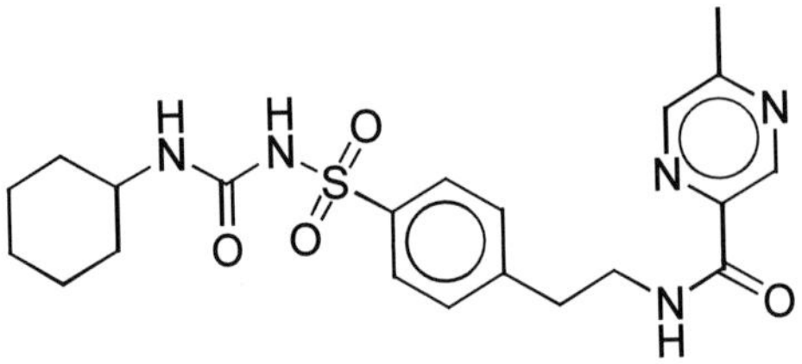

Figure 4. Chemical structure of glipizide.

Sulfonylureas are medications that increase insulin production. They lower blood sugar by directly stimulating the pancreas to release insulin, regardless of glucose levels (Thule and Umpierrez, 2014). Pancreatic beta cells normally sense blood sugar levels through a complex process involving glucose transport, phosphorylation, and metabolism, ultimately leading to insulin secretion (Prentki et al., 2013; Thorens, 2015). This process begins with glucose entering the beta cell and being converted to ATP. The increased ATP levels close potassium channels on the cell surface (Thule and Umpierrez, 2014; Prentki et al., 2013; MacDonald and Wheeler, 2003). This closure causes the cell's electrical charge to change, triggering the opening of calcium channels and a subsequent influx of calcium ions (Thule and Umpierrez, 2014; Ashcroft and Gribble, 1999). The increased calcium concentration within the cell stimulates the release of insulin-containing granules, resulting in insulin secretion (Thule and Umpierrez, 2014).

3.2.4. Meglitinides

Meglitinides are a class of medications that stimulate insulin release, similar to sulfonylureas but with a much faster onset and shorter duration of action (Tahrani et al., 2016; Guardado-Mendoza et al., 2013). Repaglinide, introduced in 1998, was the first drug in this category, followed by nateglinide in 2001. Unlike sulfonylureas, meglitinides require the presence of glucose to stimulate insulin secretion (Guardado-Mendoza et al., 2013; Tran et al., 2015). This glucose-dependent action helps reduce the risk of hypoglycemia compared to sulfonylureas, as insulin secretion decreases when blood sugar levels fall.

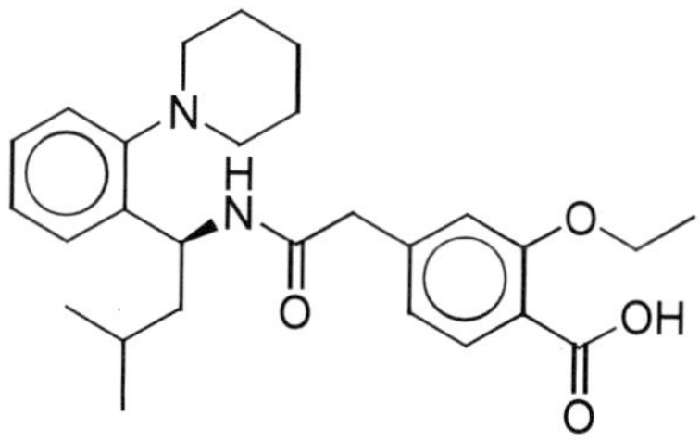

Figure 5. Chemical structure of repaglinide.

Meglitinides are rapidly absorbed into the bloodstream, reaching peak levels within an hour. They are then quickly broken down in the liver by enzymes called CYP3A4 and CYP2C8, forming inactive substances (Tahrani et al., 2016). The drug's effects last for about an hour. This rapid onset and short duration of action make meglitinides particularly effective at lowering blood sugar levels after meals (postprandial). Because of this, they should be taken 1 to 30 minutes before eating. It is essential not to take meglitinides if a meal is skipped (Guardado-Mendoza et al., 2013). Nateglinide acts more rapidly and has a shorter duration of action compared to repaglinide within the meglitinide class (Tran et al., 2015). Nateglinide induces a quicker and more pronounced initial insulin release, followed by a faster return to normal insulin levels compared to repaglinide (Guardado-Mendoza et al., 2013; Tran et al., 2015).

While both meglitinides and sulfonylureas promote insulin secretion by interacting with the sulfonylurea receptor 1 (SUR1) on beta cells, their binding sites differ (Tahrani et al., 2016; Guardado-Mendoza et al., 2013). Despite this distinction, both drug classes trigger a similar chain of events: closure of the potassium channel, cell depolarization, calcium influx, and subsequent insulin release (Tahrani et al., 2016; Guardado-Mendoza et al., 2013; Tran et al., 2015). However, the rapid onset and brief action of meglitinides make them particularly suited for managing blood sugar levels after meals (Tahrani et al., 2016; Guardado-Mendoza et al., 2013).

3.2.5. DPP-4 Inhibitors

DPP-4 inhibitors, often referred to as gliptins, are a group of oral medications approved by the U.S. Food and Drug Administration (FDA) for managing type 2 diabetes in adults. Sitagliptin, saxagliptin, linagliptin, and alogliptin are examples of FDA-approved DPP-4 inhibitors. Vildagliptin is approved for use in Europe by the European Medicines Agency (EMA) but not in the United States. These drugs work by enhancing the activity of incretin hormones, which are naturally occurring substances that help regulate blood sugar levels after eating. Beyond their blood sugar-lowering properties, DPP-4 inhibitors have demonstrated benefits for blood pressure, inflammation reduction, cell protection, and immune regulation within the heart, kidneys, and blood vessels. These effects are independent of their actions on incretin hormones. Due to this broader range of actions, some studies suggest potential applications for these drugs in individuals who develop diabetes after kidney or liver transplantation (Lim et al., 2015).

DPP-4 is a widespread enzyme that breaks down incretin hormones, primarily GLP-1 and GIP. These hormones play a crucial role in maintaining stable blood sugar levels by stimulating insulin production, reducing glucagon levels, and slowing down stomach emptying (Capuano et al., 2013).

GLP-1, produced by intestinal L cells, lowers blood sugar through these mechanisms and has a very short lifespan of less than two minutes. GIP, secreted by stomach and upper intestinal K cells, also contributes to blood sugar control. Its lifespan is around seven minutes in healthy individuals but shorter in people with type 2 diabetes (Pathak and Bridgeman, 2010; Gupta and Kalra, 2011).

Incretins are released shortly after eating and are rapidly broken down by the DPP-4 enzyme due to their short lifespan. By preventing the breakdown of these hormones, DPP-4 inhibitors increase the levels of GLP-1 and GIP in the body. This leads to increased insulin production from the pancreas, resulting in lower blood sugar levels both after meals and when fasting (Gupta and Kalra, 2011).

3.2.6. Sodium-Glucose Cotransporter 2 Inhibitors (SGLT-2 Inhibitors)

There are four primary SGLT2 inhibitors on the market: Canagliflozin (Invokana), Dapagliflozin (Farxiga), Empagliflozin (Jardiance), and Ertugliflozin (Steglatro) (Lupsa and Inzucchi, 2018). While these medications share a similar mechanism of action, there are subtle variations between them.

SGLT2 is a primary transporter in the kidney's tubules responsible for reabsorbing the majority (about 90%) of filtered glucose back into the bloodstream. In contrast, SGLT1, while also present in the kidney, plays a minor role (approximately 10%) in glucose reabsorption and is primarily found in the intestines (Tahrani et al., 2016; Thomas and Cherney, 2018).

Under normal blood sugar levels (below approximately 180 mg/dL), these transporters can reabsorb nearly all filtered glucose. However, when blood sugar exceeds this threshold, excess glucose spills into the urine (glycosuria). People with type 2 diabetes have an increased number of SGLT2 transporters in their kidney tubules, leading to enhanced glucose reabsorption and a higher blood sugar level (typically around 220 mg/dL) required for glycosuria to occur (Rahmoune et al., 2005).

3.2.7. Alpha-Glucosidase Inhibitors

Acarbose (Precose, Glucobay), miglitol (Glycet), and voglibose (Basen, Voglib) are medications belonging to a class of drugs known as alpha-glucosidase inhibitors. These drugs were introduced for the treatment of high blood sugar in the 1990s (Pathak and Bridgeman, 2010).

Alpha-glucosidase inhibitors work by blocking enzymes involved in carbohydrate digestion. These enzymes, found in the pancreas and intestinal lining, break down complex sugars into simpler sugars that the body can absorb (Tahrani et al., 2016). By inhibiting these enzymes, alpha-glucosidase inhibitors slow down the digestion and absorption of carbohydrates, leading to a gradual increase in blood sugar levels after meals. This helps to reduce blood sugar spikes. Unlike some other medications in this class, acarbose and miglitol do not affect the digestion of milk products (Tahrani et al., 2016).

In addition to their effects on carbohydrate digestion, these drugs can also influence the production of certain hormones involved in blood sugar control. SGLT2 inhibitors, a different class of diabetes medications, work by causing the kidneys to remove excess sugar from the body through urine. The amount of sugar excreted can vary based on kidney function and blood sugar levels, with an average of 60 to 90 grams of sugar lost per day (Tahrani et al., 2016).

4. Anti-Diabetic Agents and Cardiovascular Disease

Antidiabetic agents are a diverse group of medications used to manage diabetes mellitus, primarily aimed at controlling blood glucose levels. They are categorized based on their mechanisms of action, efficacy, and side effects, and are essential in both Type 1 and Type 2 diabetes management. Diabetes therapies have received a lot of study attention and have been shown to be effective in lowering blood glucose levels; however, because of unfavorable side effects such as weight gain and hypoglycemia, their efficacy is not long-lasting. (Furman, 2007).

The prevalence of diabetes mellitus has been on a steady rise globally, and with it, the risk of associated cardiovascular complications has become a significant concern for healthcare professionals. Patients diagnosed with type 2 diabetes have an increased risk of developing cardiovascular disease compared to non-diabetic patients. Type 2 diabetes significantly elevates the risk of cardiovascular disease (CVD), with patients facing a 2-4 times higher incidence of myocardial infarction and stroke compared to nondiabetic individuals. Cardiovascular complications account for approximately 75% of deaths in diabetic patients, positioning diabetes as a critical factor in CVD secondary prevention. (Kerru et al., 2018).

4.1. Evidence of Cardiovascular Benefits of Anti-Diabetic Agents

Antidiabetic agents are essential for managing blood glucose and reducing cardiovascular risks in diabetes. Some classes of these drugs provide additional cardiovascular protection, helping to lower the high incidence of CVD-related morbidity and mortality in diabetic patients. Various anti-diabetic agents, such as metformin, SGLT2 inhibitors, GLP-1 analogs, DPP-4 inhibitors, α-glucosidase inhibitors, and thiazolidinediones, exhibit cardiovascular benefits through AMPK-dependent pathways, improving heart health. (Lu et al., 2019).

Studies have shown that sodium-glucose cotransporter-2 inhibitors (SGLT2i) significantly reduce the risk of cardiovascular death, hospitalization for heart failure, and myocardial infarction compared to glucagon-like peptide-1 receptor agonists (GLP1-RA). SGLT2i lowers cardiovascular risks by promoting glycosuria and reducing blood pressure, with benefits extending to reduced cardiovascular death and hospitalization for heart failure. This effect is partly due to mechanisms such as improved endothelial function and cardiac metabolism. Several cardiovascular outcome trials (CVOTs) such as EMPA-REG OUTCOME (Steiner, 2016), CANVAS (Neal et al., 2017), and DECLARE-TIMI 58 (Wiviott et al., 2019) have been conducted to assess the cardiovascular risks of SGLT2 inhibitors—specifically empagliflozin, canagliflozin, and dapagliflozin—in patients with type 2 diabetes and cardiovascular disease (CVD). These studies demonstrated that adding

SGLT2 inhibitors to standard care significantly reduced all-cause mortality and improved cardiovascular outcomes.

In contrast, while GLP1-RA reduces cardiovascular death and all-cause mortality, its impact on heart failure hospitalization is less consistent. (Zhou et al., 2024). SGLT2 inhibitors have demonstrated significant cardiovascular benefits across various patient populations, including those with type 2 diabetes and heart failure. These medications exhibit efficacy in reducing major adverse cardiovascular events (MACE), heart failure hospitalizations, and atherosclerosis progression by modulating vascular inflammation, oxidative stress, and cellular metabolism. (Park, 2024), (Yerra & Connelly, 2024),. Additionally, GLP-1RAs and SGLT2 inhibitors have demonstrated undeniable cardiovascular benefits in large-scale trials, with GLP-1RAs decreasing atherosclerotic disease, especially stroke, and SGLT2 inhibitors mainly preventing heart failure, (Yagyu & Shimano, 2022).

Metformin, a widely used antidiabetic agent (biguanide), has shown significant cardiovascular benefits compared to other glucose-lowering therapies like insulin and sulfonylureas. It effectively reduces the risk of diabetes-related complications, including myocardial infarction (MI), stroke, heart failure, and all-cause mortality. These protective effects persist long-term, even after discontinuation of therapy, making metformin a key agent in reducing cardiovascular disease (CVD) risk in patients with diabetes. (Holman et al., 2008). Over three years, metformin treatment also resulted in decreased body weight, lower LDL cholesterol levels, and an increase in estimated glomerular filtration rate (eGFR) (Petrie et al., 2017). In patients with type 2 diabetes (T2D) and a history of coronary artery disease (CAD), metformin treatment, compared to glipizide, was linked to a reduced risk of recurrent major cardiovascular events, including myocardial infarction (MI), stroke, coronary angioplasty, coronary artery bypass grafting, cardiovascular death, and all-cause mortality. (Hong et al., 2013).

Dipeptidyl peptidase 4 (DPP4) inhibitors, used with metformin for better glycemic control in type 2 diabetes, have significant cardiovascular benefits. By inhibiting DPP4, which affects various processes including inflammation and sympathetic activation, these agents may protect against poor cardiovascular outcomes. Elevated levels of soluble DPP4 (sDPP4) are linked to worse cardiovascular outcomes and reduced heart function, indicating that DPP4 inhibition could improve cardiovascular health in patients with heart failure. (Dos Santos et al., 2013). These inhibitors have been found to regulate cardiovascular risk factors and directly impact the occurrence and progression of cardiovascular diseases through multiple mechanisms. Furthermore, DPP4 inhibitors have demonstrated therapeutic potential in preclinical models of cardiovascular diseases, including heart failure, hypertension, myocardial infarction, atherosclerosis, and stroke, regardless of the presence of diabetes. (Zakaria et al., 2022).

Thiazolidinediones (TZDs) have been extensively studied for their effects on cardiovascular health in diabetic patients. While there is some controversy regarding the cardiovascular risks associated with TZDs, recent meta-analyses, and reviews suggest potential cardiovascular benefits. TZDs have been shown to inhibit plaque progression, protect the vascular endothelium, and improve insulin sensitivity, which could contribute to a reduced risk of cardiovascular events. Researchers reviewed thiazolidinediones (TZDs), drugs used to improve insulin sensitivity in type 2 diabetes. TZDs remain viable for some patients but require careful consideration. The review also noted potential applications in other conditions, though further research is needed (Kumari et al., 2023). Another study reviewed the cardiovascular effects of pioglitazone, an antidiabetic drug, finding it reduced major adverse cardiovascular events (MACE), particularly in type 2 diabetes patients with preexisting cardiovascular disease (CVD) and in those with insulin resistance. Key studies, including PROactive and IRIS, confirmed its cardioprotective effects, especially in reducing the risk of recurrent stroke and myocardial infarction (MI). However, the benefits were less evident in patients without prior cardiovascular events or with newly diagnosed diabetes. Pioglitazone's cardiovascular protection is most significant in patients with established CVD (Tripathy et al., 2021).

4.2. Mechanisms through Which Anti-Diabetic Agents Exert Cardiovascular Effects

Anti-diabetic agents, such as sodium-glucose cotransporter-2 inhibitors (SGLT2i) and glucagon-like peptide-1 receptor agonists (GLP-1RA), exert cardiovascular effects through various mechanisms. These agents target inflammation, atherogenesis, oxidative stress-induced endothelial dysfunction, vascular smooth muscle cell reactive oxygen species production, and thrombosis, as highlighted in multiple studies. (Theofilis et al., 2022). Additionally, they modulate natriuresis/diuresis and the renin-angiotensin-aldosterone system, counteracting adverse vascular and cardiac remodeling. The activation of signaling pathways preventing adverse vascular remodeling is a specific effect of SGLT2i, while GLP-1RA directly impacts vascular cells, including endothelial and smooth muscle cells. (Khat & Husain, 2018). New antidiabetic drugs on cardiovascular outcomes in type-2 diabetes patients, particularly focusing on their effects on diabetic dyslipidemia. It was found that drugs, including SGLT2 inhibitors, DPP-4 inhibitors, and GLP-1 receptor agonists, not only lower blood sugar but also improve lipid profiles, which may help reduce cardiovascular risk. The study concludes that these medications could significantly lower cardiovascular morbidity and mortality by addressing both glucose levels and lipid abnormalities in diabetic patients. (Piccirillo et al., 2023). As mentioned previously anti-diabetic agents also exert cardiovascular effects through AMPK-dependent signaling pathways. Metformin, SGLT2 inhibitors, GLP-1 analogs, DPP-4 inhibitors, α-glucosidase inhibitors, and thiazolidinediones show AMPK-related cardiovascular benefits. (Lu et al., 2019). Traditional antidiabetic therapies primarily target hyperglycemia and have limited impact on cardiovascular disease whereas novel antidiabetic drug classes like GLP-1 agonists and DPP-4 inhibitors not only improve glycaemic control but also positively influence the underlying pathophysiological mechanisms of diabetes-related cardiovascular disease. These drugs influence various cardiovascular factors, including heart function, ischemia-reperfusion injury, endothelial function, blood pressure regulation, heart failure, lipid metabolism, and atherosclerosis. (Bistola et al., 2018). Researchers reviewed the cardioprotective mechanisms of various antidiabetic drugs, focusing on SGLT2 inhibitors (SGLT2i) and GLP-1 receptor agonists (GLP1-RA). They found that SGLT2i offers rapid cardioprotection through mechanisms like volume reduction and ketone body production, while GLP1-RA provides later-stage protection, potentially via the RISK pathway. Both drug types reduce fibrosis, arrhythmia, and apoptosis, improve mitochondrial function, and activate AMPK. (Schubert et al., 2020).

5. Anti-Diabetic Agents and Hypertension

Hypertension is prevalent among patients with both type I and type II diabetes and is closely linked to the progression of diabetic nephropathy (DN). Hyperglycaemia stimulates renal cells, leading to the production of humoral mediators, growth factors, and cytokines. These factors cause structural changes in the glomeruli of diabetic patients, such as hyaline arteriolosclerosis (mainly affecting the efferent arteriole), increased collagen deposition in the extracellular matrix, and greater permeability of the glomerular basement membrane (Basu et al., 2006). Consequently, filtration pressure rises, resulting in microalbuminuria and the activation of the renin-angiotensin system (RAAS). Chronic RAAS activation further exacerbates kidney damage in diabetic patients. The interplay between these two conditions is complex, involving a myriad of pathophysiological mechanisms. Anti-diabetic agents, primarily used for glycaemic control, also exhibit varying effects on blood pressure regulation. Understanding these effects is crucial for optimizing therapeutic strategies in patients with co-occurring diabetes and hypertension (Sun et al., 2015).

5.1. Relevant Studies and Clinical Evidence

Numerous studies have investigated the impact of anti-diabetic agents on blood pressure. Metformin, a cornerstone in type 2 diabetes management, has been shown to exhibit modest antihypertensive effects. A study by Wulffelé et al. (2004) demonstrated that metformin reduces systolic blood pressure independently of its glucose-lowering effects (Wulffelé et al., 2004).

Thiazolidinediones (TZDs), such as pioglitazone and rosiglitazone, are another class of anti-diabetic agents known for their beneficial effects on blood pressure. A randomized controlled trial indicated that pioglitazone significantly reduced both systolic and diastolic blood pressure in patients with type 2 diabetes (Basu et al., 2006). However, concerns regarding the cardiovascular safety of rosiglitazone have led to its restricted use.

Glucagon-like peptide-1 (GLP-1) receptor agonists, such as liraglutide and semaglutide, have also demonstrated promising antihypertensive effects. A meta-analysis by Sun et al. (2015) found that GLP-1 receptor agonists significantly lower systolic blood pressure, highlighting their potential dual benefits in managing diabetes and hypertension (Sun et al., 2015).

Sodium-glucose co-transporter 2 (SGLT2) inhibitors, including empagliflozin, dapagliflozin, and canagliflozin, have garnered attention for their robust antihypertensive properties. The EMPA-REG OUTCOME trial revealed that empagliflozin not only improved glycemic control but also significantly reduced systolic and diastolic blood pressure (Zinman et al., 2015). Similarly, the DECLARE-TIMI 58 trial reported significant reductions in blood pressure with dapagliflozin (Wiviott et al., 2019).

5.2. Mechanistic Insights into the Antihypertensive Effects of These Drugs

The antihypertensive effects of anti-diabetic agents are attributed to various mechanisms, some of which are independent of their glucose-lowering properties. Metformin, for instance, improves endothelial function and reduces insulin resistance, which in turn lowers blood pressure (Luque-Ramírez & Escobar-Morreale, 2010).

TZDs, by activating peroxisome proliferator-activated receptor-gamma (PPAR-γ), enhance insulin sensitivity and exert direct vascular effects, including vasodilation and improved arterial compliance (Rizos et al., 2009). These actions collectively contribute to their blood pressure-lowering effects.

GLP-1 receptor agonists exert their antihypertensive effects through multiple pathways. They promote natriuresis, reduce sympathetic nervous system activity, and improve endothelial function, all of which contribute to blood pressure reduction (Grünberger, 2017).

SGLT2 inhibitors lower blood pressure primarily through osmotic diuresis and natriuresis, leading to reduced plasma volume and arterial stiffness (Heerspink et al., 2016). Additionally, these agents may improve endothelial function and decrease sympathetic nervous system activity, further supporting their role in hypertension management.

Table 3. Impact of anti-diabetic agents on blood pressure regulation

Anti-Diabetic Agent	Mechanism of Antihypertensive Effect	Key Studies and Findings	Reference
Metformin	Improves endothelial function and reduces insulin resistance	Reduces systolic blood pressure independently of glucose-lowering effects	Wulffelé et al., 2004
Thiazolidinediones (TZDs)	Activates PPAR-γ, enhances insulin sensitivity, vasodilation	Pioglitazone reduces both systolic and diastolic blood pressure	Basu et al., 2006
GLP-1 Receptor Agonists	Promotes natriuresis, reduces sympathetic nervous system activity, improves endothelial function	Lowers systolic blood pressure, highlighting dual benefits in managing diabetes and hypertension	Sun et al., 2015
SGLT2 Inhibitors	Osmotic diuresis and natriuresis, reduced plasma volume and arterial stiffness	Empagliflozin and dapagliflozin significantly reduce systolic and diastolic blood pressure in diabetic patients.	Zinman et al., 2015; Wiviott et al., 2019

The repurposing of anti-diabetic agents for managing hypertension in diabetic patients offers a promising therapeutic strategy. The dual benefits of these agents on glycaemic control and blood pressure regulation underscore their potential in addressing the multifaceted needs of patients with diabetes. Ongoing and future studies will continue to elucidate the full scope of their cardiovascular benefits, paving the way for optimized treatment regimens.

5.3. Key Studies

5.3.1. Metformin

Study: Effects of short-term treatment with metformin on blood pressure, insulin sensitivity, and endothelial function in non-obese patients with hypertension.

Findings: Metformin reduces systolic blood pressure independently of its glucose-lowering effects.

Reference: Wulffelé et al., 2004

5.3.2. Thiazolidinediones (TZDs)

Study: Effects of pioglitazone on blood pressure and insulin sensitivity in nondiabetic subjects with elevated blood pressure.

Findings: Pioglitazone significantly reduces both systolic and diastolic blood pressure.

Reference: Basu et al., 2006

5.3.3. GLP-1 Receptor Agonists

Study: Effect of glucagon-like peptide-1 receptor agonists on weight loss: a systematic review and meta-analysis of randomized controlled trials.

Findings: GLP-1 receptor agonists significantly lower systolic blood pressure.

Reference: Sun et al., 2015

5.3.4. SGLT2 Inhibitors

Studies: EMPA-REG OUTCOME trial: Empagliflozin improves glycemic control and significantly reduces systolic and diastolic blood pressure.

DECLARE-TIMI 58 trial: Dapagliflozin significantly reduces blood pressure.

References: Zinman et al., 2015; Wiviott et al., 2019

5.4. Mechanistic Insights

Table 4. Mechanism of action of studied anti-diabetic agents

Mechanism	Anti-Diabetic Agent
Improves endothelial function	Metformin, GLP-1 Receptor Agonists
Reduces insulin resistance	Metformin
Activates PPAR-γ	Thiazolidinediones (TZDs)
Enhances insulin sensitivity	Thiazolidinediones (TZDs)
Promotes natriuresis	GLP-1 Receptor Agonists, SGLT2 Inhibitors
Reduces sympathetic nervous system activity	GLP-1 Receptor Agonists
Osmotic diuresis and natriuresis	SGLT2 Inhibitors
Reduces plasma volume and arterial stiffness	SGLT2 Inhibitors

6. Anti-Diabetic Agents and Obesity

6.1. Role of Anti-Diabetic Agents in Weight Management

Anti-diabetic agents have emerged as valuable tools in the management of obesity, particularly among individuals with type 2 diabetes. These agents not only help in controlling blood glucose levels but also have significant effects on body weight. Given the close relationship between obesity and type 2 diabetes, utilizing these medications to achieve weight reduction can have profound benefits in managing both conditions simultaneously.

I. *Metformin:* Metformin, a first-line anti-diabetic medication, has been observed to promote modest weight loss in overweight and obese individuals. This effect is primarily due to its action in reducing hepatic glucose production and improving insulin sensitivity, which can lead to a decrease in appetite and calorie intake (Lalau et al., 2019).

II. *GLP-1 Receptor Agonists:* Glucagon-like peptide-1 (GLP-1) receptor agonists, such as liraglutide and semaglutide, are known for their significant weight-reducing effects. These agents mimic the action of the incretin hormone GLP-1, which enhances insulin secretion, inhibits glucagon release, and slows gastric emptying, thereby promoting satiety and reducing food intake (Astrup et al., 2020).

III. *SGLT-2 Inhibitors:* Sodium-glucose cotransporter-2 (SGLT-2) inhibitors, including canagliflozin, dapagliflozin, and empagliflozin, facilitate glucose excretion through urine, leading to a reduction in blood glucose levels and caloric loss. This mechanism not only helps in glycaemic control but also results in weight loss due to the calorie deficit created by glucose excretion (Gallo et al., 2019).

6.2. Clinical Evidence Supporting Their Use in Obesity

The effectiveness of anti-diabetic agents in promoting weight loss has been demonstrated in various clinical trials and real-world studies. These studies provide robust evidence supporting their use in the management of obesity, especially in patients with type 2 diabetes.

I. *Metformin:* Clinical trials have shown that metformin induces weight loss in obese individuals, with an average reduction of 2-3 kg over 6-12 months of treatment. This weight loss is particularly beneficial for individuals with insulin resistance and metabolic syndrome (Knowler et al., 2002).
II. *GLP-1 Receptor Agonists:* Liraglutide and semaglutide have been extensively studied for their weight loss effects. In the SCALE Obesity and Prediabetes trial, liraglutide led to an average weight loss of 8.4 kg over 56 weeks, significantly more than the placebo (Pi-Sunyer et al., 2015). Similarly, semaglutide has demonstrated superior weight loss outcomes in the STEP trials, with participants losing an average of 15-17% of their body weight (Wilding et al., 2021).
III. *SGLT-2 Inhibitors:* SGLT-2 inhibitors have shown consistent weight loss benefits across various studies. For instance, the CANVAS program revealed that canagliflozin resulted in a mean weight reduction of approximately 2-4 kg compared to placebo (Neal et al., 2017).

6.3. Mechanistic Pathways Involved in Weight Regulation

The weight-regulating effects of anti-diabetic agents involve multiple mechanistic pathways that influence energy balance, appetite, and metabolism.

I. Metformin: The weight loss associated with metformin is primarily due to its ability to improve insulin sensitivity, reduce hepatic glucose production, and potentially decrease appetite. Metformin also modulates gut microbiota, which may contribute to its metabolic effects (Wu et al., 2017).

II. GLP-1 Receptor Agonists: GLP-1 receptor agonists enhance satiety and reduce food intake by delaying gastric emptying and acting on the central nervous system to promote feelings of fullness. They also improve insulin sensitivity and promote weight loss by increasing energy expenditure and fat oxidation (van Can et al., 2014).

III. SGLT-2 Inhibitors: The primary mechanism of weight loss with SGLT-2 inhibitors is through the excretion of glucose in the urine, leading to a caloric deficit. Additionally, these agents may promote lipolysis and fat oxidation due to the negative energy balance they induce (Ferrannini et al., 2015).

In conclusion, anti-diabetic agents play a significant role in weight management, offering valuable therapeutic options for individuals struggling with obesity, particularly those with type 2 diabetes. Clinical evidence supports their efficacy in promoting weight loss, and understanding the mechanistic pathways involved can help optimize their use in clinical practice.

7. Other Potential Therapeutic Uses

The therapeutic potential of anti-diabetic agents extends far beyond their primary use in glucose regulation. Emerging research is delving into their efficacy in managing a variety of co-occurring conditions, highlighting their versatility and potential in broader clinical applications. Notably, significant attention is being directed towards their impact on cancer, polycystic ovary syndrome (PCOS), liver disease, and neurological disorders.

1) *Cancer:* Anti-diabetic drugs, particularly metformin, have demonstrated considerable promise in oncology. Metformin is being evaluated for its ability to inhibit cancer cell proliferation and induce apoptosis. These effects are largely attributed to its role in inhibiting the mammalian target of the rapamycin (mTOR) pathway and reducing insulin levels, which in turn decreases the growth stimuli for certain cancers (Evans et al., 2019). Additionally, observational

studies have noted a lower incidence of cancer among diabetic patients treated with metformin compared to those who are not, suggesting a protective effect (Gallagher & LeRoith, 2020).

2) *Polycystic Ovary Syndrome (PCOS):* PCOS is a common endocrine disorder characterized by insulin resistance, which contributes to its metabolic and reproductive manifestations. Anti diabetic agents such as metformin are commonly used in the management of PCOS to improve insulin sensitivity, regulate menstrual cycles, and manage symptoms of hyperandrogenism (Palomba et al., 2018). Metformin's role in enhancing ovulation rates and improving fertility outcomes in women with PCOS is well-documented, making it a staple in PCOS management protocols (Costello et al., 2019).
3) *Liver Disease:* The prevalence of non-alcoholic fatty liver disease (NAFLD) and its more severe form, non-alcoholic steatohepatitis (NASH), is notably high among individuals with type 2 diabetes (Katz & Brar, 2020). Anti-diabetic medications, particularly pioglitazone, have shown efficacy in reducing liver fat accumulation and improving liver function tests (Bril et al., 2019). Pioglitazone works by enhancing insulin sensitivity and exerting anti-inflammatory effects, thereby addressing key pathophysiological components of NAFLD and NASH.
4) *Neurological Disorders:* The neuroprotective effects of anti-diabetic drugs are being explored in conditions such as Alzheimer's disease and Parkinson's disease. Insulin-sensitizing agents, including metformin and thiazolidinediones, may improve cognitive function and reduce neuroinflammation, potentially altering the progression of these neurodegenerative diseases (Athauda & Foltynie, 2018). The potential mechanisms involve improving insulin signaling in the brain, which is crucial for neuronal health and function.

However, the preliminary evidence supporting the repurposing of anti-diabetic agents for these conditions is encouraging but necessitates further investigation. For instance, ongoing clinical trials are examining the efficacy of metformin in reducing cancer recurrence and mortality rates among diabetic and non-diabetic patients (ClinicalTrials.gov, 2021). Similarly, studies are underway to evaluate the effects of pioglitazone on liver disease and the application of GLP-1 receptor agonists in neurological disorders (Garber et al., 2020). These trials aim to validate the preliminary findings and provide a robust evidence base for the broader clinical use of these drugs.

8. Future Directions and Research Opportunities

Repositioning antidiabetic drugs with good efficacy and safety in the treatment of co-occurring diseases is a promising way for drug discovery but there are notable gaps in our understanding that need to be addressed. Understanding these key areas will provide opportunities for future research and direct scientists toward more efficient repurposing of anti-diabetic drugs.

While the mechanisms of anti-diabetic drugs are well-understood in the context of glucose metabolism, how these mechanisms translate to other diseases remains unclear. For example, metformin's role in reducing cancer risk or improving cardiovascular outcomes is not fully understood at the molecular level (Olatunde et al., 2021). The ability of these drugs to target specific pathways in non-diabetic conditions without affecting glucose metabolism is another area requiring more research (Andrzejewski et al., 2018).

The long-term safety of anti-diabetic drugs when used for other conditions has not been thoroughly studied. For instance, while these drugs may be beneficial in cancer prevention, the long-term impact on patients who do not have diabetes is uncertain. Drugs like thiazolidinediones have been linked to cardiovascular risks, which could be exacerbated in non-diabetic patients (Kumar et al., 2016; Raveendran et al., 2021).

The appropriate dosing for non-diabetic conditions is not well-established. Anti-diabetic drugs are typically dosed based on their effects on blood glucose, but this may not be optimal for other conditions (Stein et al., 2013). The effects of combining these drugs with standard therapies for other diseases are not fully understood. This could be particularly relevant in cancer, where combination therapies are common.

The efficacy of these drugs in different populations, such as age groups, genders, or those with varying comorbidities, is not well-characterized. This knowledge gap makes it difficult to predict who might benefit most from repurposing. How genetic variability influences the efficacy and safety of repurposed anti-diabetic drugs is another area with limited data (Kirchheiner et al., 2005).

The regulatory pathways for repurposing drugs are not always clear, especially when the new indication is vastly different from the original (Spin et al., 2024). There are relatively few large-scale, randomized controlled trials (RCTs) that investigate the use of anti-diabetic drugs for other diseases (Mantik et al., 2023). Without this data, it is challenging to draw definitive conclusions about their efficacy and safety.

Future research on repurposing anti-diabetic drugs should focus on understanding the precise molecular mechanisms underlying their effects on co-occurring diseases, and on conducting large-scale, randomized controlled trials to assess their efficacy and safety in diverse patient populations. Additionally, exploring optimal dosing strategies and the potential benefits of combination therapies will be crucial to expanding their therapeutic applications beyond diabetes. Addressing these knowledge gaps through focused research and clinical trials will be essential to fully realize the potential of repurposing anti-diabetic drugs for other diseases.

Conclusion

The repurposing of anti-diabetic agents for managing co-occurring conditions represents a promising strategy to enhance patient care, especially for those facing complex, multifaceted health challenges. Despite significant progress in conventional drug discovery, the rising prevalence of diseases like type 2 diabetes (T2D) has outpaced the development of new drugs. Drug repurposing, which leverages existing medicines for new applications, offers a compelling solution to this challenge by significantly reducing drug development times. The bidirectional linkages between diabetes and its comorbidities—such as inflammation, oxidative stress, metabolic disruption, and obesity—highlight the potential of antidiabetic drugs to address these interconnected conditions. Substantial evidence supports the cardiovascular, antihypertensive, and weight management benefits of these drugs, with emerging research suggesting potential roles in treating conditions like cancer, PCOS, and neurological disorders. However, integrating these therapies into clinical practice demands careful consideration of safety, efficacy, and patient-specific factors, underscoring the importance of personalized medicine approaches. As research continues to advance, the strategic repurposing of anti-diabetic agents could significantly broaden therapeutic options, ultimately improving outcomes for patients with diabetes and its associated comorbidities.

References

Abdi, H., Azizi, F., & Amouzegar, A. (2018). Insulin monotherapy versus insulin combined with other glucose-lowering agents in type 2 diabetes: A narrative review. *International Journal of Endocrinology and Metabolism,* 16(2), e65600.

American Diabetes Association. (2019). Pharmacologic approaches to glycemic treatment: Standards of medical care in diabetes—2019. *Diabetes Care,* 42(Suppl 1), S90-S102.

Andrzejewski, S., Siegel, P. M., and St-Pierre, J. (2018). Metabolic Profiles Associated With Metformin Efficacy in Cancer. *Front. Endocrinol. (Lausanne).* 9, 372.

Ashcroft, F. M., & Gribble, F. M. (1999). ATP-sensitive K+ channels and insulin secretion: Their role in health and disease. *Diabetologia,* 42, 903-919.

Astrup, A., Rössner, S., Van Gaal, L., Rissanen, A., Niskanen, L., Al Hakim, M., Madsen, J., Rasmussen, M. F., Lean, M. E. J., & NN8022-1807 Study Group. (2020). Effects of liraglutide in the treatment of obesity: a randomized, double-blind, placebo-controlled study. *The Lancet*, 374(9701), 1606-1616.

Athauda, D., & Foltynie, T. (2018). Insulin resistance and Parkinson's disease: A new target for disease modification? *Progress in Neurobiology*, 145-146, 98-120.

Basu, A., Jensen, M. D., McCann, F., Mukhopadhyay, D., Joyner, M. J., & Rizza, R. A. (2006). Effects of pioglitazone on blood pressure and insulin sensitivity in nondiabetic subjects with elevated blood pressure. *Journal of Clinical Endocrinology & Metabolism*, 91(1), 156-162.

Basutkar, R. S., Sudarsan, P., Robin, S. M., Bhaskar, V., Viswanathan, B., & Sivasankaran, P. (2023). Drug repositioning of pioglitazone in management and improving cognitive function among patients with mild to moderate Alzheimer's disease: A systematic review and meta-analysis. *Neurology India, 71*(6), 1132-1141.

Baynest, H. W. (2015). Classification, pathophysiology, diagnosis, and management of diabetes mellitus. *Journal of Diabetes & Metabolism, 6, 541.*

Bistola, V., Lambadiari, V., Dimitriadis, G., Ioannidis, I., Makrilakis, K., Tentolouris, N., Tsapas, A., & Parissis, J. (2018). Possible mechanisms of direct cardiovascular impact of GLP-1 agonists and DPP4 inhibitors. *Heart Failure Reviews,* 23(3), 377-388.

Blüher, M., and Stumvoll, M. (2020). *"Diabetes and Obesity,"* In: Bonora, E., DeFronzo, R. (eds) Diabetes complications, comorbidities and related disorders. Endocrinology. Springer, Cham. 1–49.

Bril, F., Kalavalapalli, S., Clark, V. C., et al. (2019). Response to pioglitazone treatment in nonalcoholic steatohepatitis with fibrosis. *Clinical Gastroenterology and Hepatology,* 17(2), 231-238.e5.

Capuano, A., Sportiello, L., Maiorino, M. I., Rossi, F., Giugliano, D., & Esposito, K. (2013). Dipeptidyl peptidase-4 inhibitors in type 2 diabetes therapy—Focus on alogliptin. *Drug Design, Development, and Therapy,* 7, 989-1001.

Cersosimo, E., Triplitt, C., Solis-Herrera, C., Mandarino, L. J., & DeFronzo, R. A. (2021). Pathogenesis of type 2 diabetes mellitus. In: Feingold, K. R., Anawalt, B., Boyce, A., Chrousos, G., Dungan, K., Grossman, A., Hershman, J. M., et al. (Eds.), *Endotext.* South Dartmouth (MA): MDText.com, Inc.

ClinicalTrials.gov. (2021). Metformin in cancer treatment. *ClinicalTrials.gov Identifier: NCT01941953*. Retrieved from https://clinicaltrials.gov/ct2/show/NCT01941953.

Costello, M. F., Misso, M. L., Balen, A., et al. (2019). Evidence summaries and recommendations from the international evidence-based guideline for the assessment and management of polycystic ovary syndrome: Assessment and treatment of infertility. *Human Reproduction Open*, 2019(1), 21.

Davidson, M. B. (2014). Insulin analogs—Is there a compelling case to use them? No! *Diabetes Care,* 37(6), 1771-1774.

DeFronzo, R. A. (1997). Pathogenesis of type 2 diabetes: metabolic and molecular implications for identifying diabetes genes. *Diabetes Reviews*, 5(3), 177-269.

DeFronzo, R. A. (2015). Pathogenesis of type 2 diabetes mellitus. In: *International Textbook of Diabetes Mellitus* (pp. 371-400).

Delavan, B., Roberts, R., Huang, R., Bao, W., Tong, W., and Liu, Z. (2018). Computational drug repositioning for rare diseases in the era of precision medicine. *Drug Discov. Today* 23, 382–394.

DiMeglio, L. A., Evans-Molina, C., & Oram, R. A. (2018). Type 1 diabetes. *The Lancet,* 391(10138), 2449-2462.

Donner, T., & Muñoz, M. (2012). Update on insulin therapy for type 2 diabetes. *Journal of Clinical Endocrinology & Metabolism,* 97(5), 1405-1413.

Donnor, T., & Sarkar, S. (2023). Insulin—Pharmacology, therapeutic regimens, and principles of intensive insulin therapy. In: Feingold, K. R., Anawalt, B., Blackman, M. R., et al. (Eds.), South Dartmouth (MA): MDText.com, Inc.; 2000.

Dos Santos, L., Salles, T. A., Arruda-Junior, D. F., Campos, L. C. G., Pereira, A. C., Barreto, A. L. T., Antonio, E. L., Mansur, A. J., Tucci, P. J. F., Krieger, J. E., & Girardi, A. C. C. (2013). Circulating dipeptidyl peptidase IV activity correlates with cardiac dysfunction in human and experimental heart failure. *Circulation: Heart Failure,* 6(5), 1029-1038.

Evans, J. M., Donnelly, L. A., Emslie-Smith, A. M., Alessi, D. R., & Morris, A. D. (2019). Metformin and reduced risk of cancer in diabetic patients. *BMJ*, 330(7503), 1304-1305.

Ferrannini, E., Mark, M., & Mayoux, E. (2015). CV protection in the EMPA-REG OUTCOME trial: a "thrifty substrate" hypothesis. *Diabetes Care*, 39(7), 1108-1114.

Foretz, M., Guigas, B., & Viollet, B. (2023). Metformin: Update on mechanisms of action and repurposing potential. *Nature Reviews Endocrinology*, 19(8), 460-476.

Foretz, M., Guigas, B., Bertrand, L., Pollak, M., & Viollet, B. (2014). Metformin: From mechanisms of action to therapies. *Cell Metabolism,* 20, 953-966.

Furman, B. L. (2007). Antidiabetic agents. *XPharm: The Comprehensive Pharmacology Reference*, Elsevier Inc. pp1.

Gallagher, E. J., & LeRoith, D. (2020). The proliferating role of insulin and insulin-like growth factors in cancer. *Trends in Endocrinology & Metabolism*, 21(10), 610-618.

Gallo, L. A., Wright, E. M., & Vallon, V. (2019). Probing SGLT2 as a therapeutic target for diabetes: Basic physiology and consequences. *Diabetes & Metabolism*, 45(6), 404-413.

Garber, A. J., Abrahamson, M. J., Barzilay, J. I., et al. (2020). Consensus statement by the American Association of Clinical Endocrinologists and American College of Endocrinology on the comprehensive type 2 diabetes management algorithm – 2020 executive summary. *Endocrine Practice*, 26(1), 107-139.

Grunberger, G. (2014). Insulin analogs—Are they worth it? Yes! *Diabetes Care,* 37(6), 1767-1770.

Grünberger, G. (2017). GLP-1 receptor agonists: a review of head-to-head clinical studies. *Therapeutic Advances in Endocrinology and Metabolism*, 8(4), 57-74.

Guardado-Mendoza, R., Prioletta, A., Jimenez-Ceja, L. M., Sosale, A., & Folli, F. (2013). The role of nateglinide and repaglinide, derivatives of meglitinide, in the treatment of type 2 diabetes mellitus. *Archives of Medical Science,* 9, 936-943.

Gupta, V., & Kalra, S. (2011). Choosing a gliptin. *Indian Journal of Endocrinology and Metabolism, 15*(4), 298-308.

Habib, S. L., and Rojna, M. (2013). Diabetes and Risk of Cancer. *ISRN Oncol.* 2013, 1–16.

Heerspink, H. J. L., Perkins, B. A., Fitchett, D. H., Husain, M., & Cherney, D. Z. I. (2016). Sodium glucose cotransporter 2 inhibitors in the treatment of diabetes mellitus: cardiovascular and kidney effects, potential mechanisms, and clinical applications. *Circulation*, 134(10), 752-772.

Hijazi, M. A., Gessner, A., & El-Najjar, N. (2023). Repurposing of chronically used drugs in cancer therapy: A chance to grasp. *Cancers*, 15(12), 3199.

Holman, R. R., Paul, S. K., Bethel, M. A., Matthews, D. R., & Neil, H. A. W. (2008). 10-year follow-up of intensive glucose control in type 2 diabetes. *New England Journal of Medicine,* 359(15), 1577-1589.

Hong, J., Zhang, Y., Lai, S., Lv, A., Su, Q., Dong, Y., Zhou, Z., Tang, W., Zhao, J., Cui, L., Zou, D., Wang, D., Li, H., Liu, C., Wu, G., Shen, J., Zhu, D., Wang, W., Shen, W., & Ning, G. (2013). Effects of metformin versus glipizide on cardiovascular outcomes in patients with type 2 diabetes and coronary artery disease. *Diabetes Care,* 36(5), 1304-1311.

Howard-Thompson, A., Khan, M., Jones, M., & George, C. M. (2018). Type 2 diabetes mellitus: Outpatient insulin management. *American Family Physician,* 97(1), 29-37.

International Diabetes Federation International Diabetes Federation-Diabetes facts & figures. Available at: https://idf.org/about-diabetes/diabetes-facts-figures/.

Jaldin-Fincati, J. R., Pavarotti, M., Frendo-Cumbo, S., Bilan, P. J., & Klip, A. (2017). Update on GLUT4 vesicle traffic: A cornerstone of insulin action. *Trends in Endocrinology & Metabolism,* 28(8), 597-611.

Kahn, C. R. (1985). The molecular mechanism of insulin action. *Annual Review of Medicine,* 36, 429-451.

Katz, J., & Brar, P. C. (2020). Nonalcoholic fatty liver disease and metformin: What's the connection? *Journal of Clinical Endocrinology & Metabolism*, 105(4), 1164-1166.

Kearney, K., Tomlinson, D., Smith, K., Ajjan, R. (2017). Hypofibrinolysis in diabetes: A therapeutic target for the reduction of cardiovascular risk. *Cardiovascular Diabetology,* 16(1), 1-17.

Kerru, N., Singh-Pillay, A., Awolade, P., & Singh, P. (2018). Current anti-diabetic agents and their molecular targets: A review. *European Journal of Medicinal Chemistry,* 152, 436-488.

Khat, D. Z., & Husain, M. (2018). Molecular mechanisms underlying the cardiovascular benefits of SGLT2i and GLP-1RA. *Current Diabetes Reports,* 18(7), 1-15.

Khunti, K., Chatterjee, S., Gerstein, H. C., Zoungas, S., & Davies, M. J. (2018). Do sulphonylureas still have a place in clinical practice? *The Lancet Diabetes & Endocrinology,* 6, 821-832.

Kirchheiner, J., Roots, I., Goldammer, M., Rosenkranz, B., and Brockm??ller, J. (2005). Effect of Genetic polymorphisms in cytochrome P450 (CYP) 2C9 and CYP2C8 on the pharmacokinetics of oral antidiabetic drugs. *Clin. Pharmacokinet.* 44, 1209–1225.

Knowler, W. C., Barrett-Connor, E., Fowler, S. E., Hamman, R. F., Lachin, J. M., Walker, E. A., & Nathan, D. M. (2002). Reduction in the incidence of type 2 diabetes with lifestyle intervention or metformin. *New England Journal of Medicine*, 346(6), 393-403.

Kumar, R., Kerins, D. M., and Walther, T. (2016). Cardiovascular safety of anti-diabetic drugs. *Eur. Hear. J. Cardiovasc. Pharmacother.* 2, 32–43.

Kumari, D., Kainth, R., Singh, S., Singh, P., Kharyal, A., Singh, A. S. B., & Singh, J. (2023). Exploring the potential benefits and risks of 2,4-thiazolidinediones (TZDs) in the management of diabetes and related conditions. *Journal of Pharmacology and Drug Delivery,* 1(1), 101-110.

Lalau, J. D., & Arnouts, P. (2019). Metformin and other antidiabetic agents: A brief overview. *Diabetes & Metabolism*, 45(6), 494-505.

Lim, S. W., Jin, J. Z., Jin, L., Jin, J., & Li, C. (2015). Role of dipeptidyl peptidase-4 inhibitors in new-onset diabetes after transplantation. *Korean Journal of Internal Medicine,* 30(6), 759-770.

Lu, Q., Li, X., Liu, J., Sun, X., Rousselle, T., Ren, D., Tong, N., & Li, J. (2019). AMPK is associated with the beneficial effects of antidiabetic agents on cardiovascular diseases. *Bioscience Reports,* 39(2), 1-15.

Lupsa, B. C., & Inzucchi, S. E. (2018). Use of SGLT2 inhibitors in type 2 diabetes: Weighing the risks and benefits. *Diabetologia,* 61, 2118-2125.

Luque-Ramírez, M., & Escobar-Morreale, H. F. (2010). Metformin in women with polycystic ovary syndrome. *Obstetrics and Gynecology Clinics,* 37(2), 269-282.

MacDonald, P. E., & Wheeler, M. B. Voltage-dependent K(+) channels in pancreatic beta cells: Role, regulation, and potential as therapeutic targets. *Diabetologia,* 46, 1046-1062.

Mantik, K. E. K., Kim, S., Gu, B., Moon, S., Kwak, H.-B., Park, D.-H., et al. (2023). Repositioning of Anti-Diabetic Drugs against Dementia: Insight from Molecular Perspectives to Clinical Trials. *Int. J. Mol. Sci.* 24, 11450.

Mathieu, C., Gillard, P., & Benhalima, K. (2017). Insulin analogues in type 1 diabetes mellitus: Getting better all the time. *Nature Reviews Endocrinology,* 13(7), 385-399.

Moscatiello, S., Manini, R., and Marchesini, G. (2007). Diabetes and liver disease: An ominous association. *Nutr. Metab. Cardiovasc. Dis.* 17, 63–70.

Neal, B., Perkovic, V., Mahaffey, K. W., de Zeeuw, D., Fulcher, G., Erondu, N., Shaw, W., Law, G., Desai, M., & Matthews, D. R. (2017). Canagliflozin and cardiovascular and renal events in type 2 diabetes. *New England Journal of Medicine,* 377(7), 644-657.

Olatunde, A., Nigam, M., Singh, R. K., Panwar, A. S., Lasisi, A., Alhumaydhi, F. A., et al. (2021). Cancer and diabetes: the interlinking metabolic pathways and repurposing actions of antidiabetic drugs. *Cancer Cell Int.* 21, 499.

Paisey, R., Elkeles, R. S., Hambley, J., & Magill, P. (1978). The effects of chlorpropamide and insulin on serum lipids, lipoproteins, and fractional triglyceride removal. *Diabetologia*, 15, 81-85.

Palomba, S., Santagni, S., Falbo, A., & La Sala, G. B. (2018). Complications and challenges associated with polycystic ovary syndrome: Current perspectives. *International Journal of Women's Health*, 10, 625-639.

Park, J. J. (2024). SGLT2 inhibitors in cardiovascular disease. *The Journal of Korean Diabetes,* 25(1), 26-34.

Parvathaneni, V., Kulkarni, N. S., Muth, A., and Gupta, V. (2019). Drug repurposing: a promising tool to accelerate the drug discovery process. *Drug Discov. Today* 24, 2076–2085.

Pathak, R., & Bridgeman, M. B. (2010). Dipeptidyl peptidase-4 (DPP-4) inhibitors in the management of diabetes. *P&T,* 35(9), 509-513.

Petrie, J. R., Chaturvedi, N., Ford, I., Brouwers, M. C. G. J., Greenlaw, N., Tillin, T., Hramiak, I., Hughes, A. D., Jenkins, A. J., Klein, B. E. K., Klein, R., Ooi, T. C., Rossing, P., Stehouwer, C. D. A., Sattar, N., Colhoun, H. M., Nickerson, H., Lou, O., Dutta, S., & White, A. (2017). Cardiovascular and metabolic effects of metformin in patients with type 1 diabetes: A double-blind, randomized, placebo-controlled trial. *The Lancet Diabetes and Endocrinology,* 5(8), 597-609.

Petrie, J. R., Guzik, T. J., and Touyz, R. M. (2018). Diabetes, Hypertension, and cardiovascular disease: Clinical insights and vascular mechanisms. *Can. J. Cardiol.* 34, 575–584.

Piccirillo, F., Mastroberardino, S., Nusca, A., Frau, L., Guarino, L., Napoli, N., Ussia, G. P., & Grigioni, F. (2023). Novel antidiabetic agents and their effects on lipid profile: A single shot for several cardiovascular targets. *International Journal of Molecular Sciences,* 24(12), 10164.

Pinson, R. (2017). Antidiabetic agents. *Annual Reports in Medicinal Chemistry, 2*(C), 176-186.

Pi-Sunyer, X., Astrup, A., Fujioka, K., Greenway, F., Halpern, A., Krempf, M., Lau, D. C., le Roux, C. W., Violante Ortiz, R., Jensen, C. B., & Wilding, J. P. H. (2015). A randomized, controlled trial of 3.0 mg of liraglutide in weight management. *New England Journal of Medicine*, 373(1), 11-22.

Posner, B. I. (2017). Insulin signalling: The inside story. *Canadian Journal of Diabetes,* 41(1), 108-113.

Prentki, M., Matschinsky, F. M., & Madiraju, S. R. (2013). Metabolic signaling in fuel-induced insulin secretion. *Cell Metabolism,* 18, 162-185.

Pushpakom, S., Iorio, F., Eyers, P. A., Escott, K. J., Hopper, S., Wells, A., et al. (2019). Drug repurposing: progress, challenges and recommendations. *Nat. Rev. Drug Discov.* 18, 41–58.

Rahmoune, H., Thompson, P. W., Ward, J. M., Smith, C. D., Hong, G., & Brown, J. (2005). Glucose transporters in human renal proximal tubular cells isolated from the urine of patients with non-insulin-dependent diabetes. *Diabetes,* 54, 3427-3434.

Raveendran, A. V., Fernandez, C. J., and Jacob, K. (2021). Efficacy and Cardiovascular Safety of Thiazolidinediones. *Curr. Drug Saf.* 16, 233–249.

Rena, G., Hardie, D. G., & Pearson, E. R. (2017). The mechanisms of action of metformin. *Diabetologia,* 60, 1577-1585.

Rizos, C. V., Elisaf, M. S., & Mikhailidis, D. P. (2009). Liberopoulos EN. How safe is the use of thiazolidinediones in clinical practice? *Expert Opinion on Drug Safety*, 8(1), 15-32.

Roessler, H. I., Knoers, N. V. A. M., van Haelst, M. M., and van Haaften, G. (2021). Drug Repurposing for Rare Diseases. *Trends Pharmacol. Sci.* 42, 255–267.

Roglic, G., & Norris, S. L. (2018). Medicines for treatment intensification in type 2 diabetes and type of insulin in type 1 and type 2 diabetes in low-resource settings: Synopsis of the World Health Organization guidelines on second- and third-line medicines and type of insulin for the control of blood glucose levels in nonpregnant adults with diabetes mellitus. *Annals of Internal Medicine,* 169(6), 394-397. doi: 10.7326/M18-1149.

Sachedina, S. (2013). Association of comorbidities with increasing severity of peripheral neuropathy in diabetes mellitus. *World J. Diabetes* 4, 135.

Schubert, M., Hansen, S., Leefmann, J., & Guan, K. (2020). Repurposing antidiabetic drugs for cardiovascular disease. *Frontiers in Physiology,* 11, 568632.

Sharma, S., Nozohouri, S., Vaidya, B., & Abbruscato, T. (2021). Repurposing metformin to treat age-related neurodegenerative disorders and ischemic stroke. *Life Sciences*, 274, 119343.

Smith, M. J., Simmons, K. M., & Cambier, J. C. (2017). B cells in type 1 diabetes mellitus and diabetic kidney disease. *Nature Reviews Nephrology,* 13(11), 712-720.

Sørensen, A. M., & Christensen, M. B. (2018). Cardiovascular effects of antidiabetic drugs. *Drugs of Today (Barcelona, Spain: 1998),* 54(9), 547-559.

Spin, E. L., Mantel-Teeuwisse, A. K., and Pasmooij, A. M. G. (2024). International regulatory and publicly-funded initiatives to advance drug repurposing. *Front. Med.* 11.

Stein, S. A., Lamos, E. M., and Davis, S. N. (2013). A review of the efficacy and safety of oral antidiabetic drugs. *Expert Opin. Drug Saf.* 12, 153–175.

Steiner, S. (2016). Empagliflozin, cardiovascular outcomes, and mortality in type 2 diabetes. *Zeitschrift Für Gefässmedizin,* 13(1), 17-18.

Sun, F., Wu, S., Guo, S., Yu, K., Yang, Z., Li, L., & Zhang, Y. (2015). Effect of glucagon-like peptide-1 receptor agonists on weight loss: a systematic review and meta-analysis of randomized controlled trials. *Journal of Diabetes Research*, 2015, 157201.

Tahrani, A. A., Barnett, A. H., & Bailey, C. J. (2016). Pharmacology and therapeutic implications of current drugs for type 2 diabetes mellitus. *Nature Reviews Endocrinology,* 12(10), 566-592.

Tandon, S., & Sarkar, S. (2023). Glipizide ameliorates human poly (Q) mediated neurotoxicity by upregulating insulin signaling in Drosophila disease models. *Biochemical and Biophysical Research Communications*, 645, 88-96.

Teck, J. (2022). Diabetes-Associated Comorbidities. *Prim. Care Clin. Off. Pract.* 49, 275–286.

Thakur, A. K., Tyagi, S., and Shekhar, N. (2019). Comorbid brain disorders associated with diabetes: therapeutic potentials of prebiotics, probiotics, and herbal drugs. *Transl. Med. Commun.* 4, 12.

Theofilis, P., Sagris, M., Oikonomou, E., Antonopoulos, A. S., Siasos, G., Tsioufis, K., & Tousoulis, D. (2022). The anti-inflammatory effect of novel antidiabetic agents. *Life,* 12(11), 1829.

Thomas, M. C., & Cherney, D. Z. I. (2018). The actions of SGLT2 inhibitors on metabolism, renal function, and blood pressure. *Diabetologia,* 61, 2098-2107.

Thorens, B. (2015). GLUT2, glucose sensing, and glucose homeostasis. *Diabetologia,* 58, 221-232.

Thule, P. M., & Umpierrez, G. (2014). Sulfonylureas: A new look at old therapy. *Current Diabetes Reports,* 14, 473.

Tran, L., Zielinski, A., Roach, A. H., Jende, J. A., Householder, A. M., Cole, E. E., Atway, S. A., Amornyard, M., Accursi, M. L., Shieh, S. W., & Thompson, E. E. (2015). Pharmacologic treatment of type 2 diabetes: Oral medications. *Annals of Pharmacotherapy,* 49, 540-556.

Tripathy, D., Solis-Herrera, C., & Ryder, R. E. J. (2021). Cardioprotective effects of pioglitazone in type 2 diabetes. *Diabetes Spectrum,* 34(3), 243-247.

Valderas, J. M., Starfield, B., Sibbald, B., Salisbury, C., and Roland, M. (2009). Defining Comorbidity: Implications for understanding health and health services. *Ann. Fam. Med.* 7, 357–363.

Van Belle, T. L., Coppieters, K. T., & von Herrath, M. G. (2011). Type 1 diabetes: Etiology, immunology, and therapeutic strategies. *Physiological Reviews,* 91(1), 79-118.

Van Can, J. G., IJzerman, T. H., van der Wel, N. N., Brouwers, M. C., Blaak, E. E., & Saris, W. H. (2014). GLP-1 receptor activation modulates appetite- and reward-related brain areas in humans. *Diabetes*, 63(12), 4186-4196.

Wallia, A., & Molitch, M. E. (2014). Insulin therapy for type 2 diabetes mellitus. *JAMA,* 311(22), 2315-2325.

Wankumbu, S. C., Xiao-Man, J., Lukundo, S., & Xu, M. (2024). SGLT2 inhibitors extra glycemic potential: A possible novel approach to the prevention of atherosclerotic events. *Asian Journal of Health Research,* 3(1), 56-68.

Wilding, J. P. H., Batterham, R. L., Calanna, S., Davies, M., Van Gaal, L. F., Lingvay, I., McGowan, B. M., Rosenstock, J., Tran, M. T., Wadden, T. A., & Yamout, H. (2021). Once-weekly semaglutide in adults with overweight or obesity. *New England Journal of Medicine*, 384(11), 989-1002.

Willson, T. M., Lambert, M. H., & Kliewer, S. A. (2001). Peroxisome proliferator-activated receptor gamma and metabolic disease. *Annual Review of Biochemistry,* 70, 341-367.

Witham, M. D., Granic, A., Pearson, E., Robinson, S. M., & Sayer, A. A. (2023). Repurposing drugs for diabetes mellitus as potential pharmacological treatments for sarcopenia: A narrative review. *Drugs & Aging*, 40(8), 703-719.

Wiviott, S. D., Raz, I., Bonaca, M. P., Mosenzon, O., Kato, E. T., Cahn, A., Silverman, M. G., Zelniker, T. A., Kuder, J. F., Murphy, S. A., Bhatt, D. L., Leiter, L. A., McGuire, D. K., Wilding, J. P. H., Ruff, C. T., Gause-Nilsson, I. A. M., Fredriksson, M., Johansson, P. A., Langkilde, A.-M., & Sabatine, M. S. (2019). Dapagliflozin and cardiovascular outcomes in type 2 diabetes. *New England Journal of Medicine,* 380(4), 347-357.

Wu, H., Esteve, E., Tremaroli, V., Khan, M. T., Caesar, R., Mannerås-Holm, L., Ståhlman, M., Olsson, L. M., Serino, M., Planas-Fèlix, M., Xifra, G., Mercader, J. M., Torrents,

D., Burcelin, R., Ricart, W., Perkins, R., Fernàndez-Real, J. M., & Bäckhed, F. (2017). Metformin alters the gut microbiome of individuals with treatment-naive type 2 diabetes, contributing to the therapeutic effects of the drug. *Nature Medicine*, 23(7), 850-858.

Wulffelé, M. G., Kooy, A., Lehert, P., Bets, D., Ogterop, J. C., Borger van der Burg, B., Donker, A. J., & Stehouwer, C. D. (2004). Effects of short-term treatment with metformin on blood pressure, insulin sensitivity and endothelial function in non-obese patients with hypertension. *Journal of Internal Medicine*, 256(1), 23-29.

Yagyu, H., & Shimano, H. (2022). Treatment of diabetes mellitus has borne much fruit in the prevention of cardiovascular disease. *Journal of Diabetes Investigation,* 13(9), 1472-1488.

Yerra, V. G., & Connelly, K. A. (2024). Extrarenal benefits of SGLT2 inhibitors in the treatment of cardiomyopathy. *Physiology,* 39(6):0. doi: 10.1152/physiol.00008.2024.

Yki-Järvinen, H. (2004). Thiazolidinediones. *The New England Journal of Medicine,* 351, 1106-1118.

Zakaria, E. M., Tawfeek, W. M., Hassanin, M. H., & Hassaballah, M. Y. (2022). Cardiovascular protection by DPP-4 inhibitors in preclinical studies: An updated review of molecular mechanisms. *Naunyn-Schmiedeberg's Archives of Pharmacology,* 395(11), 1357-1372.

Zhang, Z., Zhou, L., Xie, N., Nice, E. C., Zhang, T., Cui, Y., et al. (2020). Overcoming cancer therapeutic bottleneck by drug repurposing. *Signal Transduct. Target. Ther.* 5, 113.

Zhou, Z., Zheng, M., Zuo, Z., & Wu, T. (2024). Comparison of cardiovascular outcomes of new antihyperglycemic agents in type 2 diabetes mellitus: A meta-analysis. *ESC Heart Failure,* 11(3), 1647-1656.

Zhu, T., Cui, J., and Goodarzi, M. O. (2021). Polycystic Ovary Syndrome and Risk of Type 2 Diabetes, Coronary Heart Disease, and Stroke. *Diabetes* 70, 627–637.

Zinman, B., Wanner, C., Lachin, J. M., Fitchett, D., Bluhmki, E., Hantel, S., Mattheus, M., Devins, T., Johansen, O. E., & Woerle, H. J. (2015). Empagliflozin, cardiovascular outcomes, and mortality in type 2 diabetes. *New England Journal of Medicine*, 373(22), 2117-2128.

Chapter 7

Drug Repurposing Rationale for Breast Cancer Therapy: Old Solutions to New Problems

Rohit Pal[1,*]
Gurubasavaraja Swamy Purawarga Matada[1,†]
Pronoy Kanti Das[1]
Nimmagadda Venkata Satya Sai Aishwarya[1]
Manjushree B. V.[1]
Haripriya E.[1]
S. Mounika[1]
and Ebna Azizal Omar[2]

[1]Integrated Drug Discovery Centre, Department of Pharmaceutical Chemistry, Acharya & BM Reddy College of Pharmacy, Bengaluru, Karnataka, India
[2]Centre for Excellence in Drug Analysis, Department of Pharmaceutical Analysis, Acharya & BM Reddy College of Pharmacy, Bengaluru, Karnataka, India

Abstract

Cancer is one of the top causes of mortality across the world, with over 29 million people expected to be diagnosed with any kind of cancer by 2040. Breast cancer is the most widespread kind of cancer in women globally, and it is linked with a high mortality rate. This high death rate is mostly due to the development of drug resistance, which is one of the most difficult obstacles in treating breast cancer. Moreover, the existing therapy options are extremely costly and have serious adverse effects.

---

* Corresponding Author's Email: rohitpal.rp096@gmail.com
† Corresponding Author's Email: gurubasavarajaswamy@gmail.com

In: Drug Repurposing
Editors: Shah Alam Khan and Md. Jawaid Akhtar
ISBN: 979-8-89530-422-8

Drug repurposing, which involves finding new uses for existing medications, has emerged as an innovative strategy for drug development. The repositioning of old, clinically approved, off-patent non-cancer drugs with established targets into other indications is analogous to employing old weapons in a new conflict. Advances in genomics, proteomics, and information computational biology have facilitated pharmaceutical repurposing. The repositioning strategy not only speeds up medicine development, but also results in more efficient, cheaper, and safer medications with fewer/known side effects. Over the last decade, drugs such as alkylating agents, anthracyclines, antimetabolites, CDK4/6 inhibitors, aromatase inhibitors, mTOR inhibitors, and mitotic inhibitors have been repositioned for breast cancer treatment. The repositioned drugs have shown effectiveness in the treatment of the most aggressive triple-negative breast cancer. The literature study suggests that chance plays an important role in medication development. This chapter provides a detailed summary of the present state of drug repurposing for breast cancer treatment. The techniques are shown alongside various instances of repurposed pharmaceuticals. The problems of pharmacological repurposing are also discussed.

Keywords: cancer, breast cancer, drug resistance, drug repurposing, genomics

1. Introduction

Epidemiological and clinical studies confirmed that the incidence of breast cancer continues to increase and remains the ubiquitous cancer in women globally (Liao et al., 2021). The second biggest cause of mortality for women is breast cancer. Globally, about 500,000 people die every year because of breast cancer (Ferlay et al., 2018). As per the report of WHO (World Health Organization) in 2022, 2.3 million women were diagnosed with breast cancer and 670,000 deaths occurred worldwide. All across the globe, breast cancer affects women at any age after adolescence, however, its prevalence rises with age ("Breast Cancer," 2022). According to the American Cancer Society, the United States is expected to see 3,10,720 new cases of invasive breast cancer in women and 2,800 in men, and deaths of about 42,250 women due to breast cancer in 2024 ("Key Statistics for Breast Cancer," 2024).

Breast carcinoma may be classified into three important subtypes: hormone receptor-positive, HER2-positive, and triple-negative breast cancer, depending on the positive or negative tests for molecular markers for human epidermal growth factor 2 (HER2), estrogen or progesterone receptors (ER or PR), and both. There are numerous signaling pathways involved in the development of breast cancer (Figure 1) (Britt, Cuzick, & Phillips, 2020). Twenty to twenty percent of instances of breast cancer are identified as triple-negative breast cancer (TNBC), which is the deadliest variant of the tumor (Giuli, Giuliani, Screpanti, Bellavia, & Checquolo, 2019; Hunakova et al., 2019). TNBC is distinguished by the lack of expression of estrogen receptor, progesterone receptor, and human epidermal growth factor receptor-2 (Tegowski, Fan, & Baldwin, 2019). For patients with estrogen receptor-positive breast cancer, endocrine therapy is the first line of treatment; however, since TNBC is less responsive to hormonal therapy and HER2-targeted therapy, chemotherapy is the primary treatment for TNBC patients (Waks & Winer, 2019). Furthermore, in cases of breast cancer with a high rate of recurrence and metastasis, neoadjuvant therapy which includes chemotherapy with targeted agents has been used extensively (Y Zhen et al., 2022). Sadly, the majority of patients relapsed after developing resistance to these medicines (Garcia-Martinez, Zhang, Nakata, Chan, & Morey, 2021).

Consequently, to overcome resistance to conventional therapies and recurrence, it is essential to explore new efficacious therapeutic methods or targets for diverse malignancies and therapy-resistant tumors. Still, developing a new drug or therapeutic is a tedious process that takes a long time, roughly ten to fifteen years, and is very expensive (Hanusova, Skalova, Kralova, & Matouskova, 2015). Additionally, because of its poor success rate, it is very challenging to discover and develop new agents. As such, a novel approach is needed to shorten the time and expense associated with discovering novel medications (Bayraktar et al., 2012; Choi, Kim, et al., 2021).

The scientific community and pharmaceutical industries have recently shown great interest in drugcombinations and drug repurposing techniques, as alternative treatment option for cancer. Drug repurposing, also known as drug repositioning, is an alternative to de novo drug development and may speed cancer patients' access to innovative treatments (Pantziarka, Sukhatme, Meheus, Sukhatme, & Bouche, 2017). The term "repurposed drug" refers to the process of discovering new therapeutic indications for drugs that have already received Food and Drug Administration (FDA) approval for clinical use (Antoszczak, Markowska, Markowska, & Huczyński, 2020). This approach has several advantages over developing new drugs, including the fact that repurposed drugs have well-defined toxicological and pharmacokinetic profiles (Verbaanderd, Meheus, Huys, & Pantziarka, 2017), which leads to a more cost-effective and quicker translational process and typically increases the likelihood that the drug will advance to clinical trials (Mohammad Jafari et al., 2018). In addition, most repurposed drugs are generic medications that are more reasonably priced than newly patented drugs. Repurposing of clinically approved or investigational drugs could be one type of effective and safe option for treating cancer patients. Several repurposed drugs have been successfully used alone or with other drugs for managing and treating cancers (Duarte, Guerreiro, & Vale, 2022). For example, metformin is an antihyperglycemic agent which is commonly used for the treatment of type-II diabetes. In the recent past, it has been demonstrated that metformin exhibits antiproliferative activity by reducing the level of N6-methyladenosin in breast cancer. Additionally, it may overcome the resistance to gemcitabine by arresting the progression of the epithelial-mesenchymal transition in pancreatic cancer by modifying the DNA (Deoxyribonucleic acid) methylation of miR-663 (microRNA-663). As a result, it seems that taking advantage of metformin's epigenetic effects might increase the drug's potential clinical use (Huang et al., 2021).

Auranofin is an antirheumatic agent selected for drug repositioning in cancer therapy. As per the results, Auranofin mainly attacks on thioredoxin reductase (TrxR) enzyme to produce anticancer activity. TrxR plays a crucial role in the thioredoxin (Trx) system, which mediates the reduction of disulfide bonds and allows Trx and NADPH (nicotinamide adenine dinucleotide phosphate hydrogen) to contribute to redox control. Auranofin could provoke cytotoxicity in cancerous cells by interfering with protein homeostasis (proteostasis). About 4~5 μM Auranofin elicits paraptosis in breast cancer cells by simultaneously targeting TrxR1 and the proteasome (Seo et al., 2023).

Loperamide acts on μ-opioid receptors to decline the GIT (gastrointestinal tract) motility which is helpful in the treatment of diarrhea. Loperamide also exerts inhibition activity against voltage-dependent Ca^{2+} channels, calcium influx, and calmodulin activity. Recently, the scientific community demonstrated that loperamide downregulates the anti-apoptosis protein Mcl-1, which causes apoptosis and DNA damage in leukemia cells. Furthermore, it was shown that loperamide causes a rise in the net uptake and a fall in the efflux of doxorubicin (DOX), suggesting that it may counteract the MCF-7/MDR1(Michigan Cancer Foundation-7/ multidrug resistance protein 1) cells' resistance to DOX (Elia, Al-Karmalawy, Nasr, & Elshal, 2022).

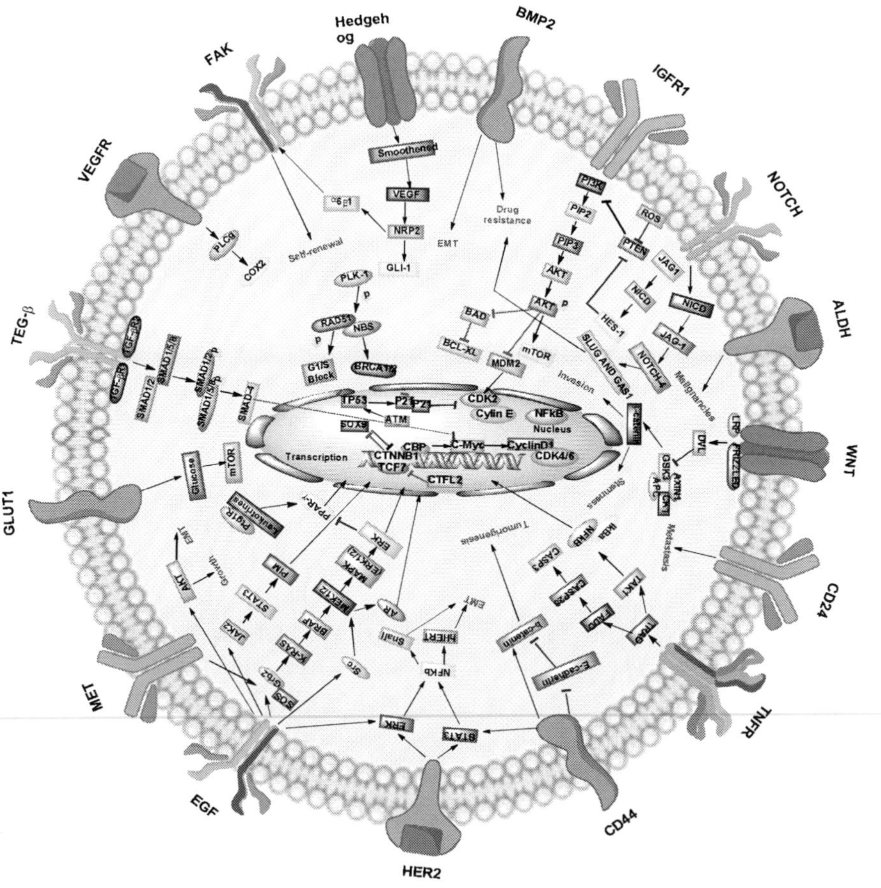

Figure 1. Signalling pathways involved in the occurrence of breast cancer.

Niclosamide is used to treat helminthic disease and was recently repurposed for the inhibition of several cancers namely breast, colorectal, lung, ovarian, etc. by downregulating the Wnt signaling pathway. Another anthelmintic drug, flubendazole also repurposed as an anticancer agent. By targeting EVA1A (eva-1 homolog A), flubendazole may influence autophagy and apoptosis, hence influencing TNBC migration and multiplication (Lohiya & Katti, 2021). Another example of repurposing drugs such as pirlindol was first licensed to treat psychotic illnesses and persistent depression, but it is now being used to treat multiple sclerosis (Shim & Liu, 2014). Originally, thalidomide was intended as an antiemetic drug but now it is used in the treatment of myeloma. Aspirin is now used to treat cardiovascular disorders, colorectal cancer, and pain but originally, it was designed as an analgesic and antipyretic agent (Banno et al., 2015; Lekka, Deftereos, Persidis, Persidis, & Andronis, 2011). As a result, repurposing medications for cancer treatment is less difficult in terms of translation, making it a viable option for treating breast cancer. Table 1 represents some of the important list of repurposed drugs for breast cancer under clinical trials.

2. Repositioning of Drugs for Breast Cancer

2.1. Repositioning of Drug Acting on CNS for Breast Cancer

Duarte et al. studied and reported the synergistic effect of CNS drugs (edaravone and quetiapine) and anti-malarial drugs combined with anticancer drugs against MCF-7 cancer cells. Edaravone showed noteworthy anti-tumor activity at every concentration (0.01–100 μM), similarly, quetiapine also showed good activity at a concentration of 100 μM. Further, the combination activity of anti-cancer drugs (DOX and paclitaxel) with edaravone and quetiapine revealed that DOX+edaravone showed similar activity as shown in the case of DOX alone at all the tested concentrations (0.01-100 μM) of edaravone with IC_{50} of DOX. Additionally, DOX (IC_{50}) +quetiapine (0.1 μM, 1 μM, 10 μM, and 100 μM) showed potential anti-breast cancer activity against the MCF-7 cell line. With paclitaxel (PTX) no significant activity was found for both drugs against MCF-7. The synergistic effect of both CNS drugs with DOX and PTX was observed. In the second stage of evaluation, the combinations of two CNS drugs (fluphenazine and sertraline), and two antimalarial drugs (mefloquine and pyronaridine) were evaluated. It was found that mefloquine (fixed) + pyronaridine (varying concentration) was not preferable when compared with the anti-cancer activity of mefloquine (> 80% cell viability) alone. The combination index (CI) for this combination indicated that 10 μM and 50 μM of pyronaridine had lower CI values (0.205 μM and 0.664 μM). Results also revealed that mefloquine (0.01 μM, 0.1 μM, and 1 μM) + pyronaridine (fixed) had moderate activity against MCF-7 cells but was not preferable for combination therapy. (Duarte, Guerreiro, et al., 2022).

Table 1. Clinical trial data for the repositioning of various drugs against the treatment of breast cancer

S. No.	Study Title	ClinicalTrials.gov Identifier	Intervention	Status
1	Bicalutamide to treat Patients With Metastatic Breast Cancer [clinicaltrials.gov].	NCT00468715	Drug: bicalutamide Other: diagnostic laboratory biomarker analysis; immunohistochemistry staining method	Phase 2
2	Palbociclib in Combination With Bicalutamide for the Treatment of AR(+) Metastatic Breast Cancer (MBC) [clinicaltrials.gov]	NCT02605486	Drug: Palbociclib Drug: Bicalutamide	Phase 1 Phase 2
3	Safety and Efficacy Study of Enzalutamide in Patients With Advanced, Androgen Receptor-Positive, Triple Negative Breast Cancer [clinicaltrials.gov]	NCT01889238	Drug: Enzalutamide	Phase 2
4	Enzalutamide and Paclitaxel Before Surgery in Treating Patients With Stage I-III Androgen Receptor-Positive Triple-Negative Breast Cancer [clinicaltrials.gov]	NCT02689427	Procedure: Axillary Lymph Node Dissection Drug: Enzalutamide Other: Laboratory Biomarker Analysis Procedure: Lymph Node Biopsy Drug: Paclitaxel Procedure: Therapeutic Conventional Surgery	Phase 2
5	Abiraterone Acetate in Molecular Apocrine Breast Cancer (AMA) [clinicaltrials.gov]	NCT01842321	Drug: Abiraterone Acetate	Phase 2
6	Neoadjuvant Zoledronate and Atorvastatin in Triple Negative Breast Cancer (YAPPETIZER) [clinicaltrials.gov]	NCT03358017	Drug: Zoledronate Drug: Atorvastatin 80mg Drug: Standard neoadjuvant cht	Phase 2
7	Zoledronic Acid as Adjuvant Treatment of Triple-negative Breast Cancer [clinicaltrials.gov]	NCT02595138	Drug: Zoledronic acid	Phase 3

Table 1. Continued

S. No.	Study Title	ClinicalTrials.gov Identifier	Intervention	Status
8	Study on Bisphosphonates Targeting Triple-negative Breast Cancer [34]	NCT04045522	Drug: Zoledronic Acid	Observational
9	L-NMMA Plus Taxane Chemotherapy in Refractory Locally Advanced or Metastatic Triple Negative Breast Cancer Patients [clinicaltrials.gov]	NCT02834403	Drug: L-NMMA Drug: Docetaxel Drug: Amlodipine Drug: Pegfilgrastim Drug: Enteric-coated aspirin	Phase 1 Phase 2
10	Orteronel as Monotherapy in Patients With Metastatic Breast Cancer (MBC) That Expresses the Androgen Receptor (AR) [clinicaltrials.gov]	NCT01990209	Drug: Orteronel	Phase 2
11	CYP17 Lyase and Androgen Receptor Inhibitor Treatment With Seviteronel Trial (INO-VT-464-006; NCT02580448) (CLARITY-01) [clinicaltrials.gov]	NCT02580448	Drug: Seviteronel	Phase 1 Phase 2
12	Oral VT-464 in Patients With Castration-Resistant Prostate Cancer Previously Treated With Enzalutamide, Androgen Receptor Positive Triple-Negative Breast Cancer Patients, and Men With ER Positive Breast Cancer [clinicaltrials.gov]	NCT02130700	Drug: VT-464: given orally twice daily in 28-day cycles Drug: VT-464: given orally once daily in 28-day cycles	Phase 2
13	Chloroquine With Taxane Chemotherapy for Advanced or Metastatic Breast Cancer After Anthracycline Failure (CAT) (CAT) [clinicaltrials.gov]	NCT01446016	Drug: Paclitaxel Drug: Docetaxel Drug: Abraxane Drug: Ixabepilone	Phase 2
14	Celecoxib in Preventing Breast Cancer in At-Risk Premenopausal Women [clinicaltrials.gov]	NCT00056082	Drug: Celecoxib	Phase 2

S. No.	Study Title	ClinicalTrials.gov Identifier	Intervention	Status
15	Pre-Surgical Trial of the Combination of Metformin and Atorvastatin in Newly Diagnosed Operable Breast Cancer [clinicaltrials.gov]	NCT01980823	Drug: Metformin Drug: Atorvastatin Procedure: Breast surgery	Early Phase 1
16	Tamoxifen Citrate, Letrozole, Anastrozole, or Exemestane With or Without Chemotherapy in Treating Patients With Invasive RxPONDER Breast Cancer [clinicaltrials.gov]	NCT01272037	Drug: Anastrozole Drug: Exemestane Other: Laboratory Biomarker Analysis Drug: Letrozole Other: Quality-of-Life Assessment Drug: Systemic Chemotherapy Drug: Tamoxifen Citrate	Phase 3

Timilsina and coworkers in 2022 demonstrated that the antidepressant agent – imipramine is an anti-breast cancer drug that acts by triggering signals of estrogen receptors and DNA repair events. The activity of imipramine was evaluated on breast cancer cell lines (MDA-MB-231, MDA-MB-468, MCF-7, and ZR75). It was determined that imipramine exhibited lower or similar IC_{50} values toward breast cancer cells than non-small cell lung cancer (NSCLC), small cell lung cancer (SCLC), and prostate cancer. Furthermore, the efficacy of imipramine was estimated on an orthotopic xenograft model by using MDA-MB-231 cells, and found that it could potentially decrease the levels of Ki-67 (cell proliferation marker) than vehicle controls. This was further confirmed through the 4T1 isograft model and it was concluded that imipramine was a potent derivative against breast cancer (Timilsina et al., 2022).

Honeybee venom was used synergistically to increase the cytotoxic activity of anticancer drug (DOX) and CNS drugs (fluphenazine, fluoxetine, sertraline, and thioridazine) by Duarte and coworkers in 2022. The results of cytotoxic activity revealed that honeybee venom reduced cell viability and cellular protein content of MCF-7 breast cancer cells. The cytotoxic activity of DOX and CNS Drugs in MCF-7 cells demonstrated that DOX ($IC_{50} = 0.17$ μM) had better anticancer activity than the CNS drugs (IC_{50}=2.68, 7.78, 5.72, and 2.22 μM). The MCF 7 breast cancer cells were further treated with combinations of DOX and CNS drugs plus increasing concentrations of honeybee venom. There was no significant difference in the anticancer activity of the combination of DOX with bee venom when compared to DOX alone. A combination of fluphenazine, fluoxetine, and sertraline with 6.25 g/mL of honeybee venom demonstrated cell viability reduction in comparison with both drugs alone. The combination of the repurposed drugs plus honeybee venom led to better anti-cancer activity over MCF-7 cells than with DOX, at lower concentrations (Duarte, Falcão, El Mehdi, Vilas-Boas, & Vale, 2022).

In the year 2021, fluvoxamine, an anti-depressant drug was repurposed in combination with tivozanib a VEGFR-TK inhibitor to target triple-negative breast cancer by Kumar and colleagues. The migration assay was performed by using MDA MB-231 cells. The results showed that inhibition with combination therapy was noteworthy compared to individual therapy. The migration analysis was also conducted on MCF-7 cells and there was a prominent decrease in cellular migration with combination treatment. (Kumar, Raza, & Sehrawat, 2021).

Badran et al., in 2020 reported aripiprazole an antipsychotic drug as a potent one against breast cancer. The cytotoxicity of the drug was tested on the MCF-7 breast cancer cell line. Aripiprazole possessed good inhibitory activity with an IC_{50} of 12.1 μM towards MCF-7 cells whereas a reference drug DOX showed an IC_{50} value of 1.25 μM. The growth inhibition activity results demonstrated that aripiprazole showed the strongest inhibition by 90% at a concentration of 50 μM. The cellular viability study results revealed that aripiprazole increased cell viability by 49.1% and 66.7% at 35 μM and 100 μM respectively. Flow cytometry study data revealed that aripiprazole caused apoptosis over the MCF-7 cell line and upregulated the number of cells in the necrotic stage by 4.12%, 8.14%, and 16.18% at 12 μM, 24 μM, and 48 μM respectively. The drug was also active in other breast cancer cells such as MDA-MB-231, AU565, and BT-474 (IC_{50} =19.83, 18.02, and 36.42 μM) (Badran et al., 2020).

In 2020 Song et al., reported an antipsychotic drug-thioridazine hydrochloride potentially inhibits triple-negative breast cancer cell progression. The anti-tumor activity was evaluated on 4T1 and MDA-MB-231 cell lines for 72 hours where Thi-hyd possessed good anti-tumor activity with IC_{50} values of 9.87 and 18.7 μM, accordingly. The colony formation assay was performed to assess the viability of the cell, while thioridazine hydrochloride treatment significantly decreased the number as well as colony size. The flow cytometry assay results revealed that thi-hyd could potentially cause G0/G1 arrest by suppressing CDK4/cyclin D1 expression levels but not CDK2/cyclin E. Additionally, p21 and p27 were increased in MDA-MB-231 cells whereas not for 4T1 cells. This in turn promoted cell apoptosis followed by MMP dysfunction that was represented in terms of decrease in Rh123 fluorescence average intensity. Thi-hyd-treated cells at 5, 10, and 15 μM decreased relative fluorescence than control by 5.49%, 11.69%, and 26.89%, accordingly. Moreover, the *in-vivo* anti-tumor activity of the drug was tested on mice with lung metastasis and found that Thi-hyd effectively inhibited tumor growth by 63.73% and prevented metastasis of the lung by 72.58% in the subcutaneous 4T1-xenograft model (Song et al., 2020).

2.2. Repositioning of Antimicrobial Drugs for Breast Cancer

Jabber et al., 2022 investigated the synergistic activity of oxytetracycline with carboplatin against breast cancer (MCF-7) cells. Antiproliferative activity was evaluated using oxytetracycline and results revealed that oxytetracycline successfully inhibited cell growth with IC_{50} values of 64.4 μM. Further, a synergistic effect was evaluated between oxytetracycline and carboplatin over the MCF-7 cell line. The *in-vitro* results demonstrated that 5% and 10% concentrations of oxytetracycline along with carboplatin (dose IC_{50} = 36.41 μM) showed potential activity with dose IC_{50} values of 13.94 and 15.62 μM respectively. Moreover, the cell viability study using the trypan blue method was examined and the result showed more cell viability in the case of combination therapy (Jabber et al., 2022).

Zhen et al., in 2022 identified flubendazole for breast cancer therapy *via* inducing damage or dysfunction of mitochondria. The effect of flubendazole on mPTP (mitochondrial permeability transition pore) opening was determined in MCF-7 and MDA-MB-231 cell lines. Through the results, it was found that the drug effectively impairs the mitochondrial outer membrane permeability and promotes dysfunction of mitochondria in the breast cancer cells, where MCF-7 and MDA-MB-231 cells, possessed decreased ATP levels, elevated superoxide and in turn decreased the levels of expression in mitochondrial components including SOD2, VDAC1, COX IV, and TOM20 (Yongqi Zhen et al., 2022).

Lohiya et al., 2021 examined and reported a synergistic effect of Niclosamide (Nic) and DOX as a combination therapy for breast cancer treatment. Effects of sequential treatment of both drugs were examined over the breast cancer cell line. The cytotoxicity assay against MDA-MB-231, SKBR3, and MCF-7 cell lines revealed that, at a high dose, both the drug niclosamide (IC_{50}= 4.23 μM, 1.09 μM, and 1.45 μM respectively) and DOX (IC_{50}=2.1 μM, 0.95 μM, and 1.79 μM respectively) caused maximum cell death. Sequential combination therapy (with niclosamide and DOX) results demonstrated that the maximum pair of niclosamide and DOX caused 50% apoptosis over MDA-MB-231 and MCF-7 cell lines and 40% apoptosis for the SKBR3 cell line. Further, a cytotoxicity assay was performed against MDA-MB-231, SKBR3, and MCF-7 cell lines and results revealed that at a high dose, both the drug niclosamide (IC_{50} = 1.24 μM, 0.44 μM, and 0.29 μM respectively) and DOX (IC_{50}= 0.64 μM, 0.31 μM, and 0.48 μM respectively) caused maximum cell death (Lohiya & Katti, 2021).

Choi et al., 2021 examined and concluded the effectiveness of benzimidazole analogs mainly mebendazole over TNBC. The colony formation assay was used for six drugs (Albendazole (ALB), Mebendazole (MBZ), Albendazole-sulfoxide (ALB-SUL), Flubendazole (FLU), Fenbendazole (FBZ), and Oxfendazole (OFZ) and the results revealed that four drugs (MBZ, ALB, FBZ, and FLU) showed potential activity against RT-R-MDA-MB-231, MCF-7, and T47D cell lines at 0.1 to 1 μM. *In-vivo* study results found that all four drugs had not caused hepatic damage, kidney toxicity, and body weight loss. In the case of the 4T1 allograft mice model, FBZ and MBZ downregulated the tumor volume, moreover, FLU and MBZ reduced the lung metastasis. Whereas in the RT-R-4T1 model, FBZ and MBZ downregulated the volume of the tumor but only MBZ reduced lung metastasis. The mechanistic study result demonstrated that MBZ nearly healed the wound in Scratch-wounded RT-R-MDA-MB-231 and MDA-MB-231 cell lines at a concentration of 0.5 μM and 1 μM. Furthermore, MBZ (0.5 μM and 1 μM) dramatically decreased the spread of both MDA-MB-231 and RT-RMDA-MB-231 cells via the EC-Matrigel-coated insert wall membrane. The cell cycle progression assay results revealed that MBZ caused arrest at the G2/M phase over MDA-MB-231 and RT-R-MDA-MB-231 cells. MBZ showed tubulin inhibitory activity in the pellet at 0.5 μM and 1 μM but was not effective in the supernatant. Additionally, MBZ enhanced the production of cyclin B1 at 0.5 and 1 μM but not cyclin D1 after 24h. The examined data revealed that MBZ cleaved caspase-3 and upregulated the level of γH2AX over MDA-MB-231 and RT-R-MDA-MB-231 cells at 0.5 μM and 1 μM. They evaluated the activity of MBZ on CSC markers and demonstrated that MBZ potentially reduced the level of CD44, OCT3/4, and ESM-1 (Choi, Ko, et al., 2021).

2.3. Repositing of the Anti-Diabetes Drugs for Breast Cancer

In the year 2021 Chequin et al., investigated the anti-breast cancer activity of the liraglutide, a drug used for type II diabetics. The colony formation assay demonstrated that liraglutide successfully inhibited the colony formation in every cell. Further, a wound healing study revealed that liraglutide had good migration inhibitory activity with a reduced migration rate of 73% and 38.9% over MDA-MB-231 and MDA-MB-436 cells respectively. To confirm the activity of the liraglutide, they investigated the individual cell migration in MDA-MB-436 using time-lapse microscopy. After liraglutide therapy, most migratory (quartile ≥25%) MDA-MB-436 cells moved 23% less than control cells, with a median difference of 60 μM ($p < 0.05$). The *in-vivo* activity (in Ehrlich mice) data revealed that Liraglutide increased the efficacy of PTX and methotrexate and reduced the chemotherapeutic dose count. In-vivo methylation assay results indicated that liraglutide therapy efficiently reached tumours and demethylated the ADAM33 gene, increasing gene and protein expression (Chequin et al., 2021).

In 2021, Huang and the group identified the non-oncological drug metformin as an anti-breast cancer agent by modulating epigenetic regulators. The antiproliferative activity was evaluated using metformin on MCF-7 with various concentrations of 0, 5, 10, 20, and 40 mmol/L and results concluded that high-concentration metformin downregulated optical density than lower concentration from 24 to 96 h. In addition, the flow cytometry study data showed that metformin upregulated the level of Bax protein expression and HRK, TNFRSF10A, and TNFRSF10B gene expression at higher concentrations (10 mmol/L and 20 mmol/L) over MCF-7 cell. Metformin caused cell cycle arrest at the G0/G1 phase by 48.62%, 58.25%, and 74.74% at a concentration of 5, 10, and 20 mmol/L respectively. According to the Western blotting and qPCR assay revealed that Metformin elevated the levels of p21 (at 20 mmol/L) and further p21 downregulated G1-CDK1. The results suggested that treatment with metformin markedly reduced the level of CDK1 and also reduced the expression of cyclin B1, cyclin B2, and cyclin D1 dose-dependently (Huang et al., 2021).

Eliaa et al., in 2020 evaluated empagliflozin and DOX against triple-negative breast cancer cells survival synergistically. DOX, empagliflozin, and DOX+empagliflozin were subjected to cytotoxic studies against the MDA-MB-231 cell line and the results revealed that DOX and DOX+empagliflozin significantly inhibited the growth of MDA-MB-231 cells with the IC_{50} value of 1.23, and 1.70µM, respectively. DOX, empagliflozin, and DOX+empagliflozin were also tested for cell cycle arrest of MDA-MB-231 cells by flow cytometry and concluded that DOX and DOX+empagliflozin diminished the G2/M phase cell population by 2.41% and 1.69% and empagliflozin significantly diminished the S phase cell population by 35.18% (Eliaa, Al-Karmalawy, Saleh, & Elshal, 2020).

2.4. Repositing of the Antimalarial Drugs for Breast Cancer

In continuation to the previous work, Duarte et al., in 2022 demonstrated the combination effect of antimalarial and CNS drugs for breast cancer. The cytotoxic effect of antineoplastic drugs (DOX, PTX, and 5-Fluorouracil) revealed that DOX and PTX showed significant cytotoxic activity against MCF-7 cells with IC_{50} values of 0.17 and 0.44µM respectively. DOX, PTX, and 5-Fluorouracil showed non-toxic characteristics against MRC-5 cells with IC_{50} values of 0.43µM, 2.61µM, and 5.78µM respectively. Next, the cytotoxic effect of antimalarial drugs (artesunate, chloroquine) against MCF-7 and

MRC-5 cell lines revealed that artesunate and chloroquine showed a cytotoxic effect against MCF-7 cells with IC_{50} values of 11.60μM and 63.98μM respectively. They found that artesunate and chloroquine had potent cytotoxic effects against MRC-5 cells with IC_{50} values of 14.70μM and 14.85μM respectively. Further, the cytotoxic effect of CNS drugs (fluoxetine, fluphenazine, benztropine, thioridazine, and sertraline) concluded that benztropine showed a significant cytotoxic effect against MCF-7 and MRC-5 cells with IC_{50} values of 21.71μM and 19.91μM respectively. The cytotoxic effect of DOX combined with artesunate and chloroquine was evaluated and results showed that both combinations had potent cytotoxicity against MRC-5 cells than MCF-7 cells. The combination of PTX with fluoxetine, fluphenazine, and benztropine was tested for anti-cancer effects and the results suggested that PTX+benztropine showed significant cytotoxic effect against MRC-5 and MCF-7 cells at low concentrations (Duarte, Nunes, Ricardo, & Vale, 2022).

Duarte et al., 2020 evaluated the anti-breast cancer activity of several anti-malarial drugs in combination with anti-cancer drugs. They evaluated the anti-cancer activity of DOX and PTX over the MCF-7 cell lines, and the result revealed that DOX had substantial action at doses over 0.1 μM. Several quinoline derivative anti-malarial agents (chloroquine, mefloquine, tafenoquine, pyronaridine, primaquine, sitamaquine, and 6-methoxy-nitroquinoline) were evaluated for cellular viability assay over the MCF-7 cell line at a concentration of 1-100 μM for 48 h. The result revealed that in between all compounds chloroquine and mefloquine showed better activity. The MTT test revealed that mefloquine had a considerable cytotoxic impact, even at concentrations as low as 1 μM. Higher doses resulted in over 50% cell loss. After that several non-quinolinic analogs (artesunate, cycloguanil, atovaquone, and lumefantrine) were evaluated for anti-breast cancer activity over the MCF-7 cell line. Artesunate treatment significantly reduced the lifespan of cells at doses over 10 μM, resulting in over 50% decrease at 50 μM or above. Cytotoxicity assay showed the activity of DOX, PTX, Mefloquine, Tafenoquine, Primaquine, Pyronaridine, Artesunate, and Cycloguanil with IC_{50} values of 0.1699, 0.003, 1.241, 2.595, 29.90, 1.388, 11.60, and 20.30 μM respectively. They evaluated the combined effect of antimalarial drugs with DOX and results found that DOX + mefloquine and DOX + chloroquine were the most effective combinations for IC_{50} levels, consistent with individual medication findings. Combining DOX with artesunate, pyronaridine, or chloroquine at doses below the IC_{50} of individuals resulted in significantly lower cell survival compared to DOX alone. At lower concentrations, the antimalarial drugs artesunate, mefloquine, and cycloguanil had the most powerful action when combined with PTX. Chloroquine displayed a more synergic effect combined with DOX than PTX. DOX (0.05 μM, 0.1 μM, 0.2 μM) and chloroquine (16 μM, 32 μM, 64 μM) generated synergic effects ranging from 0.724 to 0.851. PTX (1.5 μM, 3 μM) and chloroquine (32 μM, 64 μM) resulted in synergy, with Fa values of 0.782 and 0.831. Combining mefloquine (8 μM) with DOX (0.2 μM) and PTX (3 nM) resulted in a synergistic effect with a fractional effect of 0.845 and 0.841 respectively. DOX (0.05 μM, 0.1 μM, 0.2 μM, 0.4 μM) and artesunate (3 μM, 6 μM, 12 μM, 24 μM) generated synergic effects ranging from 0.487 to 0.821. PTX (1.5 nM, 3 nM) and artesunate (6 μM, 12 μM) resulted in synergy, with Fa values of 0.453 and 0.544 (Duarte & Vale, 2020).

2.5. Repositing of Anti-Inflammatory Drugs as Anti-Breast Cancer Drugs

In 2023, Seo et al. evaluated the paraptosis activity of auranofin in breast carcinoma cells *via* inhibition of both the proteasome and thioredoxin reductase. The antiproliferative activity over MDA-MB-435 S, MDA-MB-231, and BT549 cell lines revealed that auranofin showed excellent activity with IC_{50} values of 4.71 μM, 4.85 μM, and 4.17 μM, respectively. In the flow cytometry, at the concentration of 5 μM auranofin showed a slight raised in the ROS levels at 2–4 h and 24 h. Conversely, 2 μM AF raised ROS levels by about 2.0 fold, and 2 μM AF with 5 nM Bz amplified the rise to around 2.3 fold after 24h (Seo et al., 2023).

Li et al., 2020 utilized aspirin to treat breast cancer and defeat acquired resistance to individual treatment. The anti-proliferative activity of aspirin demonstrated that it caused MCF-7 cell death by inhibiting the PARP or caspase-3 enzyme. Flow cytometry analysis data revealed that aspirin upregulated the level of breast cancer cell apoptosis. Further, the combination treatment of aspirin with tamoxifen against the MCF-7 cell line, demonstrated that there was a significant reduction in the development of cancer cells. However, the growth of breast epithelial cells (MCF-10A) was unaffected by monotherapy with aspirin and tamoxifen, or by the combination therapy with aspirin and tamoxifen. Aspirin was then evaluated against resistance cells (MCF-7TR), breast cancer cells (MCF-7), and MCF-10A. Results revealed that aspirin showed the highest activity against MCF-7TR than MCF-7 and more activity over MCF-7 than MCF-7A (Li, Hu, Wang, Chen, & Xu, 2020).

2.6. Repositing of the Anti-Cholesterol Drugs as Anti-Breast Cancer Drugs

Recently, Korknaz and the group utilized computational studies for the repositing of ER inhibitors. The molecular docking study was done for about 2048 FDA-approved drugs including active metabolites to know the interaction among different compounds and Erα (PDB ID: 3ERT). Among all the compounds, 50 compounds exhibited good binding scores lesser than -10 kcal/mol which indicated that compounds showed excellent binding affinity towards ERα. Over 50 compounds, the top three hit compounds such as lomitapide, conivaptan, and atonement possessed the strongest binding affinity with docking scores of -13.1, -11.7, and -11.2 kcal/mol, accordingly. Tamoxifen, an ER inhibitor also showed a good binding affinity with a -10 kcal/mol docking score (Düzgün & Korkmaz, 2024).

Bhardwaj et al., 2021 determined the efficacy of fluvastatin and aspirin for breast cancer prevention. They investigated the inhibition activity of mammary tumors in the SV40C3 TAg mice model with fluvastatin and fluvastatin+aspirin. The results revealed that fluvastatin (alone) caused inhibition by 50%, but combination therapy did not show any effect. The growth inhibition assay revealed that fluvastatin showed potential apoptotic activity in mice model (5.36%; $p=0.04$) where no effectiveness was showed by aspirin (alone) and combination drug (Bhardwaj, Embury, Rojo, Albarracin, & Bedrosian, 2021).

2.7. Repositing of the Anti-Diarrheal Drugs for Anti-Breast Cancer

Elia and co-authors explored the Loperamide, to potentiate DOX sensitivity in triple-negative breast cancer cells. Combining both drugs caused G1 arrest and strengthened apoptosis more than DOX as a single treatment. Both the drugs Loperamide and DXR showed downregulated the mRNA of JNK1, MDR1, mTOR, and Bcl-2 i.e, 1.35, 2.5, 6.6, 9.5 folds respectively, and upregulated p21 gene which is 8 folds times when compared to DOX alone. The IC_{50} values for a single dose were 1.26μM for DOX and 12.26 μM for loperamide over 48 hours. Loperamide could be used to reduce DXR therapeutic doses based on the dose reduction index (DRI), reducing side effects (Elia et al., 2022).

Table 2. Summary of various repositioned drugs as anti-breast cancer

Sl. No.	Category	Structure (Name)	Summary
1.	CNS drug	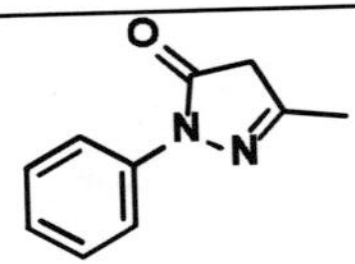 Edaravone	MTT and morphological assay results revealed that DOX (IC_{50})+edaravone (0.01-100 μM) showed similar activity as shown in case of DOX alone against the MCF-7 cell line. They performed anticancer activity and edaravone showed less anti-tumor activity at every concentration (0.01–100 μM) [Duarte et al., 2022].
2.		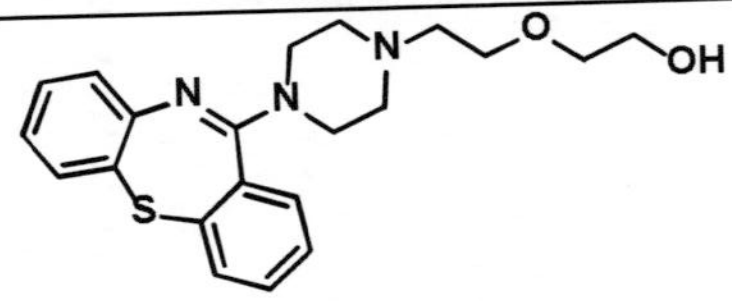 Quetiapine	The anticancer activity of quetiapine showed good anti-tumor activity at every concentration (0.01–100 μM). MTT and morphological assay results revealed that DOX (IC_{50})+quetiapine (0.1 μM, 1 μM, 10 μM, and 100 μM) showed potential anti-breast cancer activity against the MCF-7 cell line [Duarte et al., 2022].

Sl. No.	Category	Structure (Name)	Summary
3.	CNS drug	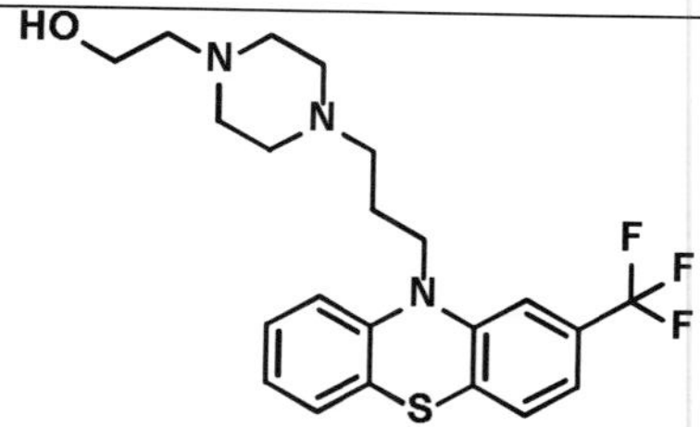 Fluphenazine	Fluphenazine (0.01 µM to 100 µM) + sertraline (fixed) and also fluphenazine (fixed) + sertraline (0.01 µM to 100 µM) showed considerable activity against the MCF-7 cell line. The CI values and the corresponding partial effects of sertraline (fixed) + fluphenazine (10 µM), and fluphenazine (fixed) + sertraline (10 µM and 50 µM) showed a synergistic effect [Duarte et al., 2022]. A combination of fluphenazine, fluoxetine, sertraline and 6.25 g/mL of honeybee venom demonstrated cell viability reduction in comparison with both drugs alone. The combination of MCF-7 cells with the repurposed drugs plus honeybee venom led to good anti-cancer efficacies than with DOX, at lower concentrations [Duarte et al., 2022]. PTX+fluphenazine upregulated PPT1 expression by 26-50% and increased P-gp expression in MCF-7 cells by 26-50% [Duarte et al., 2022].
4.		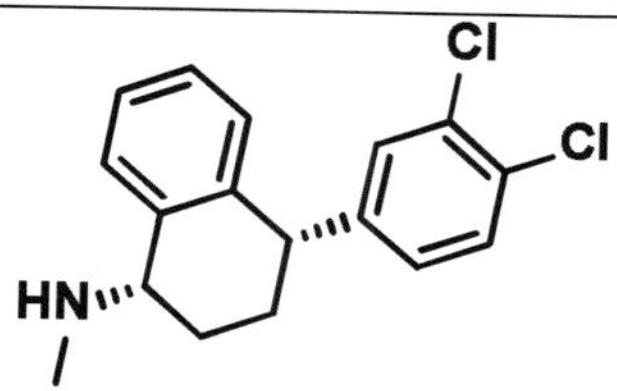 sertraline	Fluphenazine (0.01 µM to 100 µM) + sertraline (fixed) and also fluphenazine (fixed) + sertraline (0.01 µM to 100 µM) showed considerable activity against the MCF-7 cell line. The CI values and the corresponding partial effects of sertraline (fixed) + fluphenazine (10 µM), and fluphenazine (fixed) + sertraline (10 µM and 50 µM) showed a synergistic effect [Duarte et al., 2022]. A combination of fluphenazine, fluoxetine, sertraline and 6.25 g/mL of honeybee venom demonstrated cell viability reduction in comparison with both drugs alone. The combination of MCF-7 cells with the repurposed drugs plus honeybee venom led to good anti-cancer efficacies than with DOX, at lower concentrations [Duarte et al., 2022].

Table 2. Continued

Sl. No.	Category	Structure (Name)	Summary
5.	CNS drug	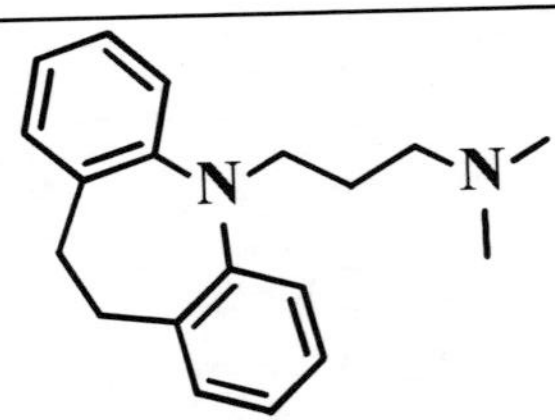 Imipramine	Imipramine decreased the levels of certain proteins that caused the transition of the G1/S phase, replication of DNA, and induced apoptosis [Timilsina et al., 2022].
6.		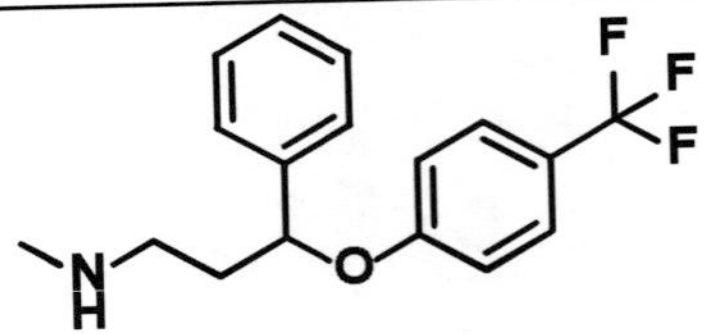 Fluoxetine	Fluoxetine with 6.25 g/mL of honeybee venom demonstrated good anti-cancer activity over MCF-7 cell [Duarte et al., 2022]. PTX+fluoxetine downregulated positive cells by 26-50% and decreased P-gp expression by 26-50% in MCF-7 [Duarte et al., 2022].
7.		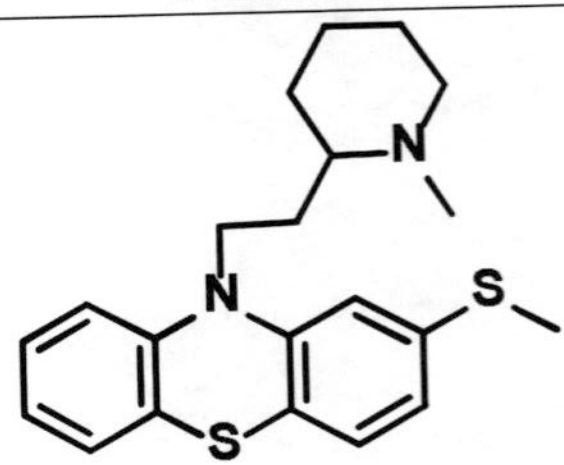 Thioridazine	The cytotoxic effect against MCF-7 cells demonstrated that thioridazine showed activity with an IC_{50} value of 2.22 μM [Duarte et al., 2022]. Thi-hyd possessed good anti-tumor activity with IC_{50} values of 9.87 and 18.7 μM, accordingly over 4T1 and MDA-MB-231 cells. The flow cytometry assay results revealed that thi-hyd could potentially cause G0/G1 phase arrest. Thi-hyd-treated cells at 5, 10, and 15 μM decreased relative fluorescence than control by 5.49%, 11.69%, and 26.89%, accordingly [Song et al., 2020].

Sl. No.	Category	Structure (Name)	Summary
8.	CNS drug	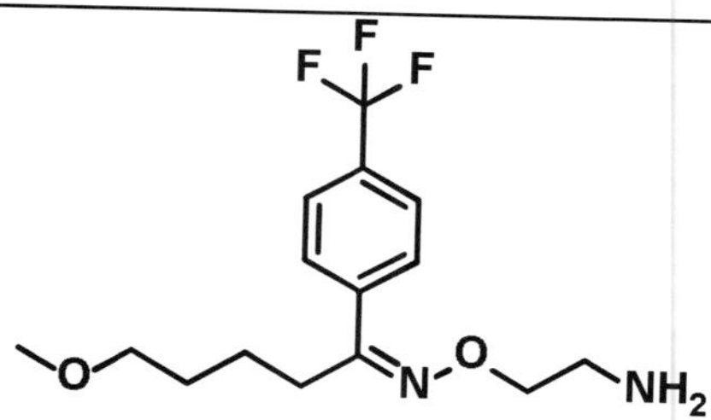 Fluvoxamine	The migration analysis results revealed that inhibition with combination therapy (Fluvoxamine + tivozanib) was noteworthy compared to individual therapy. There was prominent decrease in cellular migration, reduction of invasive cells, promising decrease in angiogenic tube formation, decrease in the expression of MENA and also the treatment led to tubulin structure deformation in MCF-7 cell lines. Combination treatment also interrupts the GM 130 level and reduction of HSP-70 expression [Kumar et al., 2021].
9.		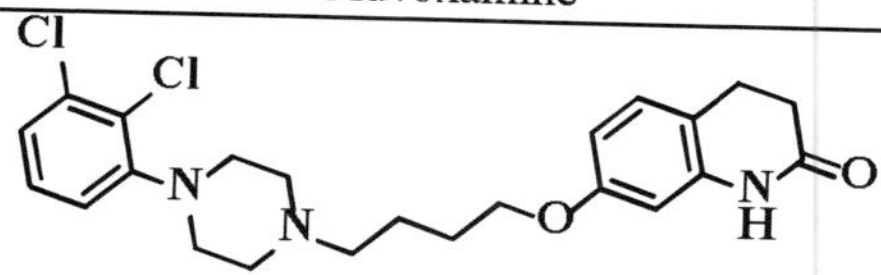 Aripiprazole	Aripiprazole possessed good inhibitory activity with an IC_{50} of 12.1 μM towards MCF-7 cells. Aripiprazole showed the strongest growth inhibition by 90% at a concentration of 50 μM over MCF-7 cell. Aripiprazole also increased the cell viability by 49.1% and 66.7% at 35 μM and 100 μM respectively [Badran et al., 2020].
10.		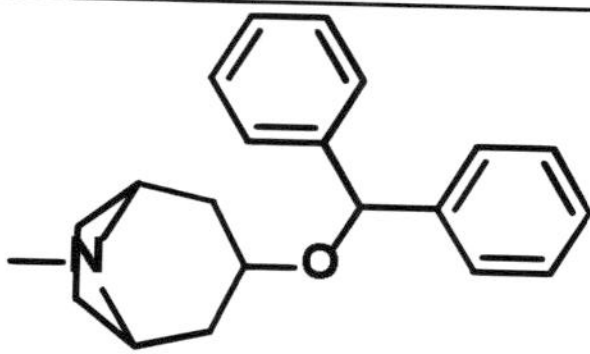 Benztropine	Benztropine showed a significant cytotoxic effect against MCF-7 and MRC-5 cells with IC_{50} values of 21.71 μM and 19.91 μM respectively. As a combination therapy PTX+benztropine showed significant cytotoxic effect against MRC-5 and MCF-7 cells at low concentrations [Duarte et al., 2022].

Table 2. Continued

Sl. No.	Category	Structure (Name)	Summary
11.	Anti-microbial	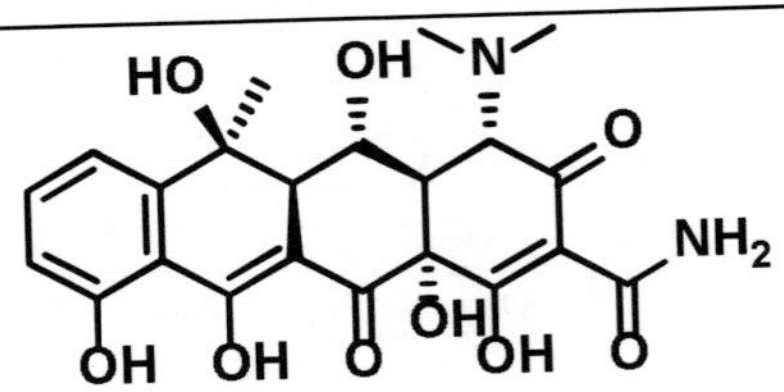 Oxytetracycline	Antiproliferative activity was performed using oxytetracycline over MCF-7 cell and results revealed that oxytetracycline successfully inhibited the cell growth with IC_{50} values of 64.4 μM. The in-vitro results demonstrated that 5% and 10% concentrations of oxytetracycline along with carboplatin (dose IC_{50} = 36.41 μM) showed potential activity with dose IC_{50} values of 13.94 μM and 15.62 μM respectively [Jabber et al., 2020].
12		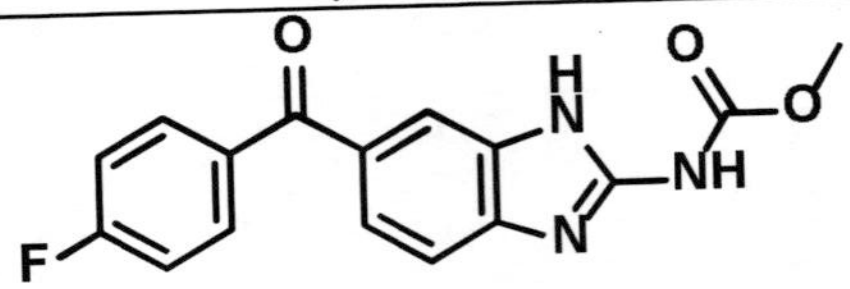 Flubendazole	Flubendazole effectively impairs the mitochondrial outer membrane permeability and promotes dysfunction of mitochondria over MCF-7 and MDA-MB-231 cell lines [Zhen et al., 2020]. Colony formation assay demonstrated that flubendazole showed potential activity against RT-R-MDA-MB-231, MCF-7, and T47D cell lines at 0.1 to 1 μM. In case of the 4T1 allograft mice model, FLU downregulated the volume of the tumor [Choi et al., 2021]

Sl. No.	Category	Structure (Name)	Summary
13.	Anti-microbial	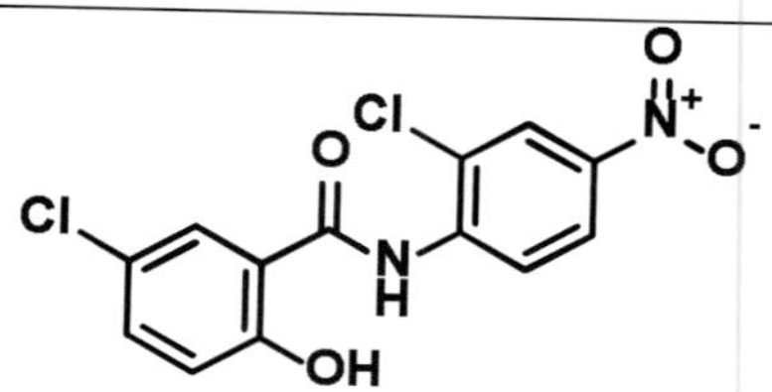 Niclosamide	In sequential combination therapy They performed a cytotoxicity assay against MDA-MB-231, SKBR3, and MCF-7 cell lines and results revealed that at a high dose, the drug Nic (IC_{50}= 4.23 µM, 1.09 µM, and 1.45 µM respectively) caused maximum cell death. Results also demonstrated that the maximum pair of Nic and Dox caused 50% apoptosis over MDA-MB-231 and MCF-7 cell lines and 40% for the SKBR3 cell line. According to the cell cycle assay results Nic caused cell cycle arrest at the G0/G1 phase. In concurrent treatment also Nic showed potential activity with IC50 values of 1.24 µM, 0.44 µM, and 0.29 µM over MDA-MB-231, SKBR3, and MCF-7 cell lines respectively [Lohiya & Katti, 2021].
14.		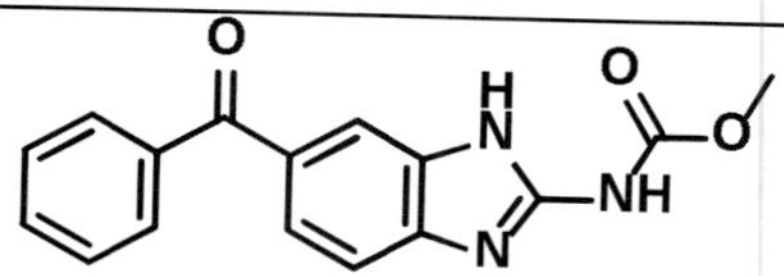 Mebendazole	Colony formation assay demonstrated that mebendazole showed potential activity against RT-R-MDA-MB-231, MCF-7, and T47D cell lines at 0.1 to 1 µM. MBZ (0.5 µM and 1 µM) dramatically decreased the spread of both MDA-MB-231 and RT-RMDA-MB-231 cells. The cell cycle progression assay results revealed that MBZ caused arrest at the G2/M phase. MBZ enhanced the production of cyclin B1 at 0.5 and 1 µM but not cyclin D1 after 24h [Choi et al., 2021].
15.		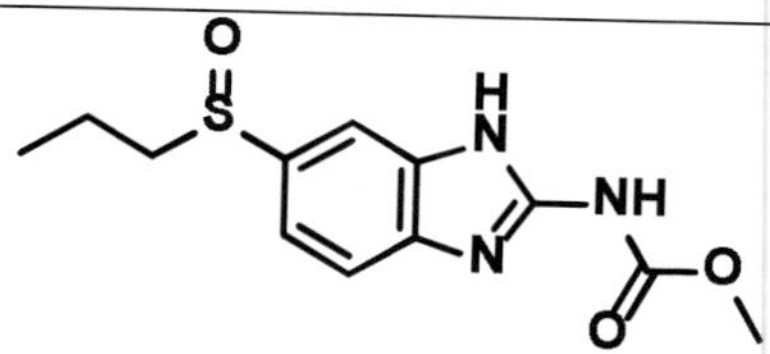 Albendazole-sulfoxide	Colony formation assay demonstrated that mebendazole showed very little activity against RT-R-MDA-MB-231, MCF-7, and T47D cell lines at 0.1 to 1 µM [Choi et al., 2021].

Table 2. Continued

Sl. No.	Category	Structure (Name)	Summary
16.	Anti-microbial	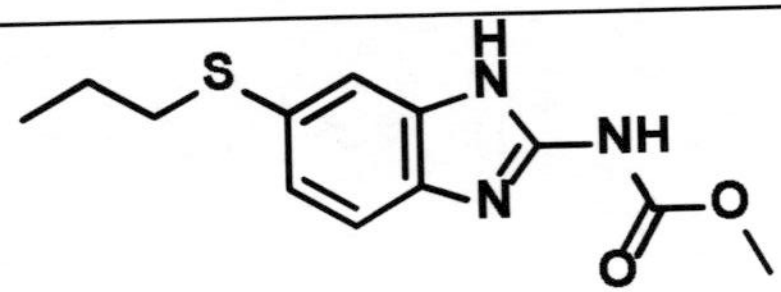 Albendazole	Colony formation assay demonstrated that albendazole showed potential activity against RT-R-MDA-MB-231, MCF-7, and T47D cell lines at 0.1 to 1 µM. In *in-vivo* showed no activity [Choi et al., 2021].
17.		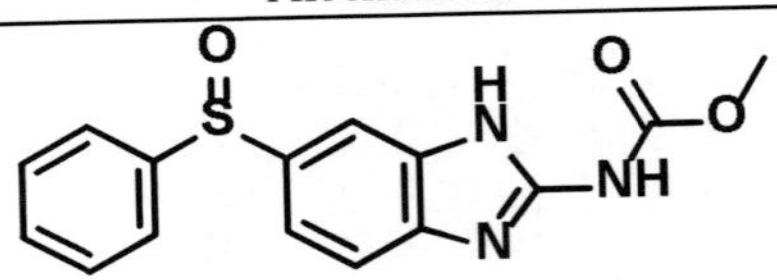 Oxfendazole	Colony formation assay demonstrated that mebendazole showed very little activity against RT-R-MDA-MB-231, MCF-7, and T47D cell lines at 0.1 to 1 µM [Choi et al., 2021].
18.		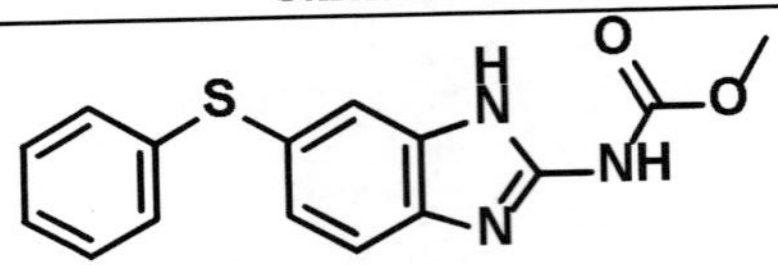 Fenbendazole	Colony formation assay demonstrated that fenbendazole showed potential activity against RT-R-MDA-MB-231, MCF-7, and T47D cell lines at 0.1 to 1 µM. In case of the 4T1 allograft mice model and RT-R-4T1, FBZ downregulated the volume of the tumor [Choi et al., 2021].
19.	Anti-diabetes	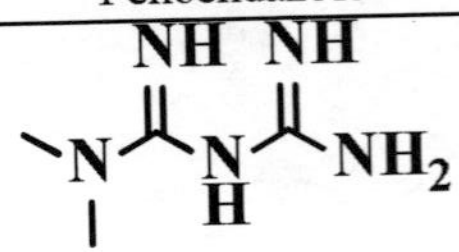 Metformin	Metformin upregulated the level of Bax protein expression and HRK, TNFRSF10A, and TNFRSF10B gene expression at higher concentrations (10 mmol/L and 20 mmol/L) over MCF-7 cell. Metformin caused cell cycle arrest at the G0/G1 phase by 48.62%, 58.25%, and 74.74% at a concentration of 5, 10, and 20 mmol/L respectively [Huang et al., 2021].

Sl. No.	Category	Structure (Name)	Summary
20.	Anti-microbial	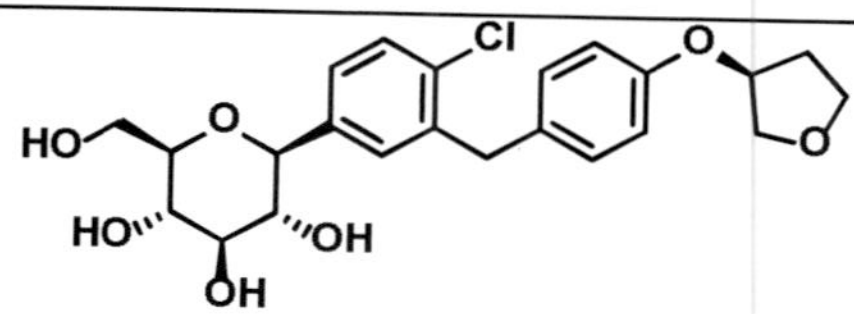 Empagliflozin	DOX+empagliflozin significantly inhibited the growth of MDA-MB-231 cells with the IC_{50} value of 1.70μM. The flow cytometry results concluded that DOX with empagliflozin diminished the G2/M phase cell population by 1.69% [Elia et al., 2022].
21.	Anti-malarial	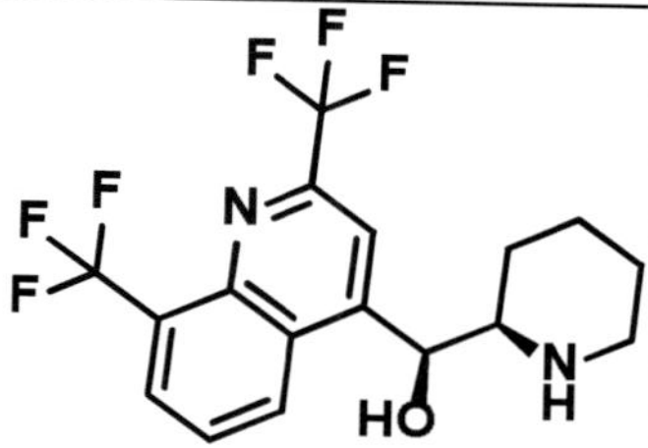 Mefloquine	Mefloquine (fixed) + pyronaridine (varying concentration) was not preferable when compared with the anti-cancer activity of mefloquine (> 80% cell viability) alone. The combination index for this combination showed that 10 μM and 50 μM of pyronaridine had lower CI values (0.205 and 0.664). Results also revealed that mefloquine (0.01 μM, 0.1 μM, and 1 μM) + pyronaridine (fixed) had moderate activity against MCF-7 cell but was not preferable for combination therapy. The combination index for this combination showed lower CI values (0.025 to 0.279) at >10 μM concentration of mefloquine [Duarte et al., 2022]. Cytotoxicity assay results revealed that tafenoquine showed activity with an IC_{50} value of 1.241 Mm over the MCF-7 cell line. Combining mefloquine (8 μM) with DOX (0.2 μM) and PTX (3 nM) resulted in a synergistic effect with a fractional effect of 0.845 and 0.841 respectively [Duarte et al., 2020].

Table 2. Continued

Sl. No.	Category	Structure (Name)	Summary
22.	Anti-malarial	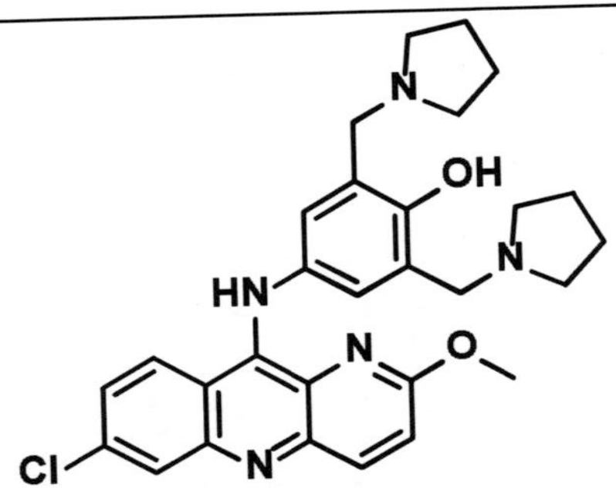 pyronaridine	Mefloquine (fixed) + pyronaridine (varying concentration) was not preferable when compared with the anti-cancer activity of mefloquine (> 80% cell viability) alone. The combination index for this combination showed that 10 μM and 50 μM of pyronaridine had lower CI values (0.205 and 0.664). Results also revealed that mefloquine (0.01 μM, 0.1 μM, and 1 μM) + pyronaridine (fixed) had moderate activity against MCF-7 cell but was not preferable for combination therapy. The combination index for this combination showed lower CI values (0.025 to 0.279) at >10 μM concentration of mefloquine [Duarte et al., 2022]. Cytotoxicity assay results revealed that pyronaridine showed activity with an IC_{50} value of 1.388 μM over the MCF-7 cell line [Duarte et al., 2020].
23.		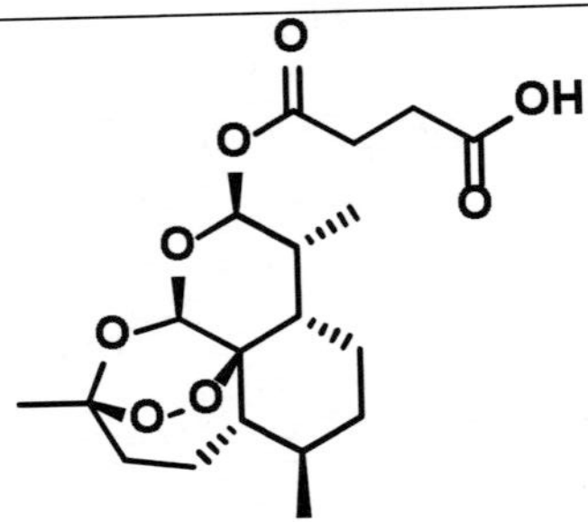 Artesunate	Artesunate treatment significantly reduced the lifespan of cells at doses over 10 μM, resulting in over 50% decrease at 50 μM or above. Cytotoxicity assay results revealed that artesunate showed activity with an IC_{50} value of 1.388 μM. And also, DOX (0.05 μM, 0.1 μM, 0.2 μM, 0.4 μM) and artesunate (3 μM, 6 μM, 12 μM, 24 μM) generated synergic effects ranging from 0.487 to 0.821. PTX (1.5 nM, 3 nM) and artesunate (6 μM, 12 μM) resulted in synergy, with Fa values of 0.453 and 0.544 [Duarte et al., 2022]. Cytotoxicity assay results revealed that pyronaridine showed activity with an IC_{50} value of 11.60 μM over the MCF-7 cell line [54].

Sl. No.	Category	Structure (Name)	Summary
24.	Anti-malarial	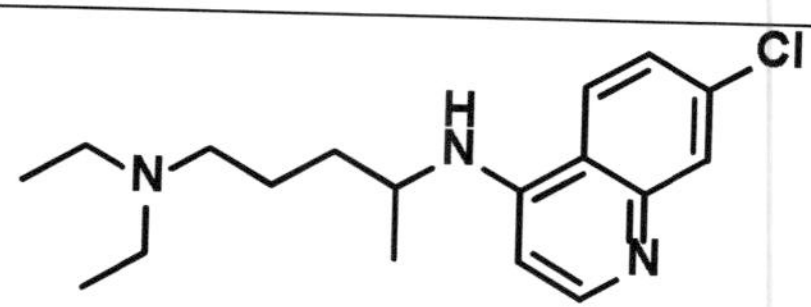 Chloroquine	Chloroquine displayed a synergic effect combined with DOX rather than PTX. DOX (0.05 μM, 0.1 μM, 0.2 μM) and chloroquine (16 μM, 32 μM, 64 μM) generated synergic effects ranging from 0.724 to 0.851. PTX (1.5 μM, 3 μM) and chloroquine (32 μM, 64 μM) resulted in synergy, with Fa values of 0.782 and 0.831 [Duarte et al., 2022]. Chloroquine showed a cytotoxic effect against MCF-7 cell with IC_{50} value of 63.98μM. Also found that chloroquine had potent cytotoxic effects against MRC-5 cells with IC_{50} value of 14.85μM [Duarte et al., 2020].
25.		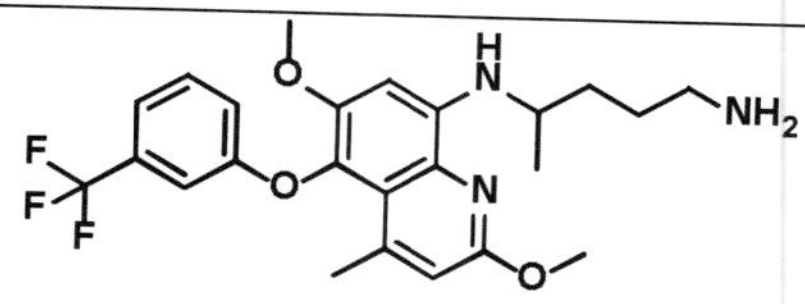 Tafenoquine	Cytotoxicity assay results revealed that tafenoquine showed activity with an IC_{50} value of 2.595 μM over the MCF-7 cell line [Duarte et al., 2020].
26.		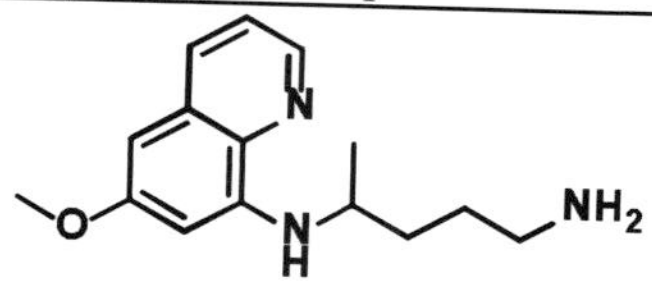 Primaquine	Cytotoxicity assay results revealed that primaquine showed activity with an IC_{50} value of 29.90 μM over the MCF-7 cell line [Duarte et al., 2020].
27.		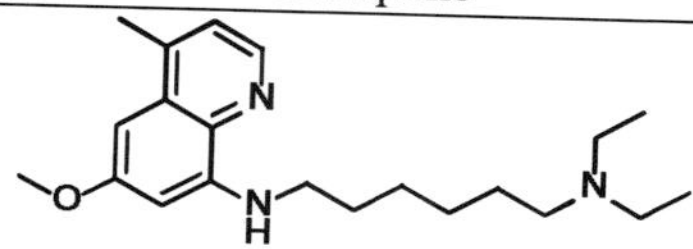 Sitamaquine	The MTT and SRB test results revealed that sitamaquine was ineffective in reducing cell viability and protein content [Duarte et al., 2020].

Table 2. Continued

Sl. No.	Category	Structure (Name)	Summary
28.	Anti-malarial	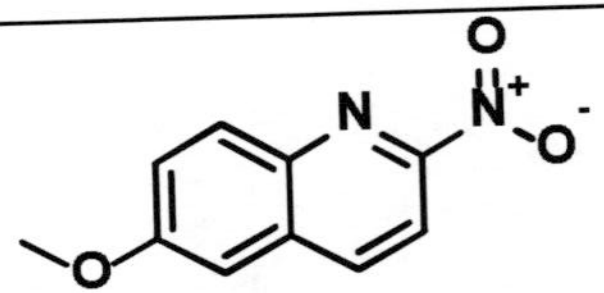 6-methoxy-nitroquinoline	The MTT and SRB test results revealed that 6-methoxy-nitroquinoline was ineffective in reducing cell viability and protein content [Duarte et al., 2020].
29.		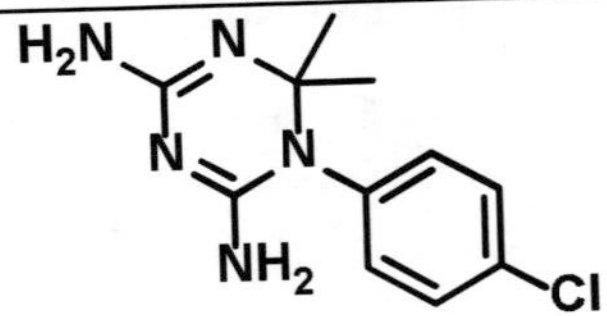 Cycloguanil	Cytotoxicity assay results revealed that cycloguanil showed activity with an IC_{50} value of 20.30 μM [Duarte et al., 2020].
30.		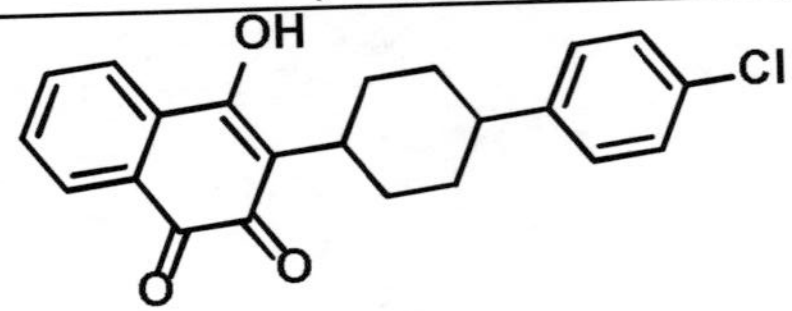 Atovaquone	In MTT and SRB tests, atovaquone did not exhibit good activity against MCF-7 cells [Duarte et al., 2020].

Sl. No.	Category	Structure (Name)	Summary
31.	Anti-malarial	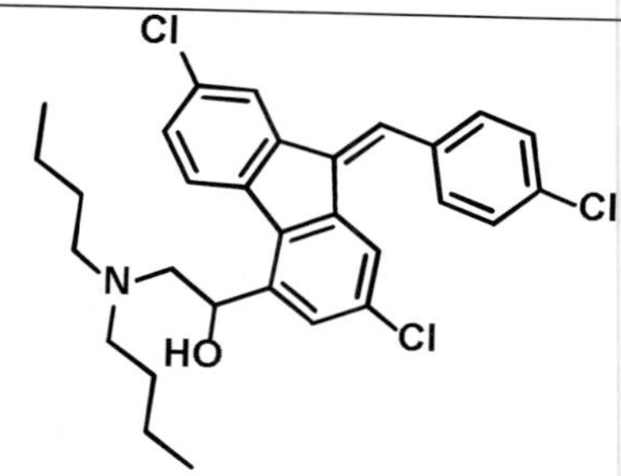 Lumefantrine	In MTT and SRB tests, lumefantre did not exhibit good activity against MCF-7 cells [Duarte et al., 2020].
32.	Anti-inflammation	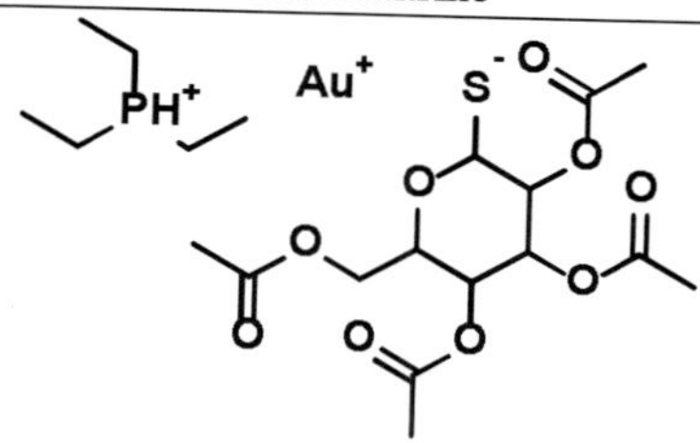 Auranofin	Auranofin was subjected to antiproliferative activity over MDA-MB-435 S, MDA-MB-231, and BT549 cell lines and results revealed that it showed inhibition activity with IC50 values of 4.71 µM, 4.85 µM, and 4.17 µM, respectively. AF primarily inhibits TrxR at its cytostatic dose of 1~2 µM, although both TrxR and the proteasome were inhibited at paraptosis-inducing dosages (4~5 µM). 2 µM AF raised ROS levels by about 2.0 fold, and 2 µM AF with 5 nM Bz amplified the rise to around 2.3 fold after 24h [See et al., 2023].
33.		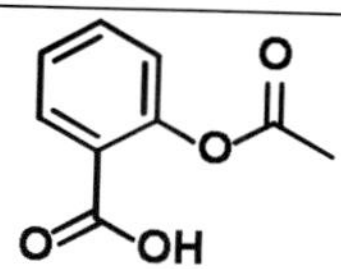 aspirin	MCF-7TR and MCF-7 cells with 1 µM or 2 µM and 2 µM or 5 µM aspirin respectively and discovered that aspirin caused cell death by raising the proportion of apoptotic cells and boosted the levels of ruptured PARP and caspase-3 enzyme [Li et al., 2020].

Table 2. Continued

Sl. No.	Category	Structure (Name)	Summary
34.	Anti-cholesterol	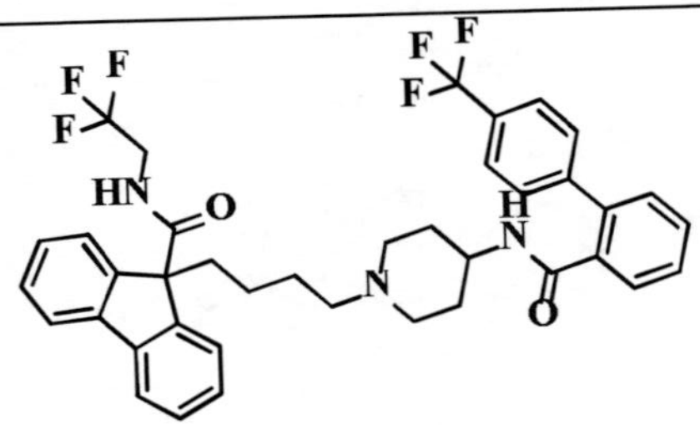 Lomitapide	Lomitapide possessed the strongest binding affinity with docking scores of -13.1 kcal/mol over PDB ID: 3ERT. Lomitapide formed H-bond with His524, halogen bond with Glu419, Gly420, and Asp351, and several alkyl interactions with Ala350, Trp383, and Lys529 amino acids [Duzgun & Korkmaz, 2024].
35.		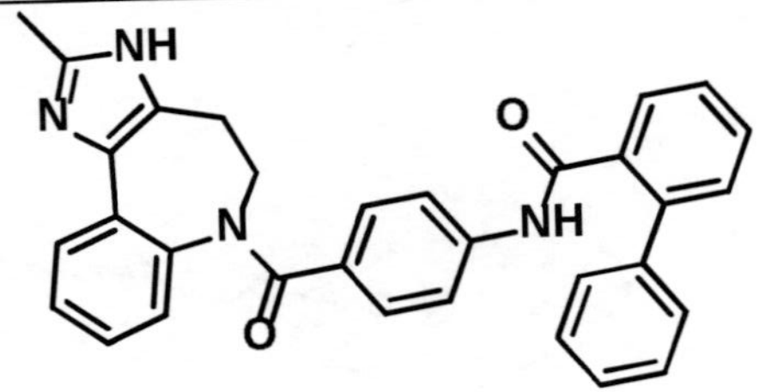 Conivaptan	Conivaptan possessed the strongest binding affinity with docking scores of -11.7 kcal/mol over PDB ID: 3ERT [Duzgun & Korkmaz, 2024].

Sl. No.	Category	Structure (Name)	Summary
36.	Anti-cholesterol	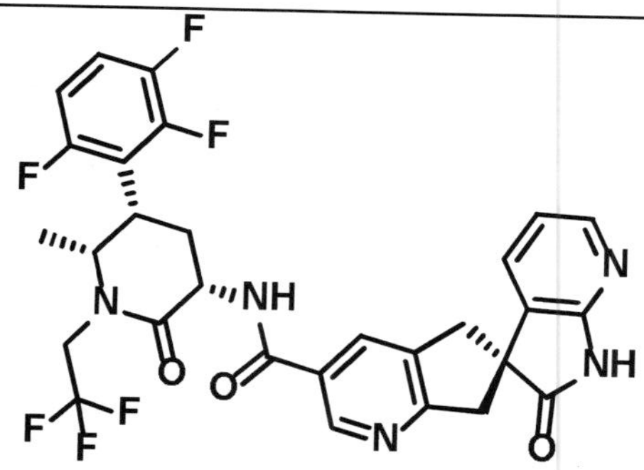 Atogepant	Atogepant possessed the strongest binding affinity with docking scores of -11.2 kcal/mol over PDB ID: 3ERT [Duzgun & Korkmaz, 2024].
37.		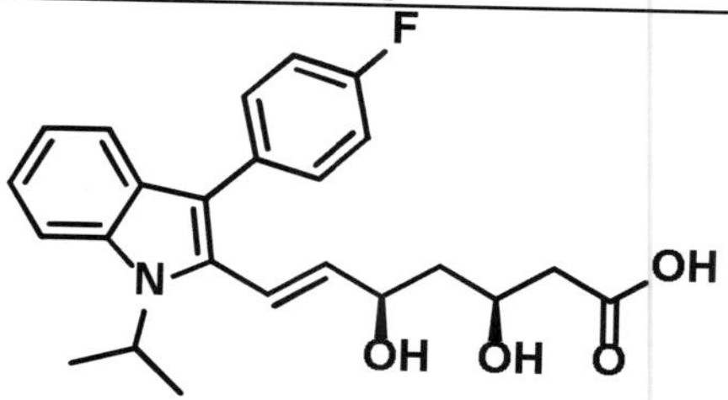 Fluvastatin	Fluvastatin showed potential apoptotic activity in mice model (5.36%; p=0.04) Fluvastatin with GGPP significantly reversed the induction of apoptosis by 1.63% [Bhardwaj et al., 2021].

Table 2. Continued

Sl. No.	Category	Structure (Name)	Summary
38.	Anti-diarrhea	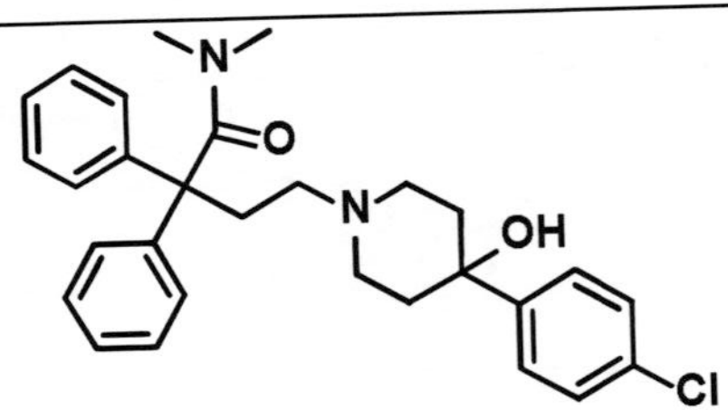 Loperamide	Both the drugs Loperamide and DXR showed downregulated the mRNA of JNK1, MDR1, mTOR, and Bcl-2 *i.e,* 1.35, 2.5, 6.6, and 9.5 folds respectively. Combining both drugs caused G1 arrest [Elia et al., 2022].
39.	Anti-hypertensive	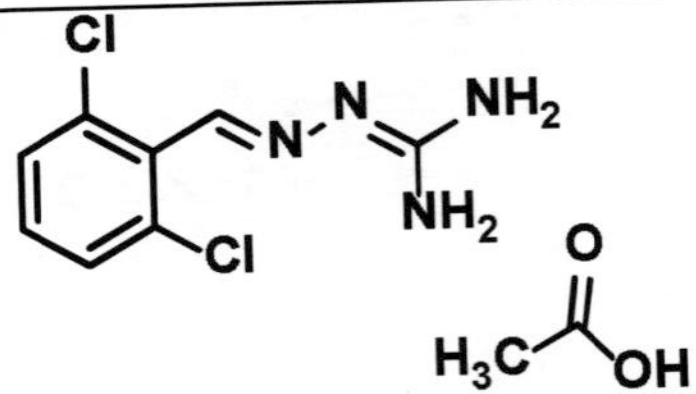 Gunanabenz acetate	The IC_{50} values of free Gunanbenz acetate and Gunanbenz acetate-loaded PS in MDA-MB-231 are 85.71 ± 5.26 and 48.78 ± 6.55 μM respectively [Hoggag et al., 2021].

2.8. Repositing of the Anti-Hypertensive Drugs for Breast Cancer

Haggag and co-researchers repurposed Gunanbenz acetate (GA) by encapsulation into long-circulating nano polymersomes for treating triple-negative breast cancer. The *in-vitro* cellular uptake and anticancer activity of Gunanbenz acetate and Gunanbenz acetate-loaded PEGylated polycaprolactone PS were tested against MDA-MB-231 and MCF-7 cell lines. The cellular uptake of GA-loaded PS results showed that confocal mapping demonstrated intracellular distribution and growth of GA-loaded PS in MDA-MB 231 breast cancer cells. GA-loaded PS (F16), with its small size, low PDI, and low negative zeta potential, was quickly absorbed by MDA-MB 231 cells, enhancing its intracellular anticancer efficacy. The IC_{50} values of free Gunanbenz acetate and Gunanbenz acetate-loaded PS in MDA-MB-231 are 85.71 and 48.78 μM respectively (Haggag et al., 2021).

Summary of various repositioned drugs as anti-breast cancer is presented in Table 2.

Conclusion

The public sector/academics and pharmaceutical companies have begun to pay more attention to drug repurposing as a quicker and less expensive way to increase the number of authorized cancer medications in circulation. Thalidomide is currently used for the treatment of myeloma cancer which was discovered by drug repurposing approaches. In this chapter, we summarized the various chemical entities that are regularly used in the treatment of different diseases such as CNS, fungal, viral infections, etc. The scientific communities are repurposing these chemical entities as anticancer agents. In this modern era, several computational techniques such as high-throughput screening, machine learning, deep learning approaches, etc. are available, which will give new perspectives into therapeutic mechanisms and enable drug repurposing. By screening the vast pharmacological data using these techniques, we enhance the clinical potential of drug repurposing in oncology and may reduce the time consumption and high-cost burden for the discoveries as potential anticancer agents.

References

Antoszczak, M., Markowska, A., Markowska, J., & Huczyński, A. (2020). Old wine in new bottles: Drug repurposing in oncology. *European Journal of Pharmacology, 866*, 172784.

Badran, A., Tul-Wahab, A., Zafar, H., Mohammad, N., Imad, R., Ashfaq Khan, M., . . . Choudhary, M. I. (2020). Antipsychotics drug aripiprazole as a lead against breast cancer cell line (MCF-7) in vitro. *PloS One, 15*(8), e0235676.

Banno, K., Iida, M., Yanokura, M., Irie, H., Masuda, K., Kobayashi, Y., . . . Aoki, D. (2015). Drug repositioning for gynecologic tumors: a new therapeutic strategy for cancer. *The Scientific World Journal, 2015*(1), 341362.

Bayraktar, S., Hernadez-Aya, L. F., Lei, X., Meric-Bernstam, F., Litton, J. K., Hsu, L., . . . Gonzalez-Angulo, A. M. (2012). Effect of metformin on survival outcomes in diabetic patients with triple receptor-negative breast cancer. *Cancer, 118*(5), 1202-1211.

Bhardwaj, A., Embury, M. D., Rojo, R. D., Albarracin, C., & Bedrosian, I. (2021). Efficacy of fluvastatin and aspirin for prevention of hormonally insensitive breast cancer. *Breast Cancer Research and Treatment, 187*(2), 363-374.

Breast Cancer. (2022, 13 March 2024). Retrieved from https://www.who.int/news-room/fact-sheets/detail/breast-cancer.

Britt, K. L., Cuzick, J., & Phillips, K.-A. (2020). Key steps for effective breast cancer prevention. *Nature Reviews Cancer, 20*(8), 417-436.

Chequin, A., Costa, L. E., de Campos, F. F., Moncada, A. D., de Lima, L. T., Sledz, L. R., . . . Goncalves, M. B. (2021). Antitumoral activity of liraglutide, a new DNMT inhibitor in breast cancer cells *in vitro* and *in vivo*. *Chemico-Biological Interactions, 349*, 109641. https://clinicaltrials.gov/

Choi, H. S., Kim, J. H., Jang, S. J., Yun, J. W., Kang, K. M., Jeong, H., . . . Jeong, B. K. (2021). Synergistic tumoricidal effects of alpha-lipoic acid and radiotherapy on human breast cancer cells via HMGB1. *Cancer Research and Treatment: Official Journal of Korean Cancer Association, 53*(3), 685-694.

Choi, H. S., Ko, Y. S., Jin, H., Kang, K. M., Ha, I. B., Jeong, H., . . . Jeong, B. K. (2021). Anticancer effect of benzimidazole derivatives, especially mebendazole, on triple-negative breast cancer (TNBC) and radiotherapy-resistant TNBC *in vivo* and *in vitro*. *Molecules, 26*(17), 5118.

Duarte, D., & Vale, N. (2020). New trends for antimalarial drugs: Synergism between antineoplastics and antimalarials on breast cancer cells. *Biomolecules, 10*(12), 1623.

Duarte, D., Falcão, S. I., El Mehdi, I., Vilas-Boas, M., & Vale, N. (2022). Honeybee venom synergistically enhances the cytotoxic effect of CNS drugs in HT-29 colon and MCF-7 breast cancer cell lines. *Pharmaceutics, 14*(3), 511.

Duarte, D., Guerreiro, I., & Vale, N. (2022). Novel strategies for cancer combat: Drug combination using repurposed drugs induces synergistic growth inhibition of MCF-7 breast and HT-29 colon cancer cells. *Current Issues in Molecular Biology, 44*(10), 4930-4949.

Duarte, D., Nunes, M., Ricardo, S., & Vale, N. (2022). Combination of antimalarial and CNS drugs with antineoplastic agents in MCF-7 breast and HT-29 colon cancer cells: Biosafety evaluation and mechanism of action. *Biomolecules, 12*(10), 1490.

Düzgün, Z., & Korkmaz, F. D. (2024). Lomitapide as a Potential Estrogen Receptor Inhibitor: A Computational Drug Repurposing Study. *Kocaeli Üniversitesi Sağlık Bilimleri Dergisi, 10*(1), 8-14.

Elia, S. G., Al-Karmalawy, A. A., Nasr, M. Y., & Elshal, M. F. (2022). Loperamide potentiates doxorubicin sensitivity in triple-negative breast cancer cells by targeting MDR1 and JNK and suppressing mTOR and Bcl-2: *In vitro* and molecular docking study. *Journal of Biochemical and Molecular Toxicology, 36*(1), e22938.

Eliaa, S. G., Al-Karmalawy, A. A., Saleh, R. M., & Elshal, M. F. (2020). Empagliflozin and doxorubicin synergistically inhibit the survival of triple-negative breast cancer cells via interfering with the mTOR pathway and inhibition of calmodulin: in vitro and molecular docking studies. *ACS Pharmacology & Translational Science, 3*(6), 1330-1338.

Ferlay, J., Colombet, M., Soerjomataram, I., Dyba, T., Randi, G., Bettio, M., . . . Bray, F. (2018). Cancer incidence and mortality patterns in Europe: Estimates for 40 countries and 25 major cancers in 2018. *European Journal of Cancer, 103*, 356-387.

Garcia-Martinez, L., Zhang, Y., Nakata, Y., Chan, H. L., & Morey, L. (2021). Epigenetic mechanisms in breast cancer therapy and resistance. *Nature Communications, 12*(1), 1786.

Giuli, M., Giuliani, E., Screpanti, I., Bellavia, D., & Checquolo, S. (2019). Notch signaling activation as a hallmark for triple-negative breast cancer subtype. *Journal of Oncology, 2019*(1), 8707053.

Haggag, Y. A., Yasser, M., Tambuwala, M. M., El Tokhy, S. S., Isreb, M., & Donia, A. A. (2021). Repurposing of Guanabenz acetate by encapsulation into long-circulating nanopolymersomes for treatment of triple-negative breast cancer. *International Journal of Pharmaceutics, 600*, 120532.

Hanusova, V., Skalova, L., Kralova, V., & Matouskova, P. (2015). Potential anti-cancer drugs commonly used for other indications. *Current Cancer Drug Targets, 15*(1), 35-52.

Huang, Y., Zhou, Z., Zhang, J., Hao, Z., He, Y., Wu, Z., . . . Zhao, Q. (2021). lncRNA MALAT1 participates in metformin inhibiting the proliferation of breast cancer cell. *Journal of Cellular and Molecular Medicine, 25*(15), 7135-7145.

Hunakova, L., Horvathova, E., Gronesova, P., Bobal, P., Otevrel, J., & Brtko, J. (2019). Triorganotin isothiocyanates affect migration and immune check-point receptors in human triple-negative breast carcinoma MDA-MB-231 cells. *Anticancer Research, 39*(9), 4845-4851.

Jabber, E. J., Alzayd, A. M., Jawad, M. J., Sulbi, I. M., Hassan, S. M., Jawad, M. J., & Jasim, A. M. (2022). Synergistic effect of oxytetracycline as a combination treatment with Carboplatin on MCF-7 breast cancer cell line. *Brazilian Journal of Veterinary Research and Animal Science, 59*, e191527-e191527.

Key Statistics for Breast Cancer. (2024). Retrieved from https://www.cancer.org/cancer/types/breast-cancer/about/how-common-is-breast-cancer.html#:~:text=The%20American%20Cancer%20Society's%20estimates,will%20die%20from%20breast%20cancer.

Kumar, N., Raza, M., & Sehrawat, S. (2021). Intuitive repositioning of an anti-depressant drug in combination with tivozanib: precision medicine for breast cancer therapy. *Molecular and Cellular Biochemistry, 476*(11), 4177-4189.

Lekka, E., Deftereos, S. N., Persidis, A., Persidis, A., & Andronis, C. (2011). Literature analysis for systematic drug repurposing: a case study from Biovista. *Drug Discovery Today: Therapeutic Strategies, 8*(3-4), 103-108.

Li, L., Hu, M., Wang, T., Chen, H., & Xu, L. (2020). Repositioning aspirin to treat lung and breast cancers and overcome acquired resistance to targeted therapy. *Frontiers in Oncology, 9*, 1503.

Liao, M., Zhang, J., Wang, G., Wang, L., Liu, J., Ouyang, L., & Liu, B. (2021). Small-molecule drug discovery in triple negative breast cancer: current situation and future directions. *Journal of Medicinal Chemistry, 64*(5), 2382-2418.

Lohiya, G., & Katti, D. S. (2021). A synergistic combination of niclosamide and doxorubicin as an efficacious therapy for all clinical subtypes of breast cancer. *Cancers, 13*(13), 3299.

Mohammad Jafari, R., Sheibani, M., Nezamoleslami, S., Shayesteh, S., Jand, Y., & Dehpour, A. R. (2018). Drug repositioning: a review. *Journal of Iranian Medical Council, 1*(1), 7-10.

Pantziarka, P., Sukhatme, V., Meheus, L., Sukhatme, V., & Bouche, G. (2017). Repurposing non-cancer Drugs in Oncology–How many drugs are out there? *bioRxiv*, 197434.

Seo, M. J., Kim, I. Y., Lee, D. M., Park, Y. J., Cho, M.-Y., Jin, H. J., & Choi, K. S. (2023). Dual inhibition of thioredoxin reductase and proteasome is required for auranofin-induced paraptosis in breast cancer cells. *Cell Death & Disease, 14*(1), 42.

Shim, J. S., & Liu, J. O. (2014). Recent advances in drug repositioning for the discovery of new anticancer drugs. *International Journal of Biological Sciences, 10*(7), 654.

Song, Y., Li, L., Chen, J., Chen, H., Cui, B., Feng, Y., . . . Luo, M. (2020). Thioridazine hydrochloride: an antipsychotic agent that inhibits tumor growth and lung metastasis in triple-negative breast cancer via inducing G0/G1 arrest and apoptosis. *Cell Cycle, 19*(24), 3521-3533.

Tegowski, M., Fan, C., & Baldwin, A. S. (2019). Selective effects of thioridazine on self-renewal of basal-like breast cancer cells. *Scientific Reports, 9*(1), 18695.

Timilsina, S., Rajamanickam, S., Rao, A., Subbarayalu, P., Nirzhor, S., Abdelfattah, N., . . . Brenner, A. (2022). The antidepressant imipramine inhibits breast cancer growth by targeting estrogen receptor signaling and DNA repair events. *Cancer Letters, 540*, 215717.

Verbaanderd, C., Meheus, L., Huys, I., & Pantziarka, P. (2017). Repurposing drugs in oncology: next steps. *Trends in Cancer, 3*(8), 543-546.

Waks, A. G., & Winer, E. P. (2019). Breast cancer treatment: a review. *JAMA, 321*(3), 288-300.

Zhen, Y., Yuan, Z., Zhang, J., Chen, Y., Fu, Y., Liu, Y., . . . Zhou, X.-L. (2022). Flubendazole induces mitochondrial dysfunction and DRP1-mediated mitophagy by targeting EVA1A in breast cancer. *Cell Death & Disease, 13*(4), 375.

Zhen, Y., Yuan, Z., Zhang, J., Chen, Y., Fu, Y., Liu, Y., . . . Zhou, X. (2022). Flubendazole induces mitochondrial dysfunction and DRP1-mediated mitophagy by targeting EVA1A in breast cancer. *Cell Death & Disease* 13: 375.

Chapter 8

Beyond Traditional Therapies: Expanding the Armamentarium Against Kidney Diseases Through Drug Repurposing

Debajyoti Roy[1,*]
Maitrayee Ghosh[2]
Rupam Bachar[3]
and Naresh Kumar Rangra[4]

[1]Faculty of Pharmacy, C.V. Raman Global University, Mahura, Bhubaneswar, Odisha, India
[2]Department of Pharmaceutics, Amity Institute of Pharmacy, Amity University Uttar Pradesh, Noida, Uttar Pradesh, India
[3]Department of Pharmacy, Bengal College of Pharmaceutical Technology, Dubrajpur, Birbhum, West Bengal, India
[4]Chitkara University School of Pharmacy, Chitkara University, Himachal Pradesh, India

Abstract

The increasing prevalence of kidney disorders presents a considerable global health challenge, prompting the search for innovative treatment methods to alleviate the strain on affected individuals and healthcare systems. A potential solution lies in the concept of drug repurposing, which involves repurposing existing medications for new therapeutic purposes. This book chapter delves into the landscape of drug repurposing concerning kidney disorders, shedding light on potential avenues for therapeutic intervention.

---

* Corresponding Author's Email: droy4896@gmail.com

In: Drug Repurposing
Editors: Shah Alam Khan and Md. Jawaid Akhtar
ISBN: 979-8-89530-422-8

Through a thorough examination of literature and clinical trials, this book chapter emphasizes the multifaceted role of drug repurposing in addressing various causes and presentations of kidney disorders, spanning chronic kidney disease, glomerulopathies, and renal fibrosis. Researchers have identified numerous candidates capable of mitigating renal inflammation, fibrosis, and oxidative stress while preserving renal function and alleviating associated comorbidities by repurposing FDA approved drugs or investigational compounds.
This chapter highlights the potential transformative impact of drug repurposing as a pivotal aspect of therapeutic advancement in kidney disorders. By harnessing existing pharmacotherapeutic agents, clinicians and researchers can pave the way for more effective, cost-efficient, and personalized treatments, ultimately enhancing outcomes and quality of life for patients on a global scale.

Keywords: chronic kidney disease, drug repurposing, FDA-approved drugs, kidney disorders, therapeutic strategies

1. Introduction

Kidney diseases turn out to be a silent killer and have gained significant momentum in public health importance globally. The International Society of Nephrology estimates that about 850 million people worldwide suffer from various kidney diseases, the majority of whom are undiagnosed. Besides, the global burden of kidney diseases remains so high that it is estimated that 2.6 million people receive dialysis treatment with many millions more needing care. The world is currently confronting an unprecedented burden of kidney diseases. The World Health Organisation (WHO) considers kidney disease a high-frequency disorder and predicts that by 2040, it will emerge as the fifth leading cause of death worldwide (Couser et al., 2011).

Despite development and research in medical technologies, therapeutic interventions for renal diseases are very limited, and most of them are ineffective. Most of the cases with end-stage renal disease undergo dialysis support for their survival. It is, however, a very costly and highly time-consuming process that greatly affects quality of life (Daenen et al., 2019). Besides this, dialysis is not that accessible, especially within low and middle-income countries, where the accessibility of health care is already at its limit. Kidney transplantation is the treatment of choice in end-stage renal disease (ESRD). Still, it is also beset by limited organ availability and the very high cost of the procedure itself. Immunosuppressive therapy is fraught with enhanced chances of infection and cancer development (Duni et al., 2019).

Nephrolithiasis, the formation of renal stones within the kidneys, is a condition. When these stones leave the renal pelvis and travel into the remaining portions of the urine collecting system, which includes the ureters, bladder, and urethra, this condition is known as urolithiasis as shown in Figure 1. Kidney stones, or renal calculi and nephroliths as known in medical terms, refer to a build-up of small, hard deposits of mineral and acid salts on the inner surfaces of the kidney (Arya et al., 2017). Renal colic is a painful condition caused by a disorder in the kidney, which blocks the flow of urine, leading to difficulty in urination. Following prostate diseases and urinary tract infections (UTIs), kidney stones rank as the third most common urinary tract problem, with a high recurrence rate, making it a significant cause of morbidity (Liu et al., 2018). About 50% of individuals may experience a recurrence within 5-10 years (Nirumand et al., 2018). Most patients with urolithiasis are treated based on supportive care, analgesics, and antiemetics. Stones presenting with obstruction, renal failure, and infection require the addition of, and urgency with, therapy. The most common stone composition includes calcium oxalate or phosphate stones at 80%. Less common forms include struvite stones at 10%, cystine stones at 1%, and uric acid stones at 9%. A host of risk factors is responsible for the development of various forms of kidney stones which include dietary practices, individual or family history of stone formation, environmental influences, pharmacological use, and patient's background in medication. Common risk factors include insufficient fluid intake, high consumption of animal proteins, excessive intake of oxalate-rich foods (such as beans, beer, coffee, berries, spinach, chocolate, certain nuts, soda, some teas, and potatoes), and high salt consumption (Ghosh et al., 2024). Consuming citrate can help prevent stone formation by preventing crystal aggregation through the formation of complexes with calcium salts in the urine. Hypocitraturia is seen in 60% of patients with calcium stones, according to research. Contrary to popular assumption, low calcium intake has been demonstrated to raise the risk of kidney stone development. Reduced oral calcium consumption will lower gastrointestinal tract (GIT) calcium levels, which would otherwise be available to bind to oxalate. The risk of stone formation will rise because of increased oxalate absorption and excretion. It has also been demonstrated that consuming fish oil and vitamin C increases the incidence of calcium stones (Bartges et al., 2015; Skolarikos et al., 2022).

Urolithiasis has been considered a public health issue with an incidence rate of about 12% worldwide (Edvardsson et al., 2013). This is more common in men than in women, and the recurrence is also much higher in men with a rate of 70-81%, compared to women who register 47-60% (Khaling et al., 2014). Urolithiasis can be observed at any age; however, most calculus incidents due to calcium oxalate are between 50 and 60 years of age (Baker et al., 1993). Some of the identified risk factors in contributing to the development of chronic kidney stones include high intake of animal protein, low fluid consumption, sedentary lifestyle, and poor dietary habits with excessive intake of oxalate foods such as beans, beer, coffee, berries, spinach, chocolate, certain nuts, and soft drinks. Other risk factors include a low-calcium diet, obesity, familial predisposition to renal stone formation, and a high-sodium diet (Ghosh et al., 2024). Several metabolic disorders can aggravate the formation of one or the other types of renal calculi. Such disorders include:

- Hyperuricosuria: An increased concentration of uric acid present in the urine.
- Hyperoxaluria: Increased excretion of oxalate in the urine.
- Hypocitraturia – urine citrate is low.
- Hypercalciuria: Excess calcium in the urine.
- Gout: A disorder associated with uric acid metabolism dysfunction (Sansores-Espana et al., 2022).

Even with the tremendous strides that have occurred in the field of medical science, the therapeutic options available for most conditions relating to the kidneys are relatively few and are often targeted more towards symptomatic management rather than approaching the fundamental aetiologies of the condition. Thus, standard ways of treating CKD control systemic blood pressure through glycaemic surveillance to retard the disease process by ACE inhibitors and ARBs. These treatments do not fully stop the progression, and most of the patients progress toward end-stage renal disease.

Drug repurposing is one of the techniques to identify new therapeutic indications for old existing drug molecules. Recently, drug repurposing has gained so much interest that one-third of recently approved drugs are repurposed for some therapeutic indication, in addition to the previously identified targets (Pantziarka et al., 2018; Sachs et al., 2017). Drug repurposing, also known as drug repositioning, is the process of finding new therapeutic uses for existing medications. This approach can markedly cut down the time and expense associated with developing new drugs, as repurposed drugs have already been extensively tested for safety and efficacy in their original applications. By building on existing knowledge of a drug's pharmacokinetics, pharmacodynamics, and safety profile, drug repurposing facilitates a faster transition from research to clinical use (Ashburn and Karl 2004).

Drug repurposing is the observation of new uses for medications already approved or under development in other indications. In contrast to conventional drug development, which starts with the discovery and further extensive testing of a new chemical entity, drug repurposing makes use of knowledge of the safety and efficacy profile of a drug obtained previously for retesting against other diseases. This often saves considerable time and costs.

The concept of drug repurposing is quite well-established. There are numerous successful instances that go ahead to prove its relevance. One of the first documented cases in this field was the case of thalidomide, which was initially prepared as a sedative but was stopped due to teratogenicity, and hence the substance was repurposed for use in treating leprosy and multiple myeloma. For instance, the well-known antihistamine preparation called diphenhydramine later turned out to have antipsychotic properties, although its use had been intended for allergies only (Roy et al., 2023).

Drug repurposing in nephrology allows an opportunity for new therapeutic intervention development in disorders with limited treatment availability. Researchers and clinical scientists alike, by following this approach, will use their findings as a new way to fight against renal inflammation, fibrosis, and oxidative stress—aiming at protecting the sole function of kidney and improving patients' outcomes. This book chapter discusses the future of drug repositioning for renal disorders, presenting emerging treatment options and how this might represent a paradigm shift in the management of patients (Chong et al., 2007).

2. Overview of Kidney Diseases

A wide range of disorders, often with systemic effects, affects kidney morphology and function. Because of different aetiologies and mechanisms of disease, these disorders have widely differed clinical presentations; however, they often share a propensity to cause progressive renal impairment with resultant significant morbidity and mortality. Understanding the different categories of renal diseases, their basic mechanisms, and the nature of their clinical manifestations remains paramount if appropriate therapeutic interventions are to be developed.

2.1. Chronic Kidney Disease (CKD)

2.1.1. Epidemiology and Burden

Chronic kidney disease is a global health issue, and the prevalence over recent decades has remained on a steady rise, currently affecting nearly 10% of adults in the world. The number of CKD cases has been continuously increasing with, in part, the increasing prevalence of major risk factors, such as diabetes and hypertension. According to worldwide trends, more than 30 million people are affected in the United States alone. CKD is considered a paramount concern in healthcare-related systems because its comorbid conditions are related to an increased risk of cardiovascular diseases, infections, and mortality.

2.1.2. Pathophysiology

CKD is defined as a progressive reduction in renal function, usually associated with lowering glomerular filtration rate (GFR) level and/or the presence of albuminuria. CKD represents a complex pathophysiological process in turn comprising several hemodynamic, metabolic, and inflammatory crosstalk between different systems. Hyperglycaemia associated with diabetes and glomerular hypertension secondary to hypertension are the main driving factors in the progression of CKD. These conditions result in glomerular injury, tubulointerstitial inflammation, and fibrosis, which culminate in the loss of nephrons and a decline in renal function.

CKD is asymptomatic in most cases during the early period of the disease process; hence, any early diagnosis and intervention are rather challenging. The signs and symptoms will include fatigue, edema, hypertension, and anaemia as the disease progresses. Advanced CKD may present with uraemia characterized by the accumulation of uremic toxins manifested clinically by nausea, vomiting, pruritus, and cognitive impairment. CKD management involves a comprehensive approach that includes management through lifestyle modifications, pharmacotherapy, and finally, in advanced stages, by dialysis or transplantation (Vivekanand et al., 2013).

2.2. Glomerulopathies

Glomerulopathies are kidney diseases related to primary glomerular structures taking part in the process of filtration of blood passing through the nephrons. These can roughly be divided into primary glomerulopathies and secondary glomerulopathies.

2.2.1. Primary Glomerulopathies

The glomeruli develop segment sclerosis typical for focal segmental glomerular sclerosis (FSGS) from a common cause to nephrotic syndrome in adults. Causes of FSGS are often idiopathic, previously described as primary or secondary, conditions identified by being overweight, high blood pressure, or viral infections. The injured podocytes in FSGS bring damage to the scarring of the glomeruli along with proteinuria and progressive decline in renal function.

Membranous nephropathy is one of the most common causes of nephrotic syndrome, especially among adults. Immune complexes deposited on the glomerular basement membrane thicken it, hence leading to its dysfunction. Membranous nephropathy can be primary or idiopathic, or associated with secondary conditions such as autoimmune diseases, infections, and malignancies. The most common clinical presentations of the disease are proteinuria, edema, and hypoalbuminemia. The course of membranous nephropathy in patients may present with a spontaneous remission to a slowly or rapidly progressive deterioration in renal function (Jha Jay et al., 2016).

2.2.2. Secondary Glomerulopathies

Diabetic nephropathy is one of the many significant burdens all over the world associated with CKD and ESRD. The disease, according to its definition, results from chronic hyperglycemia, which over time causes renal hyperfiltration, hypertrophy, and fibrosis. In the hallmark, there is persistent albuminuria, a decline in GFR, and an increased risk of cardiovascular problems. Good control of diabetic nephropathy includes tight control of blood glucose, proper evaluation of blood pressure and its management, and RAAS-targeting inhibitors (Morales et al., 2023). Lupus nephritis is the most serious complication of the autoimmune disease systemic lupus erythematosus. It is characterized by the deposition of immune complexes in the glomeruli, leading to inflammation and damage. Clinical features span a wide spectrum, from mild proteinuria to full-blown nephrotic syndrome and rapidly progressive glomerulonephritis. In general, good management includes immunosuppressive medication and close follow-up to prevent disease progression and complications (Stroie et al., 2022).

2.3. Renal Fibrosis

Renal fibrosis is a common endpoint for most chronic kidney diseases and an important aspect of renal failure progression. The deposition of extracellular matrix proteins is the main characteristic in this pathological condition, which ultimately results in scarring of the tissue followed by impairment of renal function. Renal fibrosis is usually portrayed as the common endpoint during a variety of chronic kidney diseases, representing a central driving force toward renal failure. It is characterized by abnormal, excessive deposition of extracellular matrix proteins, which cause scar tissue and subsequently result in loss of renal function (Nogueira et al., 2017).

2.4. Other Related Renal Pathologies

2.4.1. Acute Kidney Injury (AKI)

Acute kidney injury is characterized by an abrupt, usually reversible reduction in renal function that commonly results from one of three broad etiologic categories: reduced blood flow, nephrotoxic exposure, and infection. Generally, tubular injury, inflammation, and oxidative damage are the basic pathophysiologic processes underlying this disease. Generally, the illness has been associated with high mortality and could also enhance the risk for CKD. Effective management is through the early identification and treatment of the etiological factor for the injury (Kellum et al., 2021).

2.4.2. Polycystic Kidney Disease (PKD)

PKD is characterized by the progressive development of innumerable cysts within renal tissues, leading to impairment in renal function. The most common form is Autosomal Dominant PKD, caused by mutations in either PKD1 or PKD2 genes. This tends to progress to ESRD, for which therapeutic measures are mainly symptomatic and complication-oriented (Bergmann et al., 2018).

2.4.3. Kidney Stones

Nephrolithiasis is another more common term for renal calculi; it means crystalline structures are formed in the urinary system. Such stones may lead to severe pain, bleeding, and obstructive urine outflow. Factors predisposing to the formation of a stone include inadequate fluid intake, diet with certain types of content, as well as a range of metabolic states. Treatment usually includes relieving the pain, removing the stone, and using preventative measures to reduce the frequency of recurrence (Couser et al., 2011).

3. Drug Repurposing: Concept and Mechanisms

Drug repurposing would be identifying new uses of an already developed therapeutic for its subsequent uses. Some of the techniques that are under practice currently in drug repurposing involve:

3.1. Phenotypic Screening

This strategy uses high-throughput drug screening to induce a particular phenotype in a cell or animal model. Such a form of strategy could identify drugs that have unsuspected and new activities resulting in determining new disease applications (Swamidass, 2011; Doman et al., 2002).

3.2. Target-Based Mechanisms

Most pharmacological agents exert their therapeutic effect due to interaction with specific molecular targets including, among others, receptors, enzymes, or ion channels. Examples include minoxidil, developed originally as an antihypertensive agent, which was supposed to dilate vessels and lower blood pressure; it was used later for alopecia due to its unplanned effectiveness in inducing hair growth. Examples of these are metformin, which is intended to control the level of blood sugar related to diabetes, but currently finds application as an anticancer drug (Luo et al., 2017). This is associated with the activation of the AMP-activated protein kinase (AMPK) and suppression of the mTOR signaling pathway (Xu et al., 2015).

3.3. Computational Methodologies

These methodologies apply algorithms together with techniques for data analysis to find potential new uses of currently available pharmaceuticals. For instance, network-based approaches utilize computational frameworks to unearth new drug-target interactions and potential drug combinations. An example could be databases that provide information on identified drugs, their targets, and clinical uses. Data analysis allows one to discover potential new uses for an already existing drug. Sometimes, medical applications are discovered by sheer chance or accident long after a drug has been in use. A case in point is sildenafil: having been developed to treat arterial hypertension, it was later proven to be effective in treating erectile dysfunction (Pushpakom et al., 2018). Drug repurposing represents a highly varied strategy in the medical world, and its potential benefits are increasingly being recognized (Hodos et al., 2016). Therefore, drug repositioning can relatively be viewed as a realistic alternative in many cases in which few options for the therapy of rare diseases exist. A good example is thalidomide: first synthesized as a sedative and antiemetic, it was further found to treat leprosy, and a rare type of blood cancer called multiple myeloma (Chen et al., 2012).

3.4. Personalised Medicine Development

Repurposed pharmaceuticals can be used for designing personalized treatments targeting a specific need of the individual patient. This is of paramount importance in the case of oncology, because the hope to achieve personalized medicine makes it possible to understand which therapies work best for which individuals. For example, imatinib was developed for treating chronic myeloid leukemia but was later found to reduce the severity of many other forms of cancer (Druker et al., 2001).

Many repurposed pharmaceuticals have shown efficacy for a great many medical conditions, and thus provide a probable mode of treatment for several different diseases caused by one therapeutic agent. One fine example of this is metformin, a medication developed for the treatment of type 2 diabetes, through time repurposed for its cancer and polycystic ovary syndrome treatments (Rena et al., 2017).

Accelerating new drugs time-to-market saves time and resources by repurposing already existing drugs rather than the development of an entirely new drug. This can be of essence in cases of emerging infectious diseases where fast preparation of treatments is needed. A well-timed example was the COVID-19 pandemic, when several already existing drugs were rapidly repurposed for possible use against the virus (Touret et al., 2020).

4. Therapeutic Targets in Kidney Diseases

4.1. Chronic Kidney Disease (CKD)

The renin-angiotensin-aldosterone system (RAAS) plays a very important role in regulating blood pressure, maintaining fluid homeostasis, and balancing electrolytes. Inappropriate functioning of RAAS has been implicated to play a contributory role in promoting the progression of CKD. Angiotensin II is one key component in promoting vasoconstriction, sodium retention, and release of aldosterone, which subsequently promotes hypertension and thereby promotes renal injury inhibitors. Angiotensin-converting enzyme (ACE) inhibitors and angiotensin II receptor blockers (ARBs) have been convincingly shown to be effective in reducing proteinuria, retard the course of CKD and diminish the risk for cardiovascular events (Levey and Coresh., 2012).

Basically, as a cytokine, transforming growth factor-beta (TGF-β) regulates cell growth, differentiation, and fibrosis. TGF-β levels are heightened in CKD patients. It actively stimulates renal fibrosis by upregulating the synthesis of extracellular matrix (ECM) proteins and downregulating their degradation. Thus, therapeutic targeting of the TGF-β signaling pathways is aimed at preventing fibrosis and maintaining renal function. Importantly, other TGF-β inhibitors, including pirfenidone and fresolimumab, have shown effectiveness in both preclinical and clinical studies of CKD treatment.

Nuclear Factor Kappa B (NF-κB) is a powerful transcription factor that regulates the expression of a wide array of pro-inflammatory genes encoding cytokines, chemokines, and adhesion molecules. In this light, NF-κB will be centrally involved in the inflammatory pathways leading to CKD development. Suppression of NF-κB signaling would decrease inflammation and prevent further damage to the tissues. More research is likely to uncover various agents that can be modulated by this pathway in CKD, including corticosteroids, NSAIDs, and specific inhibitors of NF-κB (Levey et al., 2005; Barnes and Michael., 1997). Oxidative stress is the result of an imbalance between the generation of reactive oxygen species and antioxidant defenses in the organism. ROS are known to be at the origin of cellular injuries, inflammation, and fibrosis related to CKD. Some antioxidant-based interventions, such as N-acetylcysteine, vitamin E, and bardoxolone methyl, have shown the potential for minimizing renal damage and reducing the rate of CKD progression. Clinical outcomes from this treatment have been variable, however, and thus further research is required to establish effective antioxidant treatment modes (Levey and Coresh, 2012).

4.2. Glomerulopathies

Among epithelial cells, podocytes represent one of the most highly specialized cell types playing a major role in maintaining the integrity of the glomerular filtration barrier. Podocyte injury is one of the hallmark features of most glomerulopathies, including FSGS and membranous nephropathy. Current concepts in therapeutic protection and repair of podocytes include immunosuppressive treatment using corticosteroids and calcineurin inhibitors, and novel therapeutic strategies interfering with podocyte-specific pathways (He et al., 2023). For example, the abatacept-CTLA-4 immunoglobulin fusion protein has so far successfully stabilized the function of podocytes, reducing proteinuria in FSGS patients (Boeckhaus and Gross., 2021).

Many glomerulopathies have, as part of their pathogenesis, immune-mediated mechanisms such as the formation of immune complexes and the production of autoantibodies; thus, manipulation of an immune response plays an important role in treatment. Several autoimmune glomerulopathies have very heavily featured the monoclonal antibody rituximab, which attaches to CD20, thus depleting B cells, in the treatment of membranous nephropathy among other autoimmune glomerulopathies. Other drugs frequently used in immune-mediated glomerular disease include mycophenolate mofetil and cyclophosphamide (Duarte et al., 2021). The complement system is crucial for an innate immune response and inflammatory processes, and its malfunction has been associated with several glomerular diseases such as aHUS or C3 glomerulopathy. The efficiency of the blockade of the complement cascade with agents like eculizumab-a monoclonal antibody against C5-is already appreciated for its ability to decrease complement-induced damage to the kidney and improve clinical outcomes in those conditions (Wenzel et al., 2010).

4.3. Renal Fibrosis

Fibroblast Activation and Myofibroblast Differentiation is mainly fibroblasts that contain all the essential cellular machinery for synthesizing the ECM. During injury, fibroblasts differentiate into the myofibroblast phenotype, which is much more effective in synthesizing and depositing the ECM. In this regard, Wu et al. (2021) stated that blocking the activation and differentiation of fibroblasts into myofibroblasts could represent a potential approach by which fibrosis can be avoided. "Medications like pirfenidone and nintedanib, which are considered to hinder the proliferation of fibroblast and deposition of the matrix in most extracellular elements, demonstrate promising preclinical and clinical results in renal fibrosis treatment" (Fuchs et al., 2021; Lingampally et al., 2020). TGF-β has classically been thought to be one of the most important drivers in the control of fibrosis. When TGF-β signaling is impaired, there is less ECM production and thus less fibrosis. Fresolimumab, an anti-TGF-β monoclonal antibody, along with other TGF-β inhibitors, is being investigated as a possible treatment for renal fibrosis. Other targets by which fibrogenesis can be regulated involve downstream mediators of TGF-β signaling, such as the SMAD proteins (Sureshbabu et al., 2016).

Connective tissue growth factor CTGF is a cytokine pivotal in the events of fibroblast proliferation, synthesis of extracellular matrix, and development of fibrosis. Its levels are markedly high in various fibrotic diseases, including CKD. Inhibition of CTGF with specific monoclonal antibodies, such as pamrevlumab, was successfully achieved in preclinical and early clinical studies that manifested reduction in fibrosis with preservation of renal function References (Connolly et al., 2023; Nakayama et al., 2023). Matrix metalloproteinases mediate the degradation of extracellular matrix proteins, thereby regulating the remodeling of tissues. Dysregulated matrix metalloproteinase activity leads to fibrosis if left unchecked, as accumulated ECM stimulates fibrosis. Inhibitors that modulate matrix metalloproteinase activities have been effective in reducing renal fibrosis and restoration of function. However, the complexity and selectivity of inhibitor development pose a great challenge to therapy with a target of matrix metalloproteinase (Sureshbabu et al., 2016).

Ischemia-reperfusion injury is the major cause of acute kidney injury (AKI), particularly in the setting of surgical procedures. Blood flow impairment due to ischemia causes deprivation of oxygen to the cells, and as a result, injures them. Normally, reperfusion worsens the injury through ROS and some inflammatory mediators (Ishiuchi et al., 2020). Approaches to mitigate ischemia-reperfusion injury include antioxidants, anti-inflammatory agents, and ischemic preconditioning techniques of which have shown promise in reducing the severity of AKI (Tseng et al., 2021). In AKI, tubular epithelium is a very sensitive factor to be affected; protection of the cells, and increased proliferative regeneration is now one of the main therapeutic strategies. These other growth factors, as well as EGF and HGF, may be dissected in their facilitation of tubular cell survival and repair. However, further new horizons are now being opened about stem cell therapy inducing renal regeneration and functional recovery in AKI (Li et al., 2020).

The role of inflammation is central to AKI, not only as an initiator but also in the process of development. Anti-inflammatory therapies, including corticosteroids, NSAIDs, and selective cytokine inhibitors, might reduce the damage of inflammation in the kidneys of AKI patients (Tseng et al., 2021). In more recent research, immune modulation is included as a strategy to enhance reparative processes in tissues and to reduce injury. For example, MSCs have immunomodulatory properties and secrete anti-inflammatory cytokines. Consequently, this has placed MSCs as a very promising route for AKI treatment (Savio-Silva et al., 2020). In AKI, microvascular dysfunction and endothelial damage result in reduced blood flow and tissue hypoxia. Improving microvascular function and facilitating endothelial repair is an important therapeutic direction. Nitric oxide donors, vascular endothelial growth factor, and angiopoietins may be considered in this therapy to improve microvascular perfusion and prevent kidney injury (Duni et al., 2019).

4.4. Polygenic Kidney Diseases (PKD)

PKD is associated with mutations, most frequently in the PKD1 and PKD2 genes, which in turn encode two integral proteins involved in mechanosensation controlling calcium signaling: polycystin-1 and polycystin-2 (Sussman et al., 2020). Loss of polycystin signaling leads to the abnormal cell proliferation and cyst formation that manifest in the disease. Treatment by use of a vasopressin V2 receptor antagonist, tolvaptan, has been useful by targeting the polycystin signaling to modulate the same to slow cyst growth rate and preserve renal function in patients with autosomal dominant PKD (Zhou and Torres., 2023).

The mammalian target of the rapamycin (mTOR) pathway is a central regulator of cell growth, proliferation, and metabolism. In PKD, hyperactivation of the mTOR pathway promotes cyst growth. Several studies in the preclinical and clinical fields have proven effective in blocking excessive cyst growth and improving renal outcomes by inhibiting the mTOR pathway through agents like sirolimus (rapamycin) and everolimus. A delirious equilibrium must be brought out between various possible side effects as mentioned earlier: immunosuppression and metabolic changes (Xu et al., 2015). Cyclic adenosine monophosphate (cAMP) is a major second messenger that regulates many cellular functions, including cell growth and fluid secretion. In PKD, abnormally elevated levels of cAMP promote the formation of cysts. In PKD, several approaches have been developed for the management of cAMP signaling that have proven effective in reducing cyst growth and retarded renal function loss. This includes abrogation of either adenylyl cyclase or phosphodiesterase. Tolvaptan, for example, reduces cAMP by blocking vasopressin V2 receptors that are linked to the activity of adenylyl cyclase (Bakaj and pocai., 2023).

Diffuse fibrosis and inflammation are characteristics of PKD, both of which have been identified to play a very crucial role in the disease's progression. Antifibrotic and anti-inflammatory treatments are some of the therapies that could theoretically limit disease progression and preserve renal function. Several agents targeting fibrosis and inflammation in PKD, such as pirfenidone-a TGF-β signaling inhibitor and TNF-α inhibitors, are under study (Fogo, 2007).

4.5. Diabetic Nephropathy

In other words, the overall goal of effective management in cases of diabetic nephropathy is strict glycemic control. Good control of blood glucose—by insulin or hypoglycemic agents given orally—reduces the risk of renal damage and therefore can delay the progression of the disease. Some of the more modern treatments, such as SGLT2 inhibitors and GLP-1 receptor agonists, have also been shown to exert a nephroprotective effect, unrelated to their antiglycemic actions (Trejos et al., 2024). Activation of the RAAS represents another important part of the development and progression of diabetic nephropathy, much as it has often been in the pathogenesis of CKD. Treatment with pharmacologic RAAS-blocking agents in patients with diabetic nephropathy, especially ACE inhibitors, ARBs, and even direct renin inhibitors, results in decreased proteinuria, retarded disease progression course, and reduced occurrence of cardiovascular events. When ACE inhibitors are added to ARBs, the effect is known as dual RAAS blockade and is clinically beneficial but at the expense of hyperkalemia and renal impairment (Kovarik et al., 2021; Whitlock et al. 2023).

Oxidative stress, which results from hyperglycemia, is one of the key etiologic factors in diabetic nephropathy cell damage and inflammation. Drugs so far used to fight diabetic nephropathy with antioxidant action include vitamin E, N-acetylcysteine, and bardoxolone methyl, with favorable reports regarding lessening renal damage and boosting renal function. However, inconsistent results in the clinical variables mean that more research into effective antioxidant agents against diabetic nephropathy is required (Soliman et al., 2023; Jha et al., 2016).

Inflammation plays a very significant role in the initiation and development of diabetic nephropathy. Pharmacological intervention with anti-inflammatory agents, like corticosteroids and NSAIDs, or even targeted cytokine inhibitors, may reduce inflammation and subsequently help in reducing renal damage. Thereafter, research moved towards immune modulation approaches, which would aim at tissue repair with the least injury. For example, the use of MSCs, by its production of anti-inflammatory cytokines, is also considered one of the promising avenues for the treatment of diabetic nephropathy (Eid et al., 2021). Advanced glycation end products are the outcome of non-enzymatic glycosylation in proteins and lipids, and they contribute to major tissue damage in diabetic nephropathy. To halt the process of AGE formation and their receptor, potential agents such as aminoguanidine and soluble RAGE have shown some promise in reducing cellular injury mediated through the action of AGEs and improved renal outcomes. Other methods that can be employed to reduce the load of AGE include improved plasma glucose control and dietary modifications (Sun et al., 2024).

4.6. Novel Therapeutic Targets

Other small non-coding RNAs that regulate gene expression at the post-transcriptional level include microRNAs. Apart from CKD and glomerulopathies, these are highly involved in other kidney diseases such as diabetic nephropathy. miRNAs can be therapeutically targeted with either antagomirs—miRNA inhibitors—or miRNA mimics. This can help to modify at least some of the disease-related pathways (Garmaa et al., 2024). For example, in many forms of CKD, miR-21 is upregulated, promoting fibrosis. Therapeutic benefits by inhibition of miR-21 with antagomirs applied in renal fibrosis, including improvement in kidney function, have been provided in preclinical studies (Chau et al., 2012). Epigenetic Modifications processes have been implicated in the pathogenesis of kidney diseases and include DNA methylation, histone modification, and regulation of gene expression by non-coding RNAs (Wang et al. 2021). The use of HDAC inhibitors and DNA methyltransferase inhibitors are currently utilized in the development of epigenetic modification-targeted treatments. For instance, HDAC inhibitors have been shown in preclinical CKD models to exert an antifibrotic and anti-inflammatory effect (Neganova et al., 2020).

4.7. Autophagy and Proteostasis

Autophagy represents the procedure controlling the degradation and recycling of damaged organelles and proteins, thus helping in the maintenance of cellular homeostasis; disturbance in this process has been associated with a wide range of renal diseases, including CKD, AKI, and polycystic kidney disease (Tang et al., 2020). Induction of autophagy by pharmacological agents like rapamycin and resveratrol will be of help in maintaining cellular homeostasis and preventing renal injury. Moreover, strategies for restoring proteostasis include the chaperone-mediated treatment and the application of proteasome inhibitors as novel therapeutics for treating renal disorders (Song et al., 2020). The gut microbiota and their metabolites are thus emerging as part of general health and renal physiology. Dysbiosis or imbalance in gut microbiota also leads to the origin and development of various kidney diseases such as chronic kidney disease and diabetic nephropathy. Probiotics, prebiotics, and nutritional interventions act on the gut-kidney axis by modifying the gut microbiota and, as such, bring changes and renovation in the renal outcome. For example, SCFAs produced by the gastrointestinal microbiota bear anti-inflammatory and renoprotective properties (Levey et al., 2012).

5. Natural Drugs that have Potential Used for Drug Repurposing in Kidney Diseases

Many natural drugs have been repurposed for new therapeutic uses. Below in Table 1, we have enlisted some of the most well-known sources of drugs from natural and origin.

Table 1. List of naturally occurring drugs that have the potential used for drug repurposing in kidney diseases

Plant name	Scientific Name	Main active constituent	Pharmacological use	Drug Repurposing Use	Ref.
Turmeric	*Curcuma longa*	Curcumin	Anti-inflammatory, antioxidant	Reduces renal inflammation and oxidative stress in CKD	(Roy et al., 2021)
Green Tea	*Camellia sinensis*	Epigallocatechin gallate (EGCG)	Antioxidant, anti-inflammatory	Mitigates nephrotoxicity and oxidative stress in kidney diseases	(Sari et al., 2023)
Ginger	*Zingiber officinale*	Gingerol	Anti-inflammatory, antioxidant	Mitigates nephrotoxicity and oxidative stress in kidney diseases	(Rostamkhani et al., 2022)
Milk Thistle	*Silybum marianum*	Silymarin	Hepatoprotective, antioxidant	Reduces renal fibrosis and oxidative damage in CKD	(Abed et al., 2022)
Ginkgo	*Ginkgo biloba*	Ginkgolides	Antioxidant, neuroprotective	Reduces ischemia-reperfusion injury in acute kidney injury (AKI)	(Li et al., 2021)
Aloe vera	*Aloe vera*	Aloin	Anti-inflammatory, wound healing	Protects against nephrotoxicity and oxidative stress	(Rahman et al., 2021)
Grapes	*Vitis vinifera*	Resveratrol	Antioxidant, anti-inflammatory	Reduces renal fibrosis and inflammation in CKD	(Zhu et al., 2020)
Ashwagandha	*Withania somnifera*	Withaferin A	Adaptogenic, anti-inflammatory	Alleviates nephrotoxicity and oxidative stress	(Soliman et al., 2023)
Garlic	*Allium sativum*	Allicin	Antimicrobial, antioxidant	Reduces oxidative stress and inflammation in kidney diseases	(Albrakati., 2021)

Plant name	Scientific Name	Main active constituent	Pharmacological use	Drug Repurposing Use	Ref.
Ginseng	*Panax ginseng*	Ginsenosides	Immuno-modulatory, antioxidant	Mitigates nephrotoxicity and promotes renal function	(Fan et al., 2023)
Liquorice	*Glycyrrhiza glabra*	Glycyrrhizin	Anti-inflammatory, antiviral	Protects against nephrotoxicity and renal inflammation	(Basist et al., 2022)
Cranberry	*Vaccinium macrocarpon*	Antimicrobial, antioxidant	Antimicrobial, antioxidant	Prevents urinary tract infections and associated kidney damage	(González et al., 2020)
Bearberry	*Arctostaphylos uva-ursi*	Arbutin	Antimicrobial, diuretic	Prevents and treats urinary tract infections	(Iqbal et al., 2023)
Dandelion	*Taraxacum officinale*	Taraxasterol	Diuretic, anti-inflammatory	Promotes diuresis and reduces renal inflammation	(Pfingstgraf et al., 2021)
Bitter Melon	*Momordica charantia*	Charantin	Antidiabetic, antioxidant	Reduces renal damage in diabetic nephropathy	(Arigela et al., 2021)
Chamomile	*Matricaria chamomilla*	Apigenin	Anti-inflammatory, anxiolytic	Reduces renal inflammation and oxidative stress	(Alahmadi et al., 2021)
St. John's Wort	*Hypericum perforatum*	Hypericin	Antidepressant, anti-inflammatory	Protects against nephrotoxicity and inflammation	(Steiber et al., 2022)
Buchu	*Agathosma betulina*	Diosmin	Diuretic, anti-inflammatory	Promotes diuresis and reduces renal inflammation	(Brendlerand Abdel-Tawab., 2022)
Grape Seed Extract	*Vitis vinifera*	Proanthocya-nidins	Antioxidant, anti-inflammatory	Reduces oxidative stress and inflammation in kidney diseases	(Eid et al., 2021)
Saw Palmetto	*Serenoa repens*	Fatty acids and sterols	Anti-inflammatory, diuretic	Mitigates renal inflammation and promotes urinary health	(Mata et al., 2022)

6. FDA Approved Drugs and Investigational Compounds

A List of FDA approved drugs repurposed for kidney diseases is given below:

6.1. Kerendia

Originally approved for chronic kidney disease associated with type 2 diabetes. Kerendia reduces the risk of kidney function decline, kidney failure, and cardiovascular events such as heart attacks and strokes (Black et al., 2019).

6.2. Xphozah

Initially developed for irritable bowel disease, xphozah is now approved for controlling serum phosphate levels in chronic kidney disease patients on dialysis who do not respond well to or cannot tolerate traditional phosphate binders (Daenen et al., 2019).

6.3. Vafseo

Approved for treating anemia due to chronic kidney disease in adult patients on dialysis, Vefseo helps to reduce the need for red blood cell transfusions (Tonelli et al., 2015).

6.4. Jardiance

It is used to manage type 2 diabetes but it is also approved for chronic kidney disease. It helps to reduce the risk of kidney disease progression and cardiovascular death in chronic kidney patients (Daenen et al., 2019).

Table 2. List of FDA-approved drugs repurposed for the treatment of kidney diseases

S. No.	FDA-Approved Drug Name	Initially Used For	Name of the Risk Factor Treating Drugs	References
	Kerendia	Type 2 Diabetes	kidney function decline, kidney failure, and cardiovascular events such as heart attacks and strokes	(Black et al., 2019)
	Xphozah	Irritable bowel disease	Controlling serum phosphate level	(Daenen et al., 2019)
	Vafseo	Anemia	Reduce the need for red blood cell transfusions	(Tonelli et al., 2015)
	Jardiance	Type 2 diabetes	Kidney disease progression and cardiovascular death	(Daenen et al., 2019)

7. Challenges and Limitations in Drug Repurposing for Kidney Diseases

7.1. Biological Complexity of Kidney Diseases

Renal disorders comprise a wide array of diseases, including CKD, AKI, glomerulopathies, and PKD. Each has differing etiological causes, underlying pathophysiological mechanisms, and clinical presentations. This makes the identification of drugs that may be effectively repurposed across all cases very difficult (Levin et al., 2017).

In general, renal diseases have a multicausal origin; hence, combined with metabolic derangement, oxidative stress, inflammation, and fibrosis, strategies for treatment need personalized approaches and combination therapies because one single strategy may not hit all (Nangaku and Eckardt, 2007).

7.2. Preclinical and Clinical Challenges

Although much preclinical research on many of these repurposed drugs is highly promising, it is often difficult to translate these into successful clinical trials. Preclinical efficaciousness and clinical results can often not align as animal models cannot frequently replicate renal disorders in humans (Mak et al., 2014). Repurposed pharmaceuticals must go through rigorous regulatory reviews that will ascertain their efficacy and safety for other uses. This involves heavy documentation and evidence required for the regulatory bodies, which becomes an expensive and time-consuming affair. The research-based pharmaceutical companies and researchers face a lot of hurdles while navigating these regulatory routes (Breckenridge and Jacob., 2019).

7.3. Clinical Trial Design

Clinical trials in repurposed drugs for renal disorders represent entirely a different level of challenges, of course not excluding patient group choice, dosing, and outcome definitions, among others. Apart from that, the designs of trials are highly personalized in terms of heterogeneity in kidney diseases (Collins and Varmus., 2015).

7.4. Commercial and Economic Obstacles

Concerns are often raised over intellectual property rights on repurposed pharmaceuticals, with original patents expired. The absence of exclusivity in marketing and the competition from generic versions further prove to be disincentives for investment in repurposing projects by any pharmaceutical company (Pushpakom et al., 2019). Financing for research in studies oriented toward other uses is even more laborious to obtain than that granted to studies into new drugs. Major barriers include a lack of financial incentives, high costs of therapeutic research and the process of approval from regulatory authorities (Hernandez et al., 2017).

7.5. Scientific and Technical Challenges

New formulations or delivery routes may be required to design repurposed pharmaceuticals to target renal tissues. It is very expensive and technically challenging to develop new formulations, so efficient extensive research and development efforts are also needed (Allen et al., 2013). A repurposed drug could have unexpected interactions or off-target effects that are perhaps co-administered with the patients' widely used drugs for renal diseases. These off-target effects might create untoward outcomes and become a barrier to other therapeutic interventions for the patient.

7.6. Ethical and Patient-Related Challenges

The patient cohort is difficult to obtain informed consent from and would especially be under conditions involving concurrent illnesses, with conditions of severe renal impairment. Specific eligibility criteria and burdens associated with patients may further complicate the processes of recruiting a large enough patient cohort for clinical trials (Tatonetti et al., 2012). Equity and Access However, it is quite an uphill task to ensure that such repurposed medicines are reaching all patient populations. Socioeconomic issues, geographical barriers, and access to health care may make them accessible only to a selected few (Lantos et al., 2018).

7.7. Specific Challenges in Repurposing Naturally Originated Drugs

In some instances, informed consent from the patients is also difficult to achieve, particularly in cases of concomitant illness or renal disease that is far advanced. Additionally, specific eligibility and burden of a patient on a clinical trial further curtail appropriate recruitment of an ample patient population (Fabricant and Farnsworth., 2001). Effective repurposing of natural materials depends on the assurance of their quality and standardization. It is, therefore, crucial to establish measures for quality control that ensure such products are safe, effective, and pure (Bent., 2008). It is very tedious to go through all the regulatory processes of natural products. Obtaining full safety and efficacy data on the natural product is complicated because the composition is highly complicated and has been traditionally used. Regulatory bodies often insist on such data (Tilburt and Kaptchuk., 2008).

8. Future Directions and Potential for Personalized Medicine in Drug Repurposing for Kidney Diseases

This section elucidates future directions and opportunities for personalized medicine within this frame, with sustained current literature and research findings. The changing landscape of healthcare is an exciting area where personalized medicine might be a part of repositioning pharma in renal diseases. Precision medicine, often these days referred to as personalized medicine, tailors medical intervention treatments to the individual characteristics of each patient: their genetics, biomarkers, and lifestyle. This approach holds promise in advancing the therapeutic potential and safety of re-purposed drugs for nephrological conditions.

8.1. The Role of Genomics in Personalized Medicine

Various genomic technologies, which currently identify genetic variations and biomarkers associated with renal disorders, are being researched. Biomarkers thus can aid in the choice of the most probable repurposed drugs to be effective for certain patients, hence increasing treatment outcomes (Collins et al., 2015).

Pharmacogenomics is the study of how genetic factors influence a person's response to pharmacological treatments. A combination of this pharmacogenomic information helps healthcare providers tailor therapeutic regimens by predicting which patients would benefit from specific repurposed drugs, and those at risk for adverse reactions (Relling and Evans 2015).

8.2. Advanced Computation Methods

Artificial intelligence (AI) and machine learning (ML) can even predict the reaction of drugs and highlight regularities in huge datasets. Both technologies contribute to significantly speeding up the identification of potential candidates for drug repurposing and optimization of the treatment strategy based on personalized patient data (Esteva et al., 2019).

Big data analytics and bioinformatics technology allow the integration of multi-dimensional biological datasets, including but not limited to proteomics, metabolomics, and genomes, and their analysis. A holistic approach could find new targets for treatment and renal disease repurposing possibilities (Marx, 2013).

8.3. Personalized Therapeutic Strategies

However, combination therapies targeting multiple pathways are much more effective than monotherapy, because the pathophysiology behind renal diseases is not that simple. The application of personalized medicine to find synergistic drug combinations for the patient further improves the efficacy of treatment (Al-Lawati and Naylor., 2023).

Then, personalized medicine would apply by optimizing doses considering factors peculiar to the patients like age, weight, genetic constitution, and renal functions. It would have dual purposes: it would maximize the therapeutic efficacy of the repurposed drugs and minimize the risk of adverse toxicity or any drug interaction (Johnson, 2003).

8.4. Clinical Implementation and Integration

The incorporation of pharmacogenomic and genomic data into electronic health records can guide treatment plans. EHRs can also support the availability of all the requisite patient information for physicians which can be used to direct the use of repurposed drugs with real-time decision support (Blumenthal. And Tavenner., 2010).

CDSS can use big data and AI to make personalized recommendations for the treatment. These tools will help clinicians choose repurposed medications based on patient-specific information and, therefore, ameliorate clinical judgment and patient outcomes (Kawamoto et al., 2005).

8.5. Ethical and Social Considerations

Additionally, various initiatives that address socioeconomic disparities will ensure that repurposed pharmaceuticals and personalized medical treatment can be accessed equally, so all patients, regardless of background, are able to obtain modern options for therapeutic treatment.

Great concerns related to privacy and security are posed by the collection and use of genetic and medical information. Patient data should be kept private with sufficient security to maintain public trust in personalized medicine (Rodwin, 2009).

8.6. Future Research Areas

Therefore, further studies should be conducted with the use of an integrated approach through the multi-omics data-genomics, transcriptomics, proteomics, metabolomics-for a comprehensive view of renal diseases and identification of new repurposing opportunities. The systemic approach could discover novel therapeutic targets and complex pathologies mechanisms in different diseases (Hasin et al., 2017).

Longitudinal studies involving observations in patients are thought to have critical insights into the renal disease process and, by implication, the long-lasting consequences of pharmaceutical repurposing. It gives excellent potential for individualization of therapeutic strategy for the goal of improved patient outcomes (Belsky et al., 2020).

8.7. Prospects for Personalized Medicine in Specific Kidney Diseases

In CKD, personalization of treatment holds great promise when disease progression is informed both by environmental and genetic factors. Modification of reuse of medications based on the individual nuances in patients will lengthen disease progression and reduce injurious outcomes (KDIGO 2012).

Newer drugs are available that can prevent or reduce renal injury in the event of AKI. Biomarkers and genetic studies may help identify high-risk individuals and subpopulations at risk long before the insults occur. Thus, individualized approaches may be adopted for efficient interventions in AKI (Mehta et al., 2007). Genetic testing in patients with PKD can select therapy repurposing based on the mutated genes related to disease severity. Specific therapeutic strategies directed at the patient may prolong the life of such patients and thus protect against transplantations or dialysis (Harris and Torres, 2009).

Conclusion

Kidney disease is one fertile ground for drug repurposing, presently approved for use and therefore promising in the near future in which already-approved drugs could gain new therapeutic indications to fulfill important unmet needs within nephrology. The current state of renal disease care and the future directions are discussed considering the potentiality of drug repurposing, the challenges, and the possible future directions including personalized medicine. The repositioning of pharmaceutical agents in treating renal disorders has vast potential, especially when combined with the practice of personalized medicine. These insights illustrate a roadmap for future research and clinical applications and stress the need for out-of-the-box thinking and interdisciplinary collaboration. With these approaches comes the unprecedented opportunity for the healthcare field to foster the development of targeted, safer, and more successful treatments, improving millions of patients' lives who suffer from renal diseases. The road is hard, but the journey is worthwhile because it is a very promising way to achieve improvement in patient outcomes.

References

Abed, N. A., Musab M. K., Mohammed K. A., 2022. The protective effect of milk thistle against drug-induced renal illnesses: A review. *Iraqi J. Pharm. Sci.*19(2): 93-107.

Alahmadi, A. A., Alahmadi, B. A., Wahman, L. F. and El-Shitany, N. A., 2021. Chamomile flower extract ameliorates biochemical and histological kidney dysfunction associated with polycystic ovary syndrome. *Saudi J. Biol. Sci.* 28(11):6158-6166.

Albrakati, Ashraf., 2021. Aged garlic extract rescues ethephon-induced kidney damage by modulating oxidative stress, apoptosis, inflammation, and histopathological changes in rats. *Environ Sci Pollut Res Int.*. 28(6): 6818-6829.

Al-Lawati, H., Naylor, K., 2013. Personalized medicine and combination therapy. In: Disease Modelling and Drug Discovery. *RSC.* 162-184.

Allen, T. M., Cullis, P. R., 2013. Liposomal drug delivery systems: from concept to clinical applications. *Adv. Drug Deliv. Rev.* 65(1), 36-48.

Arigela, C. S., Nelli, G., Gan, S. H., Sirajudeen, K. N. S., Krishnan, K., Abdul Rahman, N. and Pasupuleti, V. R., 2021. Bitter gourd honey ameliorates hepatic and renal diabetic complications on type 2 diabetes rat models by antioxidant, anti-inflammatory, and anti-apoptotic mechanisms. *Foods*.10(11):2872.

Arya, P., Pandey, S., Verma. V., 2017. "Kidney stones formation and use of medicinal plants as antiurolithiatic agents." *Univers. J. Pharm. Res.* 2(4). doi:10.22270/ujpr.v2i4.RW1.

Ashburn, Ted T., Karl, B., 2004. Thor. Drug repositioning: identifying and developing new uses for existing drugs. *Nat. Rev. Drug Discov.* 3(8):673-683.

Bakaj, I., Pocai, A., 2023. Metabolism-based approaches for autosomal dominant polycystic kidney disease. *Front. Mol. Biosci.* 10:1126055.

Baker, P. W., Coyle, P., Bais, R., Rofe, A. M., 1993. Influence of season, age and sex on renal stone formation in South Australia. *Med J Aust.*. 159: 390-392.

Barnes, P. J., Michael K., 1997. Nuclear factor-κB—a pivotal transcription factor in chronic inflammatory diseases. *New England J. Med.* 336(15): 1066-1071.

Bartges, J. W., Amanda J. C., 2015. Urolithiasis. *Vet. Clin. North Am. Small Anim. Pract.* 45(4): 747-768.

Basist, P., Zahiruddin, S., Khan, M. U., Gautam, G., Jan, B., Khan, M. A., Parveen, R., Ahmad, S., 2022. Metabolite profiling and nephroprotective potential of *Glycyrrhiza glabra* L. roots against cisplatin-induced nephrotoxicity *in vitro* and *in vivo*. *Iran. J. Basic Med. Sci*. 25(11):1286.

Belsky, D. W., Caspi, A., Arseneault, L., 2020. Quantification of biological aging in young adults. *PNAS.* 117(12):660-666.

Bent, S., 2008. Herbal medicine in the United States: review of efficacy, safety, and regulation: grand rounds at University of California, San Francisco Medical Center. *J. Gen. Intern.* 23:854-859.

Bergmann, C., Guay-Woodford, L. M., Harris, P. C., Horie, S., Peters, D. J. and Torres, V. E., 2018. Polycystic kidney disease. *Nat. Rev. Dis. Pri.* 4(1):50.

Black, L. M., Lever, J. M., Agarwal, A., 2019. Renal Inflammation and Fibrosis: A Double-edged Sword. *J Histochem Cytochem.* 67(9):663-681.

Blumenthal, D., Tavenner, M., 2010. The "meaningful use" regulation for electronic health records. *New England J. Med.* 363(6):501-504.

Boeckhaus, J. and Gross, O., 2021. Sodium-glucose cotransporter-2 inhibitors in patients with hereditary podocytopathies, alport syndrome, and FSGS: a case series to better plan a large-scale study. *Cells.* 10(7):1815.

Breckenridge, A., Jacob, R., 2019. Overcoming the legal and regulatory barriers to drug repurposing. *Nat. Rev. Drug Discov.* 18(1):1-2.

Brendler, T., Abdel-Tawab, M., 2022. Buchu (*Agathosma betulina* and *A. crenulata*): rightfully forgotten or underutilized?. *Front. Pharmacol.* 13:813142.

Chau, B. N., Xin, C., Hartner, J., Ren, S., Castano, A. P., Linn, G., Li, J., Tran. P. T., Kaimal, V., Huang, X., Chang, A. N., Li, S., Kalra, A., Grafals, M., Portilla, D., MacKenna, D. A., Orkin, S. H., Duffield, J. S., 2012. MicroRNA-21 promotes fibrosis of the kidney by silencing metabolic pathways. *Sci. Transl. Med.* 4(121):121ra18.

Chen, H., Zhang, J., Luo, J., 2012.Repositioning drugs for leprosy: Thalidomide as a promising therapeutic candidate for controlling of immune reaction of multibacillary leprosy. *PLOS Negl. Trop. Dis.*. 6(11).

Chong, Curtis R., David, J., Sullivan, Jr., 2007. New uses for old drugs. *Nature.* 448(7154): 645-646.

Collins, F. S., Varmus, H., 2015. A new initiative on precision medicine. *New England J. Med.* 372(9):793-795.

Connolly, A., Zaidman, C., Brandsema, J., Phan, H., Tian, C., Zhang, X, Li J, Eisner, M. D., Carrier, E., 2023. Pamrevlumab, a fully human monoclonal antibody targeting connective tissue growth factor, for non-ambulatory patients with Duchenne muscular dystrophy. *J. Neuromuscul. Dis.* 10(4):685-699.

Couser, W. G., Giuseppe, R., Shanthi, M., Marcello, T., 2011.The contribution of chronic kidney disease to the global burden of major noncommunicable diseases. *Kidney Int.*80(12):1258-1270.

Daenen, K., Andries, A., Mekahli, D., 2019. Oxidative stress in chronic kidney disease. *Pediatr. Nephrol.* 34:975–99.

Doman, T. N., McGovern, S. L., Witherbee, B. J., Kasten, T. P., Kurumbail, R., Stallings, W. C., Connolly, D. T., Shoichet, B. K. 2002. Molecular docking and high-throughput screening for novel inhibitors of protein tyrosine phosphatase-1B. *J. Med. Chem.* 45(11):2213-2221..

Druker, B. J., Talpaz, M., Resta, D. J., Peng, B., Buchdunger, E., Ford, J. M., Sawyers, C. L., 2001. Efficacy and safety of a specific inhibitor of the BCR-ABL tyrosine kinase in chronic myeloid leukemia. *New England J. Med.* 344:1031-1037.

Duarte, I., Oliveira, J., Outerelo, C., Godinho, I., Henriques Pereira, M. S., Fernandes, P., Jorge, S. and Gameiro, J., 2021. MO316 rituximab in glomerular diseases: a cohort analysis and a literature review. *Nephrol. Dialysis Transpl.* 36(1):104-0074.

Duni, A., Liakopoulos, V., Roumeliotis, S., Peschos, D., Dounousi, E., 2019. Oxidative Stress in the Pathogenesis and Evolution of Chronic Kidney Disease: Untangling Ariadne's Thread. *Int. J. Mol. Sci.* 20(15):3711.

Edvardsson, V. O., Olafur, S., 2013. Indridason, Gudjon Haraldsson, Olafur Kjartansson, and Runolfur Palsson. Temporal trends in the incidence of kidney stone disease. *Kidney Int.* 83(1): 146-152.

Eid, R. A., Zaki, M. S. A., Al-Shraim, M., Eldeen, M. A. and Haidara, M. A., 2021. Grape seed extract protects against amiodarone-induced nephrotoxicity and ultrastructural alterations associated with the inhibition of biomarkers of inflammation and oxidative stress in rats. *Ultrastruct. Pathol. 45*(1):49-58.

Esteva, A., Robicquet, A., Ramsundar, B., Kuleshov, V., DePristo, M., Chou, K., Cui, C., Corrado, G., Thrun, S. and Dean, J., 2019. A guide to deep learning in healthcare. *Nat. Med.* 25(1):24-29.

Fabricant, D. S., Farnsworth, N. R., 2001. The value of plants used in traditional medicine for drug discovery. *Environ Health Perspect*. 109(1):69-75.

Fan, M., Xintian L., Qunling W., Mengyao S., Xiaoxue F., Yegang Z., Donglu W., Haoming L., Wenyi G., Difu Z., 2023. Renal function protection and the mechanism of ginsenosides: Current progress and future perspectives. *Front. Pharmacol.* 14:1070738.

Fogo, A. B., 2007. Mechanisms of progression of chronic kidney disease. *Pediatr. Nephrol.* 22(12): 2011-2022.

Fuchs, M. A., Broeker, K. A., Schrankl, J., Burzlaff, N., Willam, C., Wagner, C., Kurtz, A., 2021. Inhibition of transforming growth factor β1 signaling in resident interstitial cells attenuates profibrotic gene expression and preserves erythropoietin production during experimental kidney fibrosis in mice. *Kidney Int.* 100(1):122-137.

Garmaa, G., Bunduc, S., Kói, T., Hegyi, P., Csupor, D., Ganbat, D., Dembrovszky, F., Meznerics, F. A., Nasirzadeh, A., Barbagallo, C. and Kökény, G., 2024. A systematic review and meta-analysis of microRNA profiling studies in chronic kidney diseases. *ncRNA.* 10(3):30.

Ghosh, P., Ghosh, M., Roy, D., 2024. Beyond conventional treatments: exploring phytochemical remedies from lauraceae and oxalidaceae in urolithiasis. *Int. J. Pharmacog.* 11(3): 90-99.

González de Llano, D., Moreno-Arribas, M. V. and Bartolomé, B., 2020. Cranberry polyphenols and prevention against urinary tract infections: relevant considerations. *Molecules*. 25(15):3523.

Harris, P. C., Torres, V. E. 2009. Polycystic kidney disease. *Ann.l Rev. Med.* 60:321-337.

Hasin, Y., Seldin, M., Lusis, A., 2017. Multi-omics approaches to disease. *Genome Biol.* 18(1):83.

He, H. G., Huang, Y. Y., Liang, Q. Q., Ye, Q. R., Li, A. D., Ye, K., Wu, Q. X. and You, Y. W., 2023. Calcineurin inhibitors or cyclophosphamide in the treatment of membranous nephropathy superimposed with FSGS lesions: a retrospective study from China. *Renal Failure,* 45(2):.2253930.

Hernandez, J. J., Pryszlak, M., Smith, L., Yanchus, C., Kurji, N., Shahani, V. M., 2017. Giving drugs a second chance: overcoming regulatory and financial hurdles in repurposing approved drugs as cancer therapeutics. *Front. Oncol.* 7:273.

Hodos, R. A., Kidd, B. A., Shameer, K., Readhead, B. P., Dudley, J. T., 2016. In silico methods for drug repurposing and pharmacology. *WIREs Systems Biology and Medicine*. 8(3):186-210.

Iqbal, A., Ibrahim, M. and Muhammad, N., 2023. Natural Approach Used for urinary tract infections. *J. Phytopharmacol.* 2(2):1-17.

Ishiuchi, N., Nakashima, A., Doi, S., Yoshida, K., Maeda, S., Kanai, R., Yamada, Y., Ike, T., Doi, T., Kato, Y. and Masaki, T., 2020. Hypoxia-preconditioned mesenchymal stem cells prevent renal fibrosis and inflammation in ischemia-reperfusion rats. *Stem Cell Res. & Ther.*, 11:1-15.

Jha Jay, C., Chow Bryna, S. M., Cooper Mark, E., 2016. Diabetes and kidney disease: role of oxidative stress. *Antioxid Redox Signal.* 25(12) : 657-684.

Johnson, J. A., 2003. Pharmacogenetics: potential for individualized drug therapy through genetics. *Trends Genet.* 19(10):660-666.

Kawamoto, K., Houlihan, C. A., Balas, E. A., Lobach, D. F., 2005. Improving clinical practice using clinical decision support systems: a systematic review of trials to identify features critical to success. *BMJ.* 330(7494):765.

KDIGO 2012 Clinical Practice Guideline for the Evaluation and Management of Chronic Kidney Disease. KI Supplements, 3(1), 1-150.

Kellum, J. A., Romagnani, P., Ashuntantang, G., Ronco, C., Zarbock, A. and Anders, H. J., 2021. Acute kidney injury. *Nature Reviews Dis. Primers*, 7(1):1-17.

Khaling, M., Kumar S., Vandana. V., 2014. Current scenario of urolithiasis and the use of medicinal plants as antiurolithiatic agents in Manipur (North East India): A review. *International Journal of Herbal Medicine*, 2(1):1-12.

Kovarik, J. J., Kaltenecker, C. C., Domenig, O., Antlanger, M., Poglitsch, M., Kopecky, C., Säemann, M. D., 2021. Effect of mineralocorticoid receptor antagonism and ACE Inhibition on angiotensin profiles in diabetic kidney disease: An exploratory study. *Diabetes Ther.,* 12:2485-2498.

Lantos, J. D., Matlock, A. M., 2018. Informed consent: ethical and practical considerations. *JAMA*. 320(21):2255-2256.

Levey, A,S., Kai, U. E., Yusuke, T., Adeera, L., Josef, C., Jerome, R., Dick, D. E. Z., Thomas H., 2005. Hostetter, Norbert Lameire, and Garabed Eknoyan. Definition and classification of chronic kidney disease: a position statement from Kidney Disease: Improving Global Outcomes (KDIGO). *Kidney Int.* 67(6): 2089-2100.

Levey, A. S., Coresh, J., 2012. Chronic kidney disease. *Lancet*. 379(9811):165-180.

Levin, A., Tonelli, M., Bonventre, J., Coresh, J., Donner, J. A., Fogo, A. B., Fox, C .S., Gansevoort, R. T., Heerspink, H. J., Jardine, M. and Kasiske, B., 2017. Global kidney health 2017 and beyond: a roadmap for closing gaps in care, research, and policy. *The Lancet*. 390(10105):1888-1917.

Li, Ji., Jian C., Yucheng Y., Rui D., Meili W., Zhenhua Gu., 2021.Ginkgolide A attenuates sepsis-associated kidney damage via upregulating microRNA-25 with NADPH oxidase 4 as the target. *Int. Immunopharmacol.* 95:107514.

Li, X., Liao, J., Su, X., Li, W., Bi, Z., Wang, J., Su, Q., Huang, H., Wei, Y., Gao, Y. and Li, J., 2020. Human urine-derived stem cells protect against renal ischemia/reperfusion injury in a rat model via exosomal miR-146a-5p which targets IRAK1. *Theranostics*. 10(21): 9561.

Lingampally, A., Jones, M. R., Bagari, S., Chen, C., Rivetti, S., Bellusci, S., 2020. Use of the reversible myogenic to lipogenic transdifferentiation switch for the design of pre-clinical drug screening in idiopathic pulmonary fibrosis. *Front. Bioeng. Biotechnol.* 8:569865.

Liu, Yu., Yuntian, Chen., Banghua, Liao., Deyi, Luo., Kunjie, Wang., Hong, Li., Guohua, Zeng., 2018. Epidemiology of urolithiasis in Asia. *Asian J. Urol.* 5(4), 205-214.

Luo, Y., Thompson, W. K., Herr, T. M., Zeng, Z., Berendsen, M. A., Jonnalagadda, S. R., Carson, M. B., Starren, J., 2017 Natural language processing for EHR-based pharmacovigilance: a structured review. *Drug Saf.,* 40:1075-89.

Mak, I. W., Evaniew, N., Ghert, M., 2014. Lost in translation: animal models and clinical trials in cancer treatment. *Am. J. Transl. Res.* 6(2):114-118.

Marx, V., 2013. Biology: The big challenges of big data. *Nature.* 498(7453):255-260.

Mata, L. R. F. D., Motter, P. G. R., Azevedo, C., Bernardes, M. F. V. G., Chianca, T. C. M. and Vasques, C. I., 2022. Complementary therapies in the control of male lower urinary tract symptoms: A systematic review. *Rev. Latino-Am.* 30:3597.

Mehta, R. L., Kellum, J. A., Shah, S. V., Molitoris, B. A., Ronco, C., Warnock, D. G., Levin, A. and Acute Kidney Injury Network, 2007. Acute Kidney Injury Network: report of an initiative to improve outcomes in acute kidney injury. *Crit. Care.* 11:1-8.

Morales, E., Sandino, J. and Galindo, M., 2023. Lupus nephropathy beyond immunosuppression: Searching for nephro and cardioprotection. *Front. Nephrol.* 3: 1105676.

Nangaku, M., Eckardt, K. U., 2007. Pathogenesis of tubulointerstitial damage in the progression of renal disease: insights from animal models. *Contrib. Nephrol.* 156:47-53.

Neganova, M. E., Klochkov, S. G., Aleksandrova, Y. R. and Aliev, G., 2022, August. Histone modifications in epigenetic regulation of cancer: Perspectives and achieved progress. Academic Press. In *Seminars in Cancer Biology.* 83:452-471.

Nirumand, M. C., Marziyeh, H., Roja, R., Hosein Farzaei, M., Zingue, S., Seyed M. N., and Bishayee, A., 2018. Dietary plants for the prevention and management of kidney stones: preclinical and clinical evidence and molecular mechanisms. *Int. J. Mol. Sci.* 19(3):765.

Nogueira, A., Pires, M. J. and Oliveira, P. A., 2017. Pathophysiological mechanisms of renal fibrosis: a review of animal models and therapeutic strategies. *In vivo,* 31(1):1-22.

Pantziarka, P., Pirmohamed, M., Mirza, N., 2018. New Uses for Old Drugs. *BMJ.* 361:k2701.

Park K. A., 2019. Review of computational drug repurposing. *Transl. Clin. Pharmacol.* 27(2):59-63.

Pfingstgraf, I. O., Taulescu, M., Pop, R. M., Orăsan, R., Vlase, L., Uifalean, A., Todea, D., Alexescu, T., Toma, C. and Pârvu, A. E., 2021. Protective effects of *Taraxacum officinale* L.(dandelion) root extract in experimental acute on chronic liver failure. *Antioxidants*. 10(4):504.

Pushpakom, S., Iorio, F., Eyers, P. A., Escott, K. J., Hopper, S., Wells, A., Doig, A., 2018. Drug repurposing: progress, challenges and recommendations. *Nat. Rev. Drug Discovery*. 18(1):41-58.

Pushpakom, S., Iorio, F., Eyers, P. A., Escott, K. J., Hopper, S., Wells, A., Doig, A., Guilliams, T., Latimer, J., McNamee, C. and Norris, A., 2019. Drug repurposing: progress, challenges and recommendations. *Nat. Rev. Drug Discovery*. 18(1):.41-58.

Rahman, F., Mahmuda B., Chowdhury, R., 2021. Nephroprotective effect of Aloe barbadensis Miller against gentamicin induced nephrotoxicity in long evans male rats. *JBSP*. 16(1): 33-38.

Relling, M. V., Evans, W. E., 2015. Pharmacogenomics in the clinic. *Nature*. 526(7573): 343-350.

Rena, G, Hardie, D. G., Pearson, E. R., 2017. The mechanisms of action of metformin. *Diabetologia*. 60(9):1577-85.

Rodwin, M. A., 2009. The case for public ownership of patient data. *JAMA*. *302*(1):86-88.

Rostamkhani, H., Amir, H. F., Parisa, V., Alireza, R., Nooshin N., Zohreh G., 2022. The protective antioxidant activity of ginger extracts (*Zingiber officinale*) in acute kidney injury: A systematic review and meta-analysis of animal studies. *J. Funct. Foods*. 94:105-111.

Roy, R., Bhowmik, R., Seth, S., Bhattacharyya, S., Sengupta, S., 2021. Computational repurposing model of curcumin as a drug. *Int. J. Pharm. Sci. Rev. Res.*68(1).

Sachs, R. E., Ginsburg, P. B., Goldman, D. P., 2017. Encouraging new uses for old drugs. *JAMA*. 318(24):2421–2422.

Sansores-Espana, D., Pech-Aguilar, A. G., Cua-Pech, K. G., Medina-Vera, I., Guevara-Cruz, M., Gutierrez-Solis, A. L., Reyes-Garcia, J. G., Avila-Nava, A., 2022. Plants used in Mexican traditional medicine for the management of urolithiasis: A review of preclinical evidence, bioactive compounds, and molecular mechanisms. *Molecules*. 27(6): 2008.

Sari, N. W. S., Trina E. T., Beivy J. K., 2023. Benefits of green tea polyphenols for kidney health: A literature review. *GJSET*. 1(2): 60-70.

Sávio-Silva, C., Soinski-Sousa, P. E., Balby-Rocha, M. T., Lira, Á. D., Rangel, É .B., 2020. Mesenchymal stem cell therapy in acute kidney injury (AKI): Review and perspectives. *Rev. Assoc. Med. Bras*. 66:s45-54.

Soliman, M. M., Ahmed, M., Elshehawei, S. A., Samy M. S., 2023. Protective impacts of *Withania somnifera* leaf extract from Taif area against diclofenac induced hepato-renal toxicity: role of antioxidants, inflammation, apoptosis, and anti-oxidative stress biomarkers. *Toxicol. Res*. 12(4): 685-692.

Song, X., Tsakiridis, E., Steinberg, G. R. and Pei, Y., 2020. Targeting AMP-activated protein kinase (AMPK) for treatment of autosomal dominant polycystic kidney disease. *Cell Signal*. 73:109704.

Steiber, A. L., 2022. Herbal supplements in patients with kidney disease. In: *Nutritional Management of Renal Disease*. 907-919 Academic Press.

Stroie, A. M., Penescu, M. and Paul, B., 2022. Mycophenolate mofetil superior to azathioprine and as effective as calcineurin inhibitors in lupus nephritis maintenance therapy: data from a single centre prospective study. In *Nephrol. Dial. Transplant.* 37: I99-i99. Oxford univ press.

Sun, D., Zi-yi, G., Wen-yuan, L., Lin, Z., Zi-yuan, Z., Ya-ling, H., Su-fen, Li., 2024. Astragaloside IV Alleviates podocyte injury in diabetic nephropathy through regulating IRE-1α/NF-κ B/NLRP3 pathway. *Chin. J. Integr. Med.* 1-12.

Sureshbabu, A., Muhsin, S. A., Choi, M. E., 2016. TGF-β signaling in the kidney: profibrotic and protective effects. *Am. J. Physiol. Renal. Physiol.* 310(7):F596-F606.

Sussman, C. R., Wang, X., Chebib, F. T. and Torres, V. E., 2020. Modulation of polycystic kidney disease by G-protein coupled receptors and cyclic AMP signaling. *Cellular signalling*, 72:109649.

Swamidass, S. J., 2011. Mining small-molecule screens to repurpose drugs. *Brief Bioinform.* 12(4):327-35.

Tang, C., Livingston, M. J., Liu, Z. and Dong, Z., 2020. Autophagy in kidney homeostasis and disease *Nat. Rev. Nephrol.* 16(9):.489-508.

Tatonetti, N. P., Ye, P. P., Daneshjou, R., Altman, R. B., 2012. Data-driven prediction of drug effects and interactions. *Sci.Transl. Med.* 4(125):125ra31.

Tilburt, J. C., Kaptchuk, T. J., 2008. Herbal medicine research and global health: an ethical analysis. *Bulletin of the World Health Organization. 86*:94-599.

Tonelli, M., Wiebe, N., 2015. Comorbidity burden and renal outcomes in patients with chronic kidney disease. *Kidney Int.* 88(6):1316-1323.

Touret, F., Gilles, M., Barral, K., Nougairède, A., van Helden, J., Decroly, E., de Lamballerie, X., 2020. *In vitro* screening of a FDA approved chemical library reveals potential inhibitors of SARS-CoV-2 replication. *Sci. Rep.*10(1):1-8.

Trejos, J. S. S., Laura, C. N. R., Stefanya, G. B. M., Diego A. G. S., 2024. Hallazgos clínicos del uso de iSGLT2 en pacientes con nefropatía por IgA no diabéticos: una serie de casos. *Rev. Colomb. Nefrol.* 11(2). DOI:10.22265/acnef.11.2.789.

Tseng, W. C., Lee, P. Y., Tsai, M. T., Chang, F. P., Chen, N. J., Chien, C. T., Hung, S. C., Tarng, D. C., 2021. Hypoxic mesenchymal stem cells ameliorate acute kidney ischemia-reperfusion injury via enhancing renal tubular autophagy. *Stem Cell. Res. Ther.*, 12:1-22.

Vivekanand, J., Guillermo, G., Kunitoshi I., Zuo, Li., Saraladevi, N., Brett, P., Saran, R., Angela Yee-Moon, W., Chih-Wei, Y., 2013. Chronic kidney disease: global dimension and perspectives. *The Lancet.* 382(9888):260-272.

Wang, Y., Xie, Q., Tan, H., Liao, M., Zhu, S., Zheng, L. L., Huang, H. and Liu, B., 2021. Targeting cancer epigenetic pathways with small-molecule compounds: therapeutic efficacy and combination therapies. *Pharmacol. Res.* 173:105702.

Wenzel, U. O., Christian, K., Ralf B., 2010. The angiotensin II type 2 receptor in renal disease. *J Renin Angiotensin Aldosterone Syst.* 11(1): 37-41.

Whitlock, R., Leon, S. J., Manacsa, H., Askin, N., Rigatto, C., Fatoba, S. T., Farag, Y. M. and Tangri, N., 2023. The association between dual RAAS inhibition and risk of acute kidney injury and hyperkalemia in patients with diabetic kidney disease: a systematic review and meta-analysis. *Nephrol. Dial Transplant. 38*(11):2503-2516.

Wu, S. B., Hou, T. Y., Kau, H. C., Tsai, C. C., 2021. Effect of Pirfenidone on TGF-β1-induced myofibroblast differentiation and extracellular matrix homeostasis of human orbital fibroblasts in graves' ophthalmopathy. *Biomolecules.* 11(10):1424.

Xu, H., Aldrich, M. C., Chen, Q., Liu, H., Peterson, N. B., Dai, Q., Levy, M., Shah, A., Han, X., Ruan, X. and Jiang, M., 2015. Validating drug repurposing signals using electronic health records: a case study of metformin associated with reduced cancer mortality. *J Am Med Inform Assoc.*. 22(1):179-191.

Yanagihara, T., Chong, S. G., Gholiof, M., Lipson, K. E., Zhou, Q., Scallan, C., Upagupta, C., Tikkanen, J., Keshavjee, S., Ask, K., Kolb, M. R., 2020. Connective-tissue growth factor (CTGF/CCN2) contributes to TGF-β1-induced lung fibrosis. *bioRxiv.* 1:2020-07.

Zhou, J. X. and Torres, V. E., 2023. Autosomal dominant polycystic kidney disease therapies on the horizon. *Adv. Chronic Kidney Dis.* 30(3):245-260.

Zhu, J., Caigan D., 2020.Could grape-based food supplements prevent the development of chronic kidney disease? *Crit. Rev. Food Sc. Nutr.* 60(18): 3054-3062.

Index

D

N

P

Q

R

S

T

V